Catalysis by Metal Complexes

Volume 36

For further volumes:
http://www.springer.com/series/5763

Catalysis by Metal Complexes

This book series covers topics of interest to a wide range of academic and industrial chemists, and biochemists. Catalysis by metal complexes plays a prominent role in many processes. Developments in analytical and synthetic techniques and instrumentation, particularly over the last 30 years, have resulted in an increasingly sophisticated understanding of catalytic processes.

Industrial applications include the production of petrochemicals, fine chemicals and pharmaceuticals (particularly through asymmetric catalysis), hydrometallurgy, and waste-treatment processes. Many life processes are based on metallo-enzyme systems that catalyze redox and acid–base reactions.

Catalysis by metal complexes is an exciting, fast developing and challenging interdisciplinary topic which spans and embraces the three areas of catalysis: heterogeneous, homogeneous, and metallo-enzyme.

Catalysis by Metal Complexes deals with all aspects of catalysis which involve metal complexes and seeks to publish authoritative, state-of-the-art volumes which serve to document the progress being made in this interdisciplinary area of science.

Series Editors

Prof. Claudio Bianchini
Institute of Chemistry of Organometallic Compounds
Polo Scientific Area
Via Madonna del Piano 10
50019 Sesto Fiorentino
Italy

Prof. D. J. Cole-Hamilton
School of Chemistry
University of St Andrews
North Haugh
KY16 9ST St Andrews
UK

Prof. Piet W. N. M. van Leeuwen
Institute of Chemical Research of Catalonia
Av. Paisos Catalans 16
43007 Tarragona
Spain

Volume 36:
Olefin Upgrading Catalysis by Nitrogen-based Metal Complexes II
State-of-the-Art and Perspectives

Volume Editors

Giuliano Giambastiani
ICCOM, Florence Research Area
National Council of Research (CNR)
Via Madonna del Piano 10
50019 Florence
Italy

Juan Cámpora
Instituto de Investigaciones Químicas
CSIC - Universidad de Sevilla
C/ Americo Vespucio 49
41092 Sevilla
Spain

Giuliano Giambastiani · Juan Cámpora
Editors

Olefin Upgrading Catalysis by Nitrogen-based Metal Complexes II

State-of-the-art and Perspectives

Springer

Editors
Giuliano Giambastiani
ICCOM, Florence Research Area
National Council of Research (CNR)
Via Madonna del Piano 10
50019 Florence
Italy
e-mail: giuliano.giambastiani@iccom.cnr.it

Juan Cámpora
Instituto de Investigaciones Químicas
CSIC - Universidad de Sevilla
C/ Americo Vespucio 49
41092 Sevilla
Spain
email: campora@iiq.csic.es

ISSN 0920-4652

ISBN 978-94-007-3617-7 ISBN 978-94-007-0696-5 (eBook)

DOI 10.1007/978-94-007-0696-5

Springer Dordrecht Heidelberg London New York

Softcover reprint of the hardcover 1st edition 2011

Cover design: eStudio Calamar, Berlin/Figueres

Printed on acid-free paper

Springer is part of Springer Science+Business Media (www.springer.com)

Foreword

Olefin polymerization has remarkably progressed over the last two decades, mainly thanks to the contribution of organometallic chemistry to the design of innovative ligand systems and metal complexes. The irreversible decrease of fossil-resources requires continuous efforts to improve the selectivity and productivity of the industrial processes as well as to reduce the environmental impact, especially in terms of energy and waste. Due to the wealth of possible ligand structures and metal combinations, organometallic-based catalysis can indeed address many of the issues of the sustainable production of polymeric and composite materials.

These two volumes edited by Giambastiani and Campora cover a hot research subject such as that of post-metallocene nitrogen-containing complexes and their use in homogeneous catalysis for the efficient and selective olefin upgrading. These books cover the state-of-the-art of olefin polymerization by catalysts with N-donor ligands as well as hybrid ligands in conjunction with a wide range of metals across the periodic table. Particular attention has been devoted to important, still unresolved issues such as the efficient insertion polymerization of polar monomers. Advantages and limits of the known technologies have been discussed and critically addressed in the light of the most relevant contributions of the many thousand researchers active in the field.

Claudio Bianchini
Director ICCOM-CNR

Preface

Millions of tons of polyolefin-based materials are produced yearly, in most cases under relatively mild conditions mediated by transition-metal catalysts. Through a simple insertion reaction, inexpensive and abundant olefins (such as ethylene and propene) are transformed into polymeric materials for a wide range of applications, including plastics, fibers, and elastomers. The discovery of the Ziegler–Natta catalysts and the seminal works at Phillips Petroleum in the 1950s not only revolutionized polyolefin production, but also paved the way to the development of modern organometallic chemistry. Despite its long history, the polyolefin industry keeps growing steadily and remains technologically driven by the continuous discovery of new catalysts, processes, and applications.

Since Ziegler–Natta's time, important milestones in the field of homogeneous oligomerization/polymerization catalysis were set-up one after the other; from nickel complexes with phosphine donors (SHOP-type catalysts) for the highly selective and efficient production of α-olefins to Group IV metallocene polymerization catalysts and their subsequent industrial exploitation (in the early 1980s) due to the discovery of partially hydrolyzed organoaluminum compounds (MAOs) as co-catalysts/activators. All these scientific successes have shown how discrete "single-site" molecular catalysts could offer unmatched opportunities, compared with heterogeneous systems, towards the tailored synthesis of new polymeric architectures as well as the in-depth understanding of complex reaction mechanisms.

Until a few years ago there have been relatively few reports on late transition metal complexes capable of catalyzing the polymerization of ethylene and α-olefins efficiently. A distinct feature of the latter systems is a high rate of chaintransfer which favors their application as oligomerization catalysts. The discovery of new ligands and activators has been fundamental to fill the gap and make late transition metal catalysts as efficient (and in some cases even more versatile) as metallocene-based systems for the oligomerization and polymerization catalysis.

In 1995, Brookhart and co-workers synthesized a new class of $Ni^{II}$and Pd^{II} complexes stabilized by bulky α-diimine ligands (Schiff bases) which represented a real breakthrough into the development of late transition metal catalysts for the efficient olefin polymerization/oligomerization.

Since then, an almost infinite variety of imine-based ligands or, more generally, nitrogen-containing ligands in combination with either *d*- and *f*- block metals have been explored as efficient and selective oligomerization and polymerization catalysts. The major advantages of this ligand class are represented by the facile control of their stereoelectronic properties, their simple preparation from available and cheap building blocks and their easy handling and storage. All these considerations, together with the capability of most of their metal derivatives to impart high activity and selectivity in olefin upgrading processes, have contributed to make nitrogen-containing catalysts highly desirable for industry and academy.

The aim of these books is to provide an overview on the state-of-the-art and the perspectives in the field of oligomerization/polymerization catalysis mediated by metal complexes (spanning from early to late and lanthanide series) stabilized by ligands containing nitrogen donor groups. Rather than a systematic revision of the major breakthroughs achieved over the last decades, these two volumes offer to the readership the critical point of view of researchers active in specific fields of polymerization catalysis. The amplitude and rigor of each contribution also provide an exhaustive account on the topic: from the synthesis of ligands and related complexes to mechanistic details, the investigation of the catalyst performance and future perspectives. Although the chapters' extension has made the book division into two separate volumes necessary, this partition is merely due to editorial reasons.

Finally, the editors are extremely grateful to all book co-authors for their enthusiasm in participating to this editorial project and for writing up their contribution at their best. A special thank is also due to Sonia Ojo, Claudia Culierat and Ilaria Tassistro from the London Springer office, whose precious assistance in facing technical and logistic details has been essential for the success of the Editors' efforts.

Giuliano Giambastiani
Juan Cámpora

Contents

List of Contributors

Mikael Brassé
Instituto de Investigaciones Químicas, CSIC-Universidad de Sevilla, 41092, Sevilla, Spain.

Juan Cámpora
Instituto de Investigaciones Químicas, CSIC-Universidad de Sevilla, 41092, Sevilla, Spain.

Terunori Fujita
Research Center, Mitsui Chemicals Inc., Tokyo, Japan.

Pedro T. Gomes
Departamento de Engenharia Química e Biológica, Instituto Superior Técnico, Centro de Química Estrutural, Universidade Técnica de Lisboa, Av. Rovisco Pais, 1049-001, Lisbon, Portugal.

Akihiko Iwashita
Research Center, Mitsui Chemicals Inc., Tokyo, Japan.

Lidong Li
Departamento de Engenharia Química e Biológica, Instituto Superior Técnico, Centro de Química Estrutural, Universidade Técnica de Lisboa, Av. Rovisco Pais, 1049-001, Lisbon, Portugal.

Haruyuki Makio
Research Center, Mitsui Chemicals Inc., Tokyo, Japan.

Brian C. Peoples
Departamento de Ingeniería Química, Facultad de Ciencias Físicas y Matemáticas, Centro para la Investigación Interdisciplinaria Avanzada en Ciencias de los Materiales (CIMAT), Universidad de Chile, Santiago, Chile.

René S. Rojas
Departamento de Química Inorgánica, Facultad de Química, Pontificia Universidad Católica de Chile, Casilla 306, Santiago, Chile.

Chapter 1
Phenoxy–Imine Group 4 Metal Complexes for Olefin (co)Polymerization Including Polar Monomer Copolymerization

Akihiko Iwashita, Haruyuki Makio, and Terunori Fujita

Abstract About 50 years after the discovery of Ziegler–Natta catalysts, phenoxy–imine-based group 4 transition metal complexes (FI catalysts) emerged as the next frontier catalysts for the controlled (co)polymerization of olefinic monomers. FI catalysts are highly versatile catalysts capable of producing a wide range of novel polymer architectures. The inherent electronic and structural features of FI catalysts and the accessibility and variability of the phenoxy–imine ligands offer precise control over olefin polymerization. This chapter deals with the key features of FI catalysts, homopolymerization and copolymerization by FI catalysts, and the value-added olefin-based materials that can be produced with FI catalysts.

1.1 Introduction

With the discovery of the Ziegler–Natta catalyst in the 1950s, the polyolefin industry was launched into a period of unprecedented growth [1]. One reason for this growth lies in the fact that polyolefins are the most versatile of polymers in that they possess superior mechanical and physical properties, excellent chemical inertness, good processability, and easy recyclability. Another reason lies with the development of new catalysts that helped initiate new production processes and new product lineups. As a result, polyolefin resin is the most produced resin today, and polyolefinic materials are ubiquitous in everyday life (Fig. 1.1).

While the majority of commercially available polyolefins are still produced with the heterogeneous, multi-site Ziegler–Natta catalysts (as represented by

A. Iwashita · H. Makio · T. Fujita (✉)
Research Center, Mitsui Chemicals, Inc., 580-32 Nagaura,
Sodegaura, Chiba 299-0265, Japan
e-mail: Terunori.Fujita@mitsui-chem.co.jp

G. Giambastiani and J. Cámpora (eds.), *Olefin Upgrading Catalysis by Nitrogen-based Metal Complexes II*, Catalysis by Metal Complexes, 36,
DOI: 10.1007/978-94-007-0696-5_1,

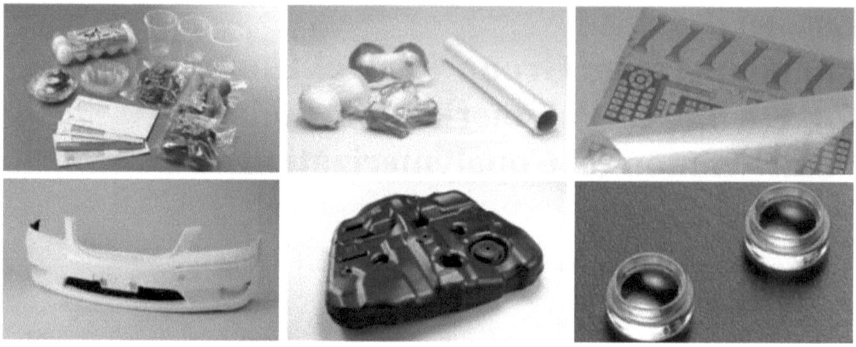

Fig. 1.1 Examples of the application of polyolefinic materials

$MgCl_2$-supported $TiCl_4$ catalysts), single-site metallocene catalysts have become more prevalent in the polyolefin industry. Originally discovered by Kaminsky and coworkers [2], single-site metallocene catalysts offered people a better understanding of reaction mechanisms, introduced new catalyst design possibilities [3–5], and allowed for the production of differentiated materials such as high-performance linear low-density polyethylenes (LLDPEs), isotactic polypropylenes (iPPs), syndiotactic polypropylenes (sPPs), ethylene/1-butene amorphous copolymers, and ethylene/propylene diene elastomers.

Following the development of metallocene catalysts, a lot of research has been undertaken to develop novel molecular catalysts based on non-cyclopentadienyl-based ligands, in other words, new post-metallocene catalysts based on both early and late transition metals [6–12]. One of the earliest examples were the bis(phenoxy–imine) group 4 metal complexes (now known as FI catalysts) that were discovered [13–15] based on the "ligand oriented catalyst design concept", a concept that is founded on the belief that the flexible electronic nature of a ligand is a key requirement for achieving high activity [16]. And due to their extremely high activity, unique selectivity, and remarkable versatility, it is not an exaggeration to say that FI catalysts represent one of the most successful post-metallocene catalysts to date.

In this chapter, unique olefin polymerization with FI catalysts and the polymerization mechanisms involved will also be discussed, and the resulting value-added polyolefin materials will be introduced.

1.2 Key Features of FI Catalysts

1.2.1 General Synthetic Schemes of FI Complexes

As shown in Scheme 1.1, a phenoxy–imine group 4 metal complex is generally synthesized by reacting MCl_4 ($M = Ti, Zr, Hf$) with a lithium or sodium salt of a phenoxy–imine ligand precursor which is derived from the Schiff-base condensation reaction of an *ortho*-hydroxy aromatic aldehyde or ketone and a primary amine. The initial phenol and amine derivatives with various substituents are easily synthesized and some of them are commercially available. This ease of synthesis allows FI catalysts to possess a wide range of catalyst design possibilities. Thus, catalyst efficiency and resulting polymer properties can be tuned by efficiently and systematically examining diverse ligands, which have sterically and electronically varied substituents, including O, S, N, P and halogen-based functional groups at strategic positions. Substituents R^1 to R^4 are defined according to Scheme 1.1 throughout this chapter.

1.2.2 Structural Features

In contrast to group 4 metallocenes that have a tetrahedral framework, bis(phenoxy–imine) group 4 complexes adopt an octahedral geometry around the metal furnished with two imine nitrogens, two phenolic oxygens, and two non-spectator ligands (X). Figure 1.2 depicts the theoretically possible five structural isomers arising from the coordination modes of FI ligands in an octahedral configuration.

Crystallographically determined structures of phenoxy–imine group 4 metal complexes (catalyst precursors) have revealed that in the solid state the complexes

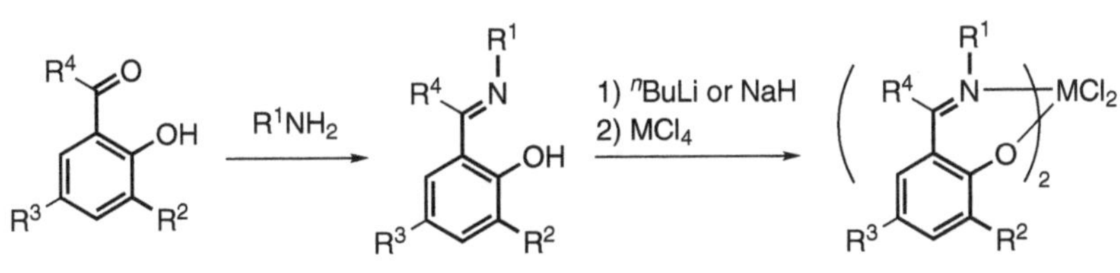

Scheme 1.1 General synthetic scheme of phenoxy–imine group 4 metal complexes

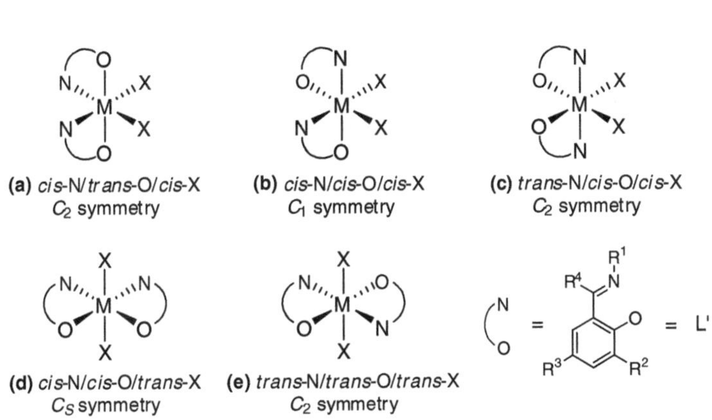

Fig. 1.2 Possible structural isomers for the FI catalyst

most often exist as isomer (a) in a *cis*-N/*trans*-O/*cis*-X arrangement and thus possess C_2 symmetry [17, 18]. The crystal structures of prototypical complexes **1** and **2** are shown in Fig. 1.3. Because of the steric repulsions of the substituents on the imine nitrogen (R^1) of the FI ligands, two nitrogens and two oxygens cannot exist on a plane including the metal center unlike isomers (d) and (e), and this inevitably results in one of those *cis*-X geometries (isomers (a)–(c)), among which isomer (a) is most commonly found. It is important to note that the *cis*-X geometry, which is similarly observed in metallocenes or any other high-performance olefin polymerization catalyst, is considered to be essential for the efficient olefin insertion reaction [3–5, 19–21]. When schematically looking at the ordinary C_2 symmetric phenoxy–imine complexes (isomer (a)), the imine nitrogens are located at the backside of the X–M–X moiety, whereas phenoxy-oxygens are situated above and below the X–M–X moiety. Therefore, the R^1 substituents on the imine nitrogen and the R^2 substituents ortho to the phenoxy–oxygen are located at specific positions near the X ligands. Since the X–M–X moiety becomes an olefin polymerization site upon activation, these R^1 and R^2 substituents located in close proximity to the X ligands heavily influence the polymerization behavior of the FI catalysts in specific ways relative to each substituent (Fig. 1.4).

NMR studies of FI catalysts in solution sometimes exhibit a minor isomer with C_1 symmetry (the two FI ligands are chemically non-equivalent, most likely a *cis*-N/*cis*-O/*cis*-X isomer in Fig. 1.2) in addition to the major C_2 symmetric isomer [22–27]. These isomers are often fluxional and interchangeable with each other on

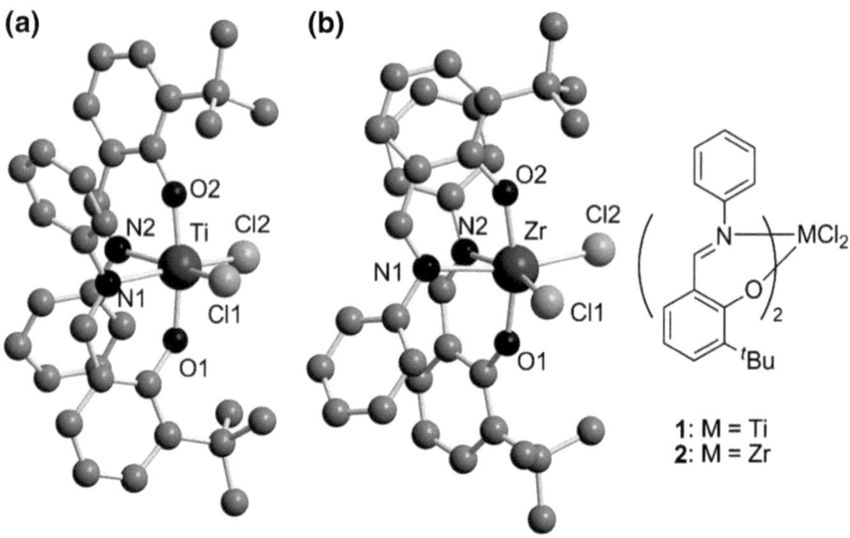

Fig. 1.3 Molecular structures of Ti– and Zr–FI complexes **1** and **2**. All hydrogens are omitted for clarity. Selected bond lengths (Å) and angles (°): **a** Ti–O1 1.852(4), Ti–N1 2.236(4), Ti–Cl1 2.305(2), O1–Ti–O2 171.6(2), N1–Ti–N2 76.4(2), Cl1–Ti–Cl2 103.10(8); **b** Zr–O1 1.985(2), Zr–N1 2.355(2), Zr–Cl1 2.4234(9), O1–Zr–O2 165.5(1), N1–Zr–N2 74.0(1), Cl1–Zr–Cl2 100.38(5)

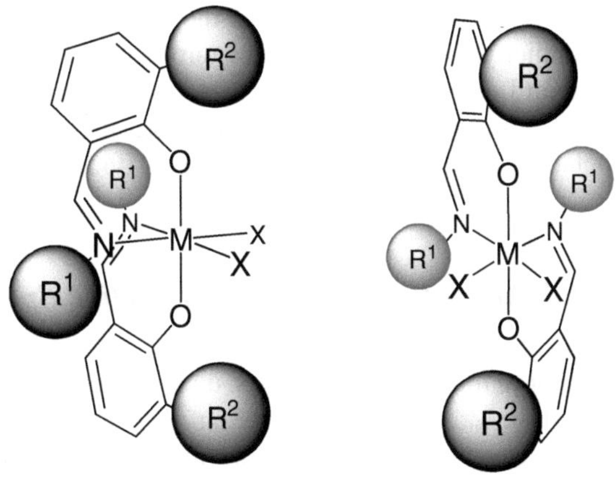

Fig. 1.4 Schematic image of a C_2 symmetric FI catalyst

an NMR time scale [18, 23, 28]. Some of the unique polymerization characteristics of FI catalysts may stem from this structurally dynamic behavior between these possible isomers. For example, for the C_2 symmetric FI catalysts, fluxional isomerization between two enantiomers (Λ/Δ racemization) was proposed by calculations for a syndioselective propylene polymerization with Ti–FI catalysts (see Sect. 1.3.2.1) [29], and has been suggested by low temperature VT-NMR experiments for a cationic monobenzyl Hf–FI catalyst [30] and other related complexes [31]. Thus, the fluxionality of FI catalysts will be an important feature in understanding their polymerization characteristics, and it is believed that the fluxional isomerization probably takes place via M–N bond dissociation due to the labile nature of the imine-N donors.

The labile and dynamic nature of the phenoxy–imine ligands, (especially of the imine-N donors) was also supported by DFT calculations, which demonstrated that the bond distances of the Zr metal center and the imine-N's of a methyl cationic FI catalyst derived from complex **2** were lengthened significantly (0.02–0.1 Å) by the coordination of an ethylene molecule, while the Zr–O bond lengths remained virtually unchanged [32].

1.2.3 Electronic Features

Another distinctive feature of FI catalysts vis-à-vis metallocene catalysts is their heteroatom ligation when compared to the sp^2-hybridized carbon-based cyclopentadienyl (Cp) anion of metallocene compounds. The Mulliken charge at the metal center calculated by DFT methods for three cationic monomethyl titanium species clearly shows the trend between these two types of catalyst. The cationic properties of the metal center increase in the following order: $(C_5H_5)_2TiMe^+$ (1.417 au) < $Me_2Si(C_5Me_4)(^t Bu–N)TiMe^+$ (1.599 au) < $(Ph–N{=}CH–C_6H_3–2–O–3–^t Bu)_2TiMe^+$ (1.741 au), implying that the presence of the heteroatom-coordinating FI ligands makes the complex more electrophilic [33]. In general, one of the reasons for the high activity of the FI catalysts is believed to be this high electrophilicity. In addition, heteroatom ligation renders the M–L bonding properties of the FI catalysts more ionic or polarized relative to that of metallocenes, which may cause the stronger affinity to inorganic surfaces and the higher tolerance to polar functionalities of the FI catalysts (vide infra).

1.2.4 Activation Methods and Active Species

Similar to metallocene catalysts, upon appropriate activation, neutral FI complexes (L'_2MX_2, L': a generic form of an FI ligand) are transformed into coordinatively unsaturated, highly electrophilic monoalkyl cationic complexes, (L'_2M^+–R) (R: alkyl group), which mediate olefin polymerizations via a coordination-insertion mechanism. The monoalkyl cationic species of some Ti–FI catalysts (and related complexes [31, 34]) were actually observed by NMR upon activation of L'_2TiCl_2 with MAO [24, 35], L'_2TiMe_2 [36] or $L'_2Hf(benzyl)_2$ [30] with $B(C_6F_5)_3$ or $Ph_3CB(C_6F_5)_4$, and L'_2ZrCl_2 with $AlMe_3/Ph_3CB(C_6F_5)_4$ [35, 37]. Since these cationic species of FI catalysts exhibited only one set of ligand signals, the species still possess C_2 symmetric structures in solution at room temperature within an NMR time scale.

Activation of metallocene ($Cp'MX_2$, Cp': a generic form of a cyclopentadienyl ligand) by triisobutylaluminum ($^iBu_3Al$) and $Ph_3CB(C_6F_5)_4$ generated a cationic alkyl species, and thus this activation method is considered to be chemically equivalent to the activation with MAO. However, when FI catalysts are activated by $^iBu_3Al$ and $Ph_3CB(C_6F_5)_4$, the imine moiety is reduced to an amine by $^iBu_3Al$ (or contaminant $^iBu_2AlH$) accompanied by isobutene formation, (FI catalysts and $^iBu_3Al$ are mixed for 10 min before polymerization) [32, 38, 39], resulting in phenoxy–amine complexes. The reduced species exhibited a number of interesting polymerization characteristics owing to their unusual N donors, $^iBu_2Al$–N, which will be bulkier and weaker as a coordinating donor than the imine-N′s (See Sects. 1.3.1.2, 1.3.2.2, and 1.3.3).

Meanwhile, it was demonstrated that $AlMe_3$ included in MAO can cause deactivation of FI catalysts [24, 35]. The deactivation is probably initiated by an attack of $AlMe_3$ on a phenoxy-O group. Subsequent C–H bond activation in $AlMe_3$ by a cationic L'_2M^+–Me species yields an Al–FI complex ($L'AlMe_2$)

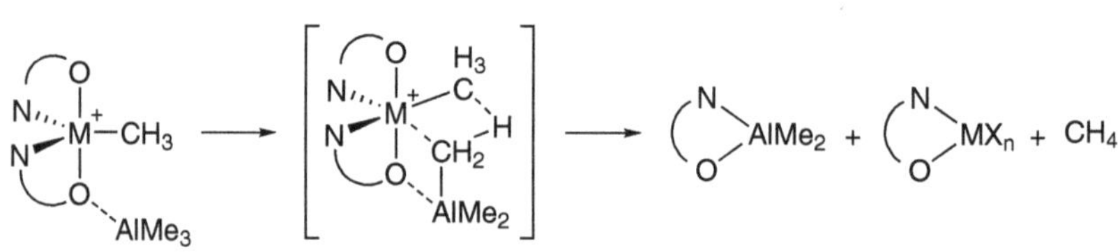

Scheme 1.2 A deactivation path of FI catalysts caused by $AlMe_3$

together with a methane molecule and a paramagnetic group 4 metal compound (Scheme 1.2). As another deactivation pathway, the formation of dinuclear μ-complexes [40, 41] or heterodinuclear complexes with alkyl aluminums [37] have also been proposed.

As described above, FI catalysts have distinctive features derived from a pair of non-symmetric phenoxy–imine ligands, which inevitably results in; (a) enormous structural diversity stemming from ligand accessibility and amenability to modification; (b) isomers arising from ligand coordination arrangements in an octahedral framework, among which *cis*-X configurations are favored; and (c) potential fluxionality among these isomers. The phenoxy–imine ligands also make FI catalysts highly electrophilic, chemically absorbable (possible chemisorption on a solid surface), and functional-group tolerant, presumably due to the heteroatoms that are present and their M–L bonding characteristics. Additionally, the high reactivity of the imine moieties allows in situ ligand modification by the choice of activation methods. The following sections reveal the basic polymerization behavior of FI catalysts and how they are able to upgrade olefins.

1.3 Homopolymerization by FI Catalysts

1.3.1 Ethylene Polymerization and Oligomerization

1.3.1.1 Exploration of High Activity

The catalytic activity of polymerization is generally measured by weight of polymers per a specific catalyst (metal) amount, time, and sometimes monomer concentration and stated in, for example, g-polymer/(mmol-M h) or g-polymer/(mmol-M h atm). Activity is one of the most important performance parameters of any catalyst. The group 4 metallocene catalysts were once known as the most active catalysts among olefin polymerization catalysts (10^1–10^2 kg-polymer/(mmol-M h)), until FI catalysts eclipsed them with an activity of the order of 10^3 kg-polymer/(mmol-M h) [15]. Even the prototypical zirconium FI catalyst **2** showed a strikingly high ethylene polymerization activity of 519 kg-polymer/(mmol-M h), which is one of the highest activities among known olefin polymerization catalysts including metallocenes.

Taking advantage of the straightforward synthesis and modular property of phenoxy–imine ligands (Sect. 1.2), systematic studies on catalyst efficiency have been carried out by introducing substituents with different steric and electronic properties [32, 42–44]. As summarized in Table 1.1, a keen relationship between the ethylene polymerization activity and the steric bulkiness of the R^2 substituents can be observed for a series of Zr–FI catalysts **2**–**11**. When a particularly bulky tertiary R^2 group, CPh_2Me, is introduced, the activity of complex **11** reaches an unprecedented level, 6552 kg-polymer/(mmol-M h). This activity corresponds to a

Table 1.1 Effects of the substituents of the Zr–FI complexes on ethylene polymerization activity

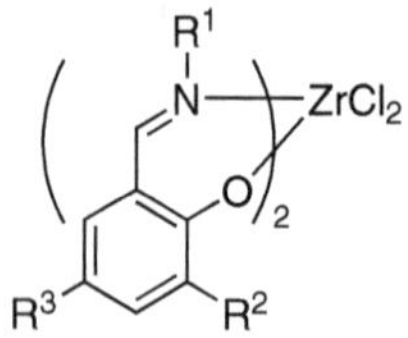

Entry	Complex	R^1	R^2	R^3	Activity [kg-polymer/(mmol-M h)]
1	**3**	Ph	Me	H	0.4
2	**4**	Ph	^{i}Pr	H	0.9
3	**2**	Ph	^{t}Bu	H	519
4	**5**	Ph	^{t}Bu	Me	331
5	**6**	Ph	1-adamantyl	Me	714
6	**7**	Ph	cumyl	Me	2096
7	**8**	Cy	^{t}Bu	Me	82
8	**9**	Cy	1-adamantyl	Me	434
9	**10**	Cy	cumyl	Me	4315
10	**11**	Cy	CPh_2Me	H	6552
11[a]	**12**	Cy	cumyl	OMe	7224

Polymerization conditions toluene 250 mL, MAO 1.25 mmol, Al/Zr = 250–625,000, ethylene 0.1 MPa, 25 °C, 5 or 10 min
[a] Polymerization conditions: *n*-heptane 500 mL, complex 0.005 μmol, MAO 1.25 mmol, ethylene pressure 0.9 MPa, 75 °C, 15 min

TOF of 64,900 s^{-1} atm^{-1}, standing as one of the most efficient catalysts not only for olefin polymerization but also for any catalytic reaction. The complexes having methyl at para to the phenolic oxygen (R^3 substituents) consistently show slightly lower activity than the unsubstituted complexes (R^3 = H) (complex **5** vs. **2**), although the reason is not clear.

The introduction of an electron-donating methoxy group at the R^3 position (para to the phenoxy-O′s) was found to increase the thermal stability of the FI catalysts, making them available for polymerizations in an industrially practical higher temperature range (e.g., **12**: 7224 kg-polymer/(mmol-M h), 75 °C, 0.9 MPa ethylene pressure) [45].

With regard to Ti–FI catalysts, the activity enhancement by bulky R^2 substituents is rather modest, but the catalytic efficiency of Ti–FI catalysts can be electronically increased by electron-withdrawing R^1 substituents (Chart 1.1). The activity increase from 3.58 kg-polymer/(mmol-M h) for prototypical Ti–FI catalyst **1** to 34.8 kg-polymer/(mmol-M h) for **13** (R^1 = 3,5-$F_2C_6H_3$) has been demonstrated. Likewise, activity values of 43.3 kg-polymer/(mmol-M h) for **14** (R^1 = 3,4,5-$F_3C_6H_2$) and 40.3 kg-PE/mmol-M h for **15** (R^1 = 3,5-$(CF_3)_2C_6H_3$) are the typical examples [46, 47].

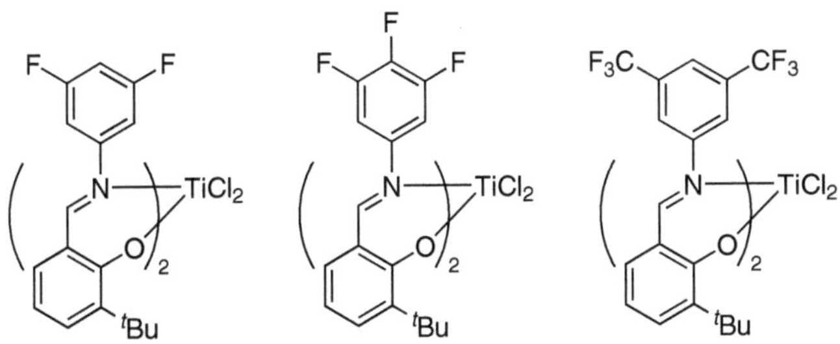

13 **14** **15**

Chart 1.1 Structures of complexes **13–15**

1.3.1.2 Control of Molecular Weight

The controllability of the molecular weight of polymers is one of the most important requirements for a catalyst vis-à-vis the end uses of polyolefin materials. Profound understanding of chain transfer reactions is necessary to gain the controllability, which may also result in gaining a control of the chain-end structures. FI catalysts permit an exceptionally wide range of molecular weight from 10^3 to 10^6 as a result of varying catalyst structures or changing activation methods (vide infra). As will be seen in Sect. 1.3.1.3, controllability of chain transfer reactions of FI catalysts can lead to living polymerizations with practically no chain-terminating processes taking place.

With regard to the molecular weight control with FI catalysts, activation methods and the size of the R^1 substituents are important. As discussed above, activation of FI catalysts with $^i Bu_3Al$ and $Ph_3CB(C_6F_5)_4$ causes formation of phenoxy–amine complexes, which consistently shows lower activities but furnishes higher molecular weight polymers than the phenoxy–imine complexes irrespective of the Ti or Zr complexes.

When R^1 substituents were varied in a series pertaining to Zr–FI catalysts as shown in Table 1.2, the molecular weights of polyethylenes (PEs) show a clear dependence on the size of the R^1, that is, $M_w = 2000$ g/mol with **16** (R^1 = cyclobutyl), $M_w = 3600$ g/mol with **17** (R^1 = cyclopentyl), $M_w = 14{,}000$ g/mol with **18** (R^1 = cyclohexyl), and $M_w = 290{,}000$ g/mol with **19** (R^1 = 2-methylcyclohexyl) [44, 48]. The terminating chain-end structures were revealed to be almost exclusively a vinyl group for the low molecular weight PEs (>90 mol%), suggesting that the main chain transfer reaction of these FI catalysts is a β-hydrogen elimination.

Further studies have shown that the molecular weights were apparently independent of the monomer concentration, while the activities increased linearly at higher monomer concentration, showing that both propagation and β-H transfer are first order in monomer concentration. Therefore, the chain transfer involved in these polymerizations is the β-H transfer to a coordinating ethylene.

Table 1.2 Effects of the R^1 substituents of the phenoxy–imine Zr complexes on ethylene polymerization behavior

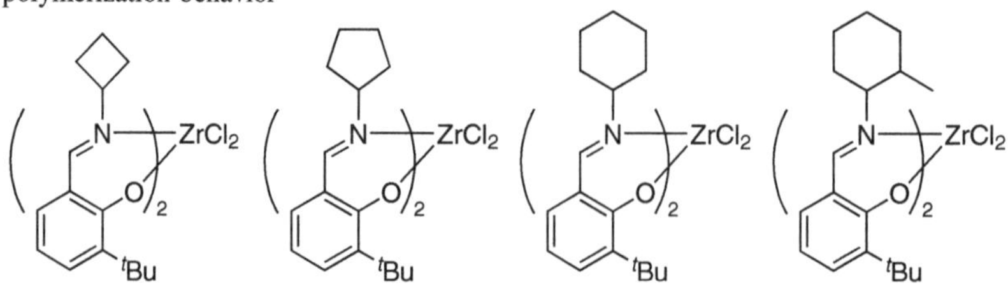

16 17 18 19

Entry	Complex	Activity [kg-polymer/(mmol-M h)]	M_w[a] (g/mol)	M_w/M_n[a]	Vinyl terminated chain end[b] (mol%)
1	**16**	31.6	2000	2.0	91
2	**17**	67.2	3600	2.1	95
3	**18**	87.7	14,000	1.7	96
4	**19**	93.0	290,000	4.9	71

Polymerization conditions toluene 250 mL, complex 0.5 μmol, dried MAO 0.625 mmol, ethylene feed 100 L/h, 25 °C, 5 min
[a]Determined by GPC (PE calibration)
[b]Determined by ^{1}H NMR

This bimolecular *β*-H transfer reaction is known to proceed via a six-centered transition state, which is sterically more encumbered than the compact four-centered transition state for the propagation (ethylene insertion) reaction or the unimolecular *β*-hydrogen transfer to a metal. Within a framework of isomer (a), *cis*-N/*trans*-O/*cis*-X in Fig. 1.2, it is quite possible that the R^1 substituents on the imine nitrogens have a non-bonding interaction with the X ligands (see Fig. 1.4). This was confirmed by the X-ray structures of Zr–FI catalysts, which demonstrated that the Cl–Zr–Cl bond angles were narrowed as R^1 substituents became larger (R^1 = cyclobutyl (**16**: 100.64(3)°), cyclohexyl (**18**: 99.8(2)°), 2-methylcyclohexyl (**19**: 98.00(6)°)) [48]. The narrower bond angles observed for the bulkier R^1 groups probably destabilize the six-centered transition state of the chain transfer more than the four-centered transition state of the propagation, and the higher rate of the propagation relative to the chain transfer results in higher molecular weight polymers. DFT calculations have confirmed this scenario, as well as the extremely unstable nature of the Zr hydrides, which means that *β*-H transfer to a metal is disfavored. These vinyl-terminated polymers can be used as a distinctive precursor for the chain-end functionalized polymers (see Sect. 1.3.1.6).

The relationship between the R^1 and the polymer molecular weight holds true for a series of Zr–FI complexes bearing *N*-aryl groups. The molecular weights of PE increase dependently upon the *ortho*-substituent in the order of H < Me < ^{i}Pr at the expense of activity, which decreases in the same order. However, even though the large R^1 substituents virtually shut down the bimolecular *β*-hydrogen transfer path, a chain transfer with main group metal alkyls via transmetallation seems to be still available [42]. Upon activation with MAO or MAO/trimethylaluminum, a Zr–FI complex that has bulky 2-isopropyl phenyl groups as R^1 substituents afforded Al-terminated PEs (M_w = 10–720 kg/mol, M_w/M_n = 2.0–2.6), demonstrating that the chain transfer to alkylaluminums predominantly takes place [49]. The exclusive formation of the Al-terminated PE was confirmed by the ^{13}C NMR study of the PE obtained with deuterolytic quenching. The Al-terminated PEs are expected to serve as useful intermediates for functionalized polyolefins.

FI catalysts can also produce ultra-high-molecular-weight PE (UHMWPE). UHMWPE in general refers to PE that possesses the molecular weights higher than 1.5 million. UHMWPE possesses excellent tribological property, abrasion resistance, impact resistance, and chemical resistance, and has a number of material applications.

As described above, Zr–FI catalysts bearing bulky R^1 substituents can produce PEs with a high molecular weight of over 10^6 g/mol. When Zr–FI catalyst **2** was activated with $^iBu_3Al/Ph_3CB(C_6F_5)_4$, the molecular weight of the obtained polymer reached 5×10^6 g/mol under atmospheric ethylene at 50 °C [32]. The UHMWPE can be produced with supported FI catalysts in an industrially practical manner, which will be discussed in Sect. 1.3.1.5.

Molecular weight distribution is another important parameter particularly relevant to processability as well as properties (strength and modulus) of the polymers. Single-site catalysts like metallocene and post-metallocene catalysts usually afford the most probable molecular weight distributions of around $M_w/M_n = 2$, contrary to conventional heterogeneous Ziegler–Natta catalysts, which are multi-site catalysts consisting of complex mixtures of multiple components and produce polymers with broader molecular weight distributions (M_w/M_n 4–10).

As discussed in Sect. 1.2.2, FI catalysts potentially possess five isomers stemming from the coordination modes of ligands in the octahedral geometry, among which isomer (a), *cis*-N/*trans*-O/*cis*-X, is the most thermodynamically stable in most cases. However, ^{15}N NMR of Zr–FI catalyst **20** having ^{15}N enriched imine functions (–C=^{15}N–Ph) unexpectedly revealed that **20** existed almost exclusively as a C_1 symmetric isomer in solution (isomer (b)), and that the two chemically non-equivalent imine-N′s rapidly interchanged with each other above room temperature [23]. Interestingly, **20** activated with MAO produced PEs with temperature-dependent trimodal molecular weight distributions, which appeared to be formed with three chemically-distinctive active species (Fig. 1.5). Due to the fluxionality of FI catalysts, each isomer can generate multiple single-site active species, which may be the reason for this well-defined multimodality of **20**/MAO.

1.3.1.3 Monodisperse Polymers

An extreme end of the spectrum with regard to control of chain transfer and chain termination reactions is a living polymerization that has no chain transfer or chain termination reactions by definition. Living polymerizations are a tool to regulate the molecular weight and the molecular weight distribution of polymers, and sometimes the means to produce block copolymers, and chain-end functionalized polymers. Living olefin polymerizations via a coordination-insertion mechanism have been achieved in the past but with significant limitations in terms of polymerization conditions, polymerization activity, applicable monomers, and the achievable molecular weight of living polymers. In 2000, scientists at Mitsui Chemicals discovered unprecedented living olefin polymerizations with Ti–FI catalysts that exhibit robust livingness even at temperatures much higher than room temperature [50–54]. The living Ti–FI catalysts possess at least one *ortho*-fluorine in the aryl groups on the imine nitrogens (R^1 substituent). As shown in Table 1.3, the molecular weight distributions (M_w/M_n) are extremely narrow for complexes **21**, **22**, **23**, and **24** that have the *ortho*-Fs, whereas the other complexes **14**, **13**, and **25** without the *ortho*-Fs show ordinary M_w/M_n values of around 2.

One rationale for the robust livingness brought by the *ortho*-F was proposed based on DFT calculations on the cationic species bearing an *n*-propyl group as a model of a propagating alkyl group [43, 52, 55–57]. In the calculated structure of the propagating model, one of the *ortho*-F(s) was observed in proximity to the β-hydrogen atoms ($\sim$2.3 Å) of the growing chain (Van der Waals radii of H and F are 1.20 and 1.47 Å, respectively) and weak attractive and electrostatic interactions between the *ortho*-F and β-H were expected (the electrostatic energy $\sim$30 kJ/mol), which can stabilize the chain-transfer-prone β-agostic state of the cationic polymeryl-titanium and avert unwanted β-H transfer reactions. Such hypothetical C–F···H–C interactions were experimentally observed by Chan and coworkers for some phenoxy–pyridine Zr complexes (for example, **26**) bearing a cyclometallated aryl group by NMR spectroscopy and X-ray and neutron crystallography (Fig. 1.6) [58–62]. Furthermore, Mecking and co-workers revealed that a structurally related Ti enolatoimine complex bearing an *ortho*-F *N*-aryl group was an excellent ethylene living polymerization catalyst, whereas the

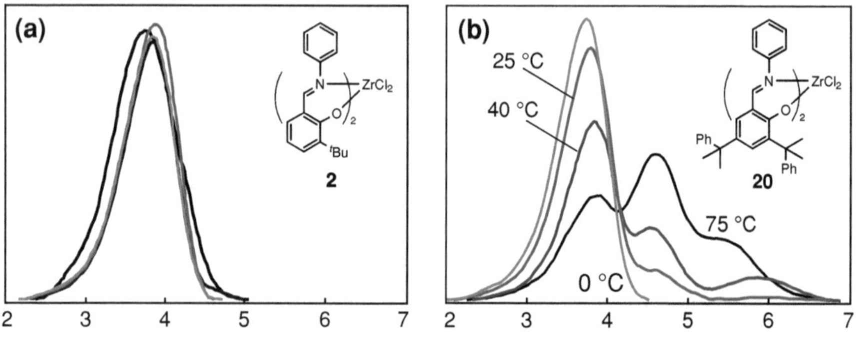

Fig. 1.5 GPC elution curves for the PEs formed by polymerization at 0–75 °C with FI catalysts **a 2** and **b 20** with MAO

corresponding complexes having *ortho*-methyl substituted or non-substituted *N*-aryl groups were not [63, 64].

Even though the *ortho*-F effects must be further clarified experimentally, the proposed concept may lead to the prospect of a new catalyst design strategy to manipulate olefin polymerizations together with conventional steric/repulsive interactions [51, 56, 57].

1.3.1.4 Selective Ethylene Trimerization

Oligomerization of ethylene to produce linear α-olefins (LAOs) is of great interest because LAOs are widely used as polyethylene comonomers, detergent alcohols, oil-field chemicals, and lubricant additives. The oligomerization catalysts employed in industry produce a range of LAOs following a Schulz–Flory distribution. In order

Table 1.3 Effects of the R^1 substituents of the phenoxy–imine Ti complexes on ethylene polymerization behavior

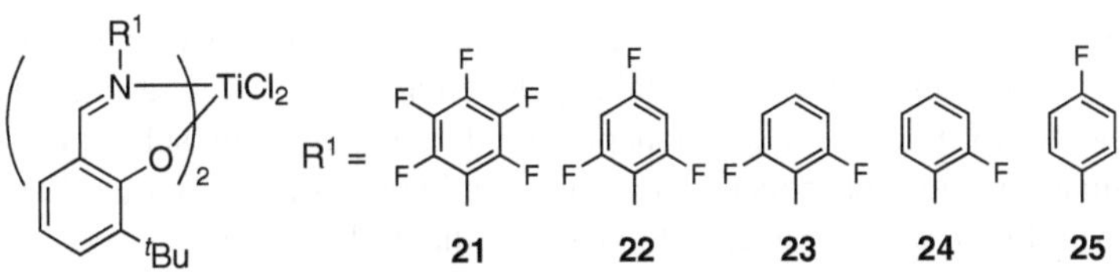

Entry	Complex	Activity [kg-polymer/(mmol-M h)]	M_n[a] (kg/mol)	M_w/M_n[a]
1	**21**	18.12	424	1.13
2	**22**	2.43	145	1.25
3	**23**	0.828	64	1.05
4	**24**	0.127	13	1.06
5	**14**	44.6	98	1.99
6	**13**	32.0	129	1.78
7	**25**	5.31	128	2.18

Polymerization conditions toluene 250 mL, MAO 1.25 mmol, complex 0.4–5.0 μmol, ethylene 0.1 MPa, 100 L/h, 1–5 min, 50 °C

[a] Determined by GPC (PE calibration)

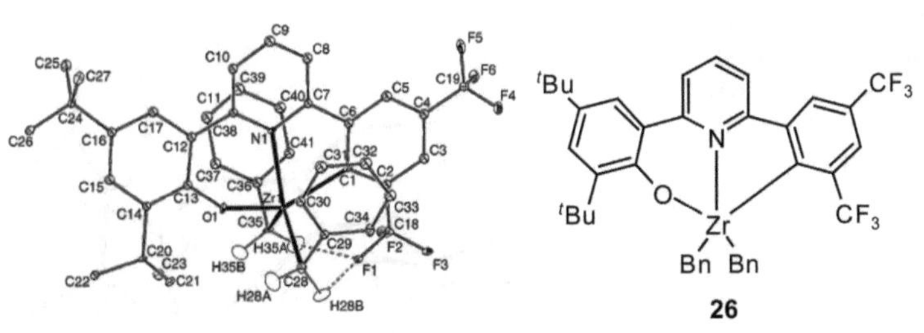

Fig. 1.6 Molecular structure of **26** from the neutron diffraction study (50% probability ellipsoids) showing selected hydrogen atoms. Reprinted with permission from [59]. Copyright 2006 Wiley–VCH Verlag GmbH & Co. KGaA

to meet the fluctuating market demands among LAOs and to reduce the fractions of unprofitable LAOs, investigation on the selective oligomerization catalysis to yield the desired LAOs has been intense, particularly on ethylene trimerization to produce 1-hexene. Most of the catalysts studied are Cr-based, although there are a few examples of Ti-, V-, or Ta-based catalysts.

Recently, scientists at Mitsui Chemicals developed tridentate phenoxy–imine ligated titanium complexes, which served, upon activation with MAO, as selective ethylene trimerization catalysts with extremely high activities [65]. The tridentate ligands possessed the 2′-alkoxybiphenyl-2-yl groups as the R^1 substituent (complexes **27**, **28**, and **29**) and the tridentate ligands wrapped around the Ti center in a facial fashion (Fig. 1.7). The oligomerization results are summarized in Table 1.4. Complex **29** exhibited exceptionally high activities and selectivity, in which branched decenes (cotrimerization products of 1-hexene and two molecules of ethylene) and a small amount of PE were the only side products. The activities were two orders of magnitude higher than commercial Cr-based catalysts (Phillips catalyst) under comparable conditions.

A metallacyclic propagation mechanism was proposed for the selective formation of 1-hexene as shown in Scheme 1.3, consisting of the oxidative addition of two ethylene molecules to form a metallacyclopentane species, insertion of another ethylene to form a metallacycloheptane intermediate, and subsequent β-hydride elimination (and subsequent reductive elimination or 3,7-H transfer). The pendant OMe donor was considered to stabilize the Ti(II) species as seen in similar cyclopentadienyl-arene titanium complexes. However, the reaction demonstrated second order dependence on ethylene pressure, indicating that formation of the metallacyclopentane intermediate can be the rate-determining step (RDS). This result contrasts with the cyclopentadienyl-arene titanium complexes which showed apparent first order dependence on [ethylene] and therefore ethylene insertion to the metallacyclopentane species is considered to be the RDS [66].

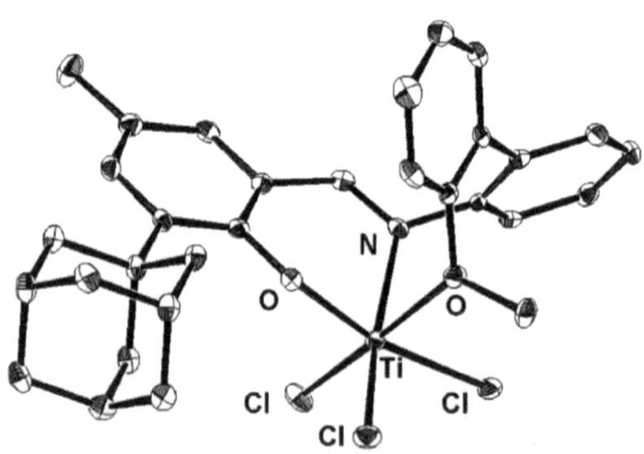

Fig. 1.7 Molecular structure of complex **29**. Reprinted with permission from [65]. Copyright 2010 American Chemical Society

Table 1.4 Catalytic ethylene conversion with the tridentate phenoxy–imine Ti complexes

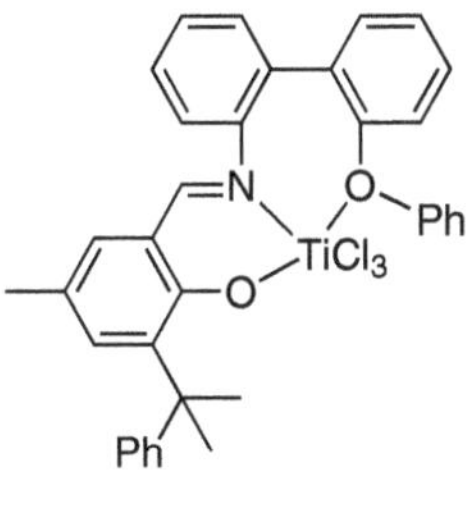

27

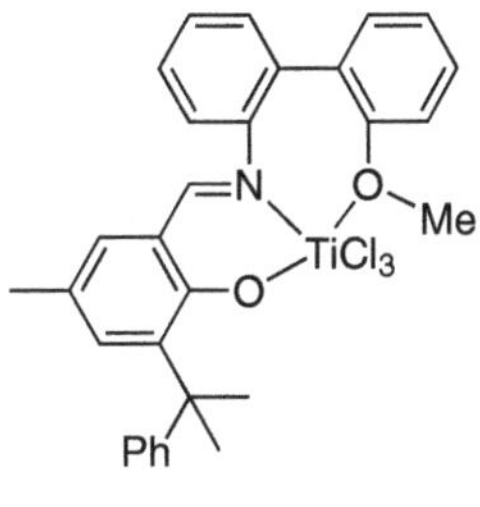

28

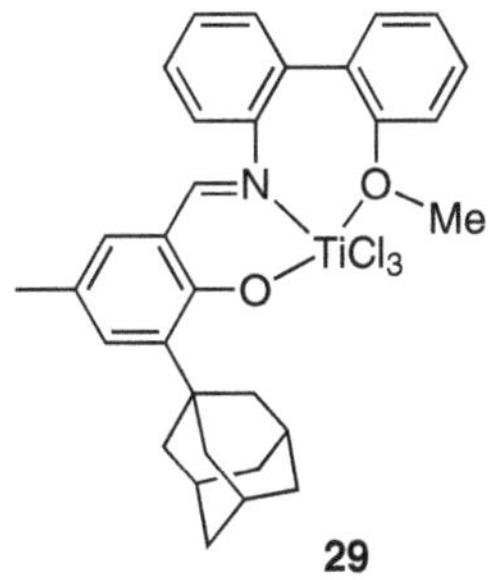

29

Entry	Complex	C_2H_4 pressure (MPa)	Activity [kg-hexene/(mmol-M h)]	Selectivity (wt%)		
				C_6	C_{10}	PE
1	**27**	0.8	0.19	76.6	0	23.4
2	**28**	0.8	5.7	86.2	12.2	1.6
3	**29**	0.8	7.4	91.4	6.4	2.1
4[a]	**29**	5.0	315	92.3	7.3	0.4

Conditions cyclohexane 30 mL, complex 0.5 μmol, MAO 5.0 mmol, 30 °C, 1 h
[a]Cyclohexane 150 mL

1.3.1.5 Ultra-Fine Non-Coherent Polyethylene Particles

$MgCl_2$ is widely used in industry as a support for Ziegler–Natta catalysts. It was demonstrated that the introduction of $MgCl_2$ increased the number of active sites and that unexpectedly and significantly it also enhanced the rate of olefin insertion reactions by about two orders of magnitude on average relative to classical Ziegler–Natta catalysts.

Inspired by the active roles that $MgCl_2$ plays in olefin polymerization, $MgCl_2$ was examined in the polymerization catalyzed by FI catalysts. As discussed in Sect. 1.2, FI catalysts have more ionic and polarized metal–ligand bonds compared to metallocene catalysts, and probably due to this feature, FI catalysts are more strongly adsorbed on Lewis acidic surfaces of inorganic compounds and generate an active supported catalytic system in combination with, for example, $MgCl_2$, hetero-poly compounds, and clays [17, 67–75].

Scheme 1.3 A plausible mechanism for the selective ethylene trimerization catalyzed by the tridentate Ti–FI catalysts

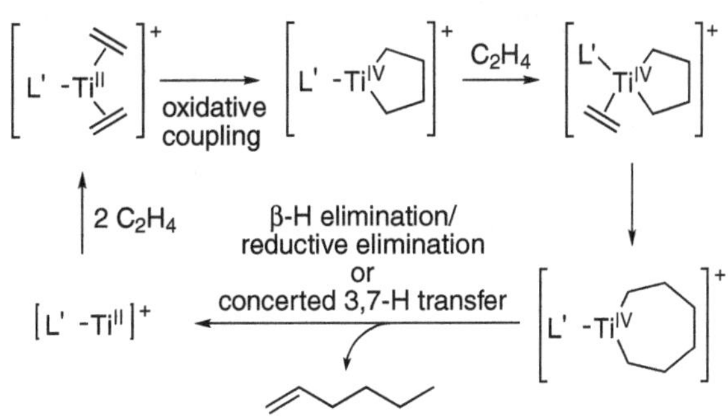

$MgCl_2$ catalyst supports can be prepared from adducts of $MgCl_2$ and alcohol (e.g., 2-ethyl-1-hexanol), which are dissolved in *n*-decane. Upon addition of R'_3Al, the $MgCl_2$ supports are precipitated out via dealcoholysis as a mixture of $MgCl_2/R'_nAl(OR)_{3-n}$. The ethylene polymerization activities of Ti–FI catalysts supported on the $MgCl_2/R'_nAl(OR)_{3-n}$ (R′: ^{i}Bu, Et) were very high and roughly comparable with those activated with MAO. With Zr–FI catalysts, ethylene polymerization activity reached 1820 kg-polymer/(mmol-M h) at 0.9 MPa ethylene pressure without using conventional MAO or boron-based activators, demonstrating that the $MgCl_2/R'_nAl(OR)_{3-n}$ serves as a highly efficient activator in a similar way to the classical Ziegler catalysts.

In addition, these FI catalysts supported on the $MgCl_2/R'_nAl(OR)_{3-n}$ keep the single-site polymerization characteristics and afford polymers with good morphology, demonstrating simultaneously the advantages of homogeneous and heterogeneous catalysts (high-performance heterogeneous single-site catalysts).

Under carefully controlled conditions, $MgCl_2/R'_nAl(OR)_{3-n}$ catalyst supports can be prepared into nearly perfectly spherical particles free from agglomerations having a wide range of controllable particle sizes (1–15 μm) and narrow size distributions. On the other hand, FI catalysts can produce UHMWPE by introducing bulky R^1 substituents as discussed in Sect. 1.3.1.2. By combining these two technologies, Zr–FI complexes bearing sterically demanding R^1 (= 2-methylcyclohexyl) with $MgCl_2/^iBu_nAl(OR)_{3-n}$ gave UHMWPE particulates (M_v 2–5 × 10^6 g/mol) at an extremely high activity of 750–1800 kg-polymer/(mmol-M h) under industrially applicable conditions (50 °C, 0.9 MPa ethylene pressure) [67–69]. Because the morphology of catalysts is replicated to that of the obtained polymers, the obtained UHMWPE particulates possess noncoherent and uniform spherical shapes. The obtained polymer shows a very high bulk density of up to 0.50 g/mL with variable diameters (Fig. 1.8).

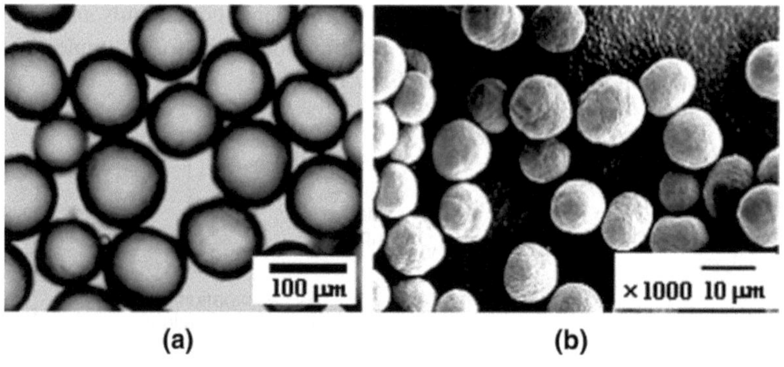

Fig. 1.8 **a** Photograph and **b** SEM images of the ultra-fine non-coherent PE particles formed with $MgCl_2$-supported Zr–FI catalyst. Average particle size: **a** 120 μm, **b** 10 μm. Reprinted with permission from [33]. Copyright 2009 American Chemical Society

These UHMWPE particulates are expected to have a wide array of applications such as sintered sheets and filters, light diffusion films, high-performance resin modifiers, and cosmetics.

1.3.1.6 End Functionalization of Polyethylenes

The addition of functionality to polyolefins that are otherwise non-polar can greatly enhance the range of attainable properties (e.g., adhesion, wettability, dyeability, printability, and compatibility). The functionalized polyolefins can also be used as versatile precursors when preparing hybrid polyolefins with non-polyolefinic materials possessing complementary functionalities. In the latter context, chain-end functionalized (telechelic) polymers serve as a building block for the hybrid polymers possessing well-defined architectures such as (multi-) block, graft and polymer networks. In this section, chain-end functionalized polyolefins prepared by FI-based catalysts are discussed.

We have already discussed in Sect. 1.3.1.2 that Zr–FI catalysts having relatively small R^1 groups afforded vinyl-terminated PE, which can be chemically modified according to various methods.

The chain-end vinyl groups can be efficiently converted into epoxy groups by treatment of the vinyl-terminated PE (M_w 2000 g/mol, M_w/M_n 2.4, vinyl functionality 95%, T_m 122 °C) with hydrogen peroxide (30% water solution) in toluene at 90 °C in the presence of Na_2WO_4 as an oxidation catalyst and methyl-tri-*n*-octylammonium hydrogen sulfate as a phase transfer catalyst [76]. In fact, epoxy-terminated PE (M_w 2000 g/mol, M_w/M_n 1.8, epoxy functionality 96%, T_m 121 °C) was obtained in almost quantitative yield. The epoxy-terminated PEs were highly reactive and could be transformed into a diol-terminated PE by the in situ hydrolysis using aqueous 2-propanol or a triol-terminated PE via a reaction with diethanolamine [44, 48, 77]. The obtained diol- or triol-terminated PEs can serve as precursors for well defined hybrid materials of PE and polyethylene glycol (PEG) having AB^2 and AB^3 type block structures (Scheme 1.4) [76, 77]. These hybrid materials formed stable nano-scale dispersion in water (up to 40 wt% of the hybrid materials) without adding any surface-active agents. The transmission electron microscopy (TEM) revealed that the AB^3 hybrid (PE, M_n 1100 g/mol; PEG, average M_n 400 × 3 g/mol, T_m 120 °C) formed a semi-transparent dispersion, consisting of nanoparticles (~18 nm on average) with a narrow size distribution. The nanoparticles appeared to have very lipophilic PE cores with hydrophilic shells of PEG because they can encapsulate large organic molecules such as 2,7,12,17-tetra-*tert*-butyl-5,10,15,20-tetraazaporphyrinato copper (II) (a water-insoluble dye) or 8-anilino-1-naphthalene sulfonic acid (a probe showing no fluorescence in a hydrophilic environment) as shown in Fig. 1.9.

As the second example of chemical modification of the vinyl chain-end PE, the Alder-ene reaction with maleic anhydride can give succinic anhydride-terminated PE. The reaction of the vinyl terminated PE (M_w 1400 g/mol, M_w/M_n 2.0, vinyl group 95%, T_m 116 °C) with maleic anhydride was carried out in the presence of

2,6-di-*tert*-butyl-4-methylphenol as a radical quencher at 195 °C for 16 h, which furnished the corresponding succinic anhydride-terminated PE (sa-t-PE: M_w 1700 g/mol, M_w/M_n 1.7, succinic anhydride functionality 102%, T_m 117 °C) (Fig. 1.10) [78].

The sa-t-PE exhibited higher thermal stability than its parent vinyl-terminated PE although the PE having epoxy-, diol-, triol-, and diamino-functionalities introduced above displayed no such improvement regarding thermal stability. The sa-t-PE also exhibited higher melt viscosity. The sa-t-PE possesses a high potential as a dispersant of hydrophilic materials (e.g., pigment) and as a modifier for polar polymers, in particular, engineering plastics.

Fig. 1.9 **a** 2,7,12,17-tetra-*tert*-butyl-5,10,15,20-tetraazaporphyrinato copper (II) (0.07 mmol) or **b** 8-anilino-1-naphthalene sulfonic acid (0.01 mmol) with water (10 mL) in the absence (A, C) or in the presence of AB[3] nanoparticles (0.2 wt%; B, D)

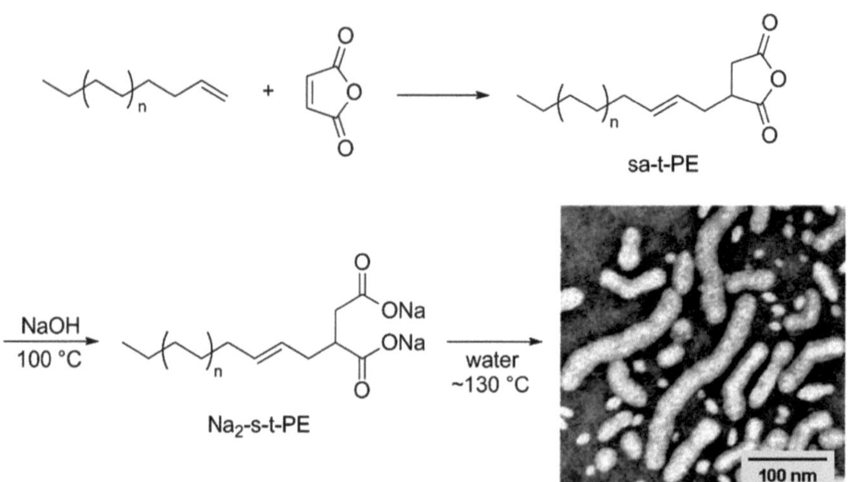

Fig. 1.10 Synthetic scheme and TEM image of a Na_2-s-t-PE water dispersion. Reprinted with permission from [78]. Copyright 2009 American Chemical Society

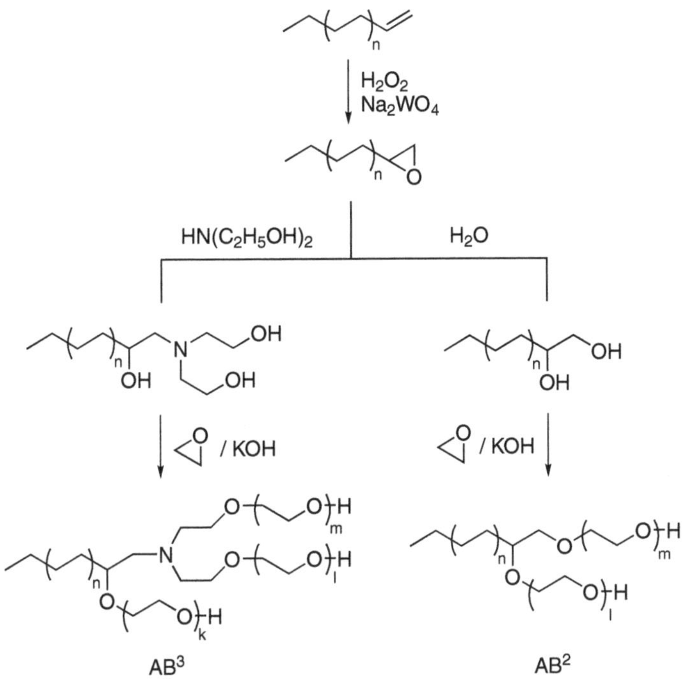

Scheme 1.4 Chain-end functionalization of vinyl-terminated PEs produced by FI catalysts

Interestingly, the Na salt of this succinic anhydride-terminated PE (Na_2-s-t-PE) are self-assembled into nano-sized particles (diameter 10–30 nm) and worm-like aggregates in water (diameter 20–30 nm, length 60–270 nm) (Fig. 1.10).

The end-functionalized PEs introduced herein possess unique features and are expected to be used in many applications (e.g., macromonomers, cationic surfactants, heat-resistant waxes and additives, film-surface modifiers, and anti-wear agents for inks and coatings), depending on the functional groups that were introduced.

A robust and highly active living Ti–FI catalyst **21** can be used for the synthesis of chain-end functionalized polyolefins, including telechelic polymers (Scheme 1.5) [79]. Activation of **21** with MAO furnishes monomethyl cationic species **30**, which reacts with ω-functionalized α-olefins (e.g., $H_2C{=}CH(CH_2)_9OSiMe_3$ (**31**)) to produce complex **32** in nearly quantitative yields, without forming multiple insertion products. The reason for this strict stoichiometry is that cationic species **30** exhibits high activity for the first monomer insertion, which is primary in its regiochemistry, resulting in species **32** bearing higher β-branched alkyls. The second monomer insertion to species **32**, which tends to be secondary, is extremely slow relative to the first insertion due to the steric repulsion. Species **32** can work as a highly active functionalized initiator for living polymerization of ordinary olefins like ethylene or propylene. Upon termination, living propagating species **33** can afford polyolefins **34** that have a functionality at the initiating chain

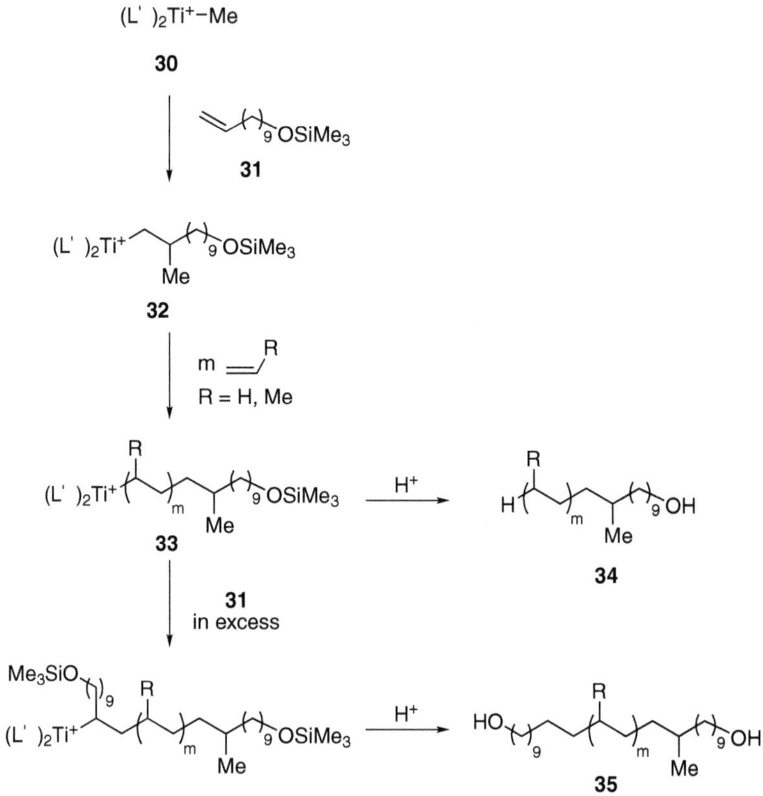

Scheme 1.5 Synthesis of mono- and difunctional telechelic polyolefins using a living Ti–FI catalyst

end. Chain-end capping at the terminating chain end can also be accomplished by treatment of an excess amount of **31** with **33**, resulting in the production of telechelic polyolefins **35** after quenching. Again the functionalization is stoichiometric and quantitative due to the slow consecutive insertions of **31**. This method is extendable to other functional groups if they have appropriate protecting groups. Heterotelechelic polyolefins having different functional groups at each chain end are also achievable simply using different monomers at initiating and terminating steps.

1.3.2 Propylene Polymerization

Propylene polymerization involves the issues of regio- and stereochemistry for monomer insertion, which is heavily dependent upon the catalysts employed. The regio- and stereoregularity has decisive effects on the physical properties of the polypropylene (PP) obtained and therefore, control of regio- and stereoselectivity of the catalysts is extremely important.

The "symmetry rule" established for metallocene-mediated propylene polymerization is well documented and understood, and it says that C_2 symmetric complexes that have homotopic reaction sites yield isotactic polypropylenes (iPPs), while complexes with C_s symmetry whose active sites are enantiotopic afford syndiotactic polypropylenes (sPPs) [5, 21]. In this scheme, each propylene monomer inserts in a primary fashion, and the face selection of the prochiral propylene monomer is believed to be made by the steric non-bonding interaction between the methyl group of the propylene monomer and the β-carbon of the metal-bound alkyl in the assumed four center metallacyclic transition state. Therefore, the methyl group and the β-carbon are always anti across the plane of the metallacycle, which is also assumed for the isotactic polymerization with conventional heterogeneous Ziegler–Natta catalysts.

Interestingly, FI catalysts can polymerize propylene in a highly isoselective and syndioselective manner by appropriate combinations of metal, ligands, and activators. These unique propylene polymerization characteristics have their base on the unique features of the FI catalysts described in Sect. 1.2 and in the following subsections we will discuss the propylene polymerization behavior with FI catalysts.

1.3.2.1 Syndiotactic Polypropylene

As described, FI catalyst precursors generally adopt a C_2-symmetric *cis*-N/*trans*-O/*cis*-Cl configuration as a predominant isomer. Additionally, DFT calculations suggest that catalytically active species favor a C_2-symmetric configuration with a *cis*-N/*trans*-O/*cis*-(polymer chain and coordinated olefin) disposition. If the symmetry rule is applied, FI catalysts are expected to work as catalysts capable of forming iPPs via a site-control mechanism. However, Fujita and Coates independently revealed that Ti–FI catalysts upon activation with MAO unexpectedly displayed moderate to extremely high syndioselectivity toward propylene polymerizations, which is mediated via a chain-end control mechanism [50, 80]. Significantly, propylene polymerizations with Ti–FI catalysts having at least one *ortho*-fluorine on the *N*-aryl R^1 groups exhibit robust living polymerization behavior in the same way as seen for ethylene polymerization with the same catalysts. These *ortho*-F Ti–FI complexes consistently exhibit higher syndioselectivity than the corresponding non-F complexes. Cavallo and coworkers proposed that the difference in syndioselectivity can be explained by "buried volume theory" based on the calculations, which can be applied to a wide range of complexes in a unified manner [29].

Subsequent research has further revealed that the size of the R^2 substituents of living Ti–FI catalysts **21**, **36**, **37**, **38**, and **39** has a direct effect on syndioselectivity (Table 1.5) [81–83]. Thus, the syndioselectivity is linearly correlated with the volume of R^2 substituents, meaning higher selectivity for larger R^2. To this end,

Table 1.5 Effects of the R^2 substituents of the phenoxy–imine Ti complexes on propylene polymerization behavior

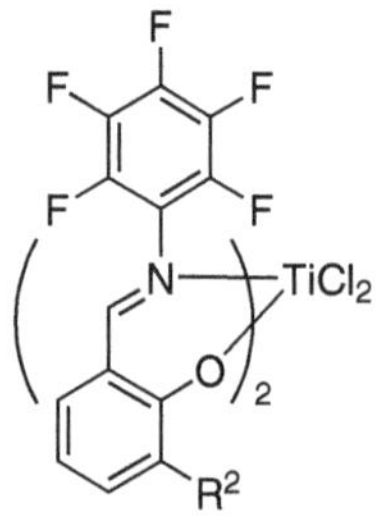

Entry	Complex	R^2	Activity [g-polymer/(mmol-M h)]	M_n[a] (kg/mol)	M_w/M_n[a]	T_m[b] (°C)	[*rr*] (%)
1	**36**	H	30.7	189.0	1.51	n. d.[c]	43
2	**37**	Me	68.8	260.2	1.22	n. d.[c]	50
3	**38**	^{i}Pr	31.1	153.7	1.16	n. d.[c]	75
4	**21**	^{t}Bu	3.7	28.5	1.11	137	87
5	**39**	TMS	5.9	47.0	1.08	152	93

Polymerization conditions toluene 250 mL, complex 10 μmol, MAO 2.5 mmol, propylene 0.1 MPa, 25 °C, 5 h
[a] Determined by GPC (PP calibration)
[b] Determined by DSC
[c] Not detected

Ti–FI catalyst **39**, which possesses a trimethylsilyl group as an R^2 substituent, was found to produce a highly syndiotactic PP (*rr* 93%, 25 °C polymerization) with a very high T_m of 152 °C, representing one of the highest T_m's for sPPs ever synthesized.

Since the syndioselectivity and the livingness are regulated independently by R^1 and R^2 substituents, living PPs with varied syndioselectivity are attainable.

Research on the chain-end structures of living and non-living polymers, sequence distributions of propylene polymers with a small amount of (^{13}C-labeled) ethylene units, and cyclopolymerization of 1,6-heptadiene, all established the peculiar regiochemistry involved in propylene polymerization mediated by Ti–FI catalysts activated with MAO [84–88]. This can be described as: (1) exclusive 1,2-insertion to Ti–H (assumed after the *β*-H transfer) and Ti–Me species; (2) preferential 1,2-insertion to Ti–CH_2CH_2–R; (3) regio-random insertion to Ti–$CH_2CH(R')$–R and (4) highly regulated 2,1-insertion to Ti–$CH(R')CH_2$–R. This selectivity is consistent with the observed regio-block structures obtained by this class of catalyst [83, 89].

Based on the theoretical calculations performed by Cavallo and coworkers, and in association with our experimental results, the observed chain-end controlled syndiospecificity could be explained by fluxional interconversion between Δ and Λ isomers of the octahedral Ti–FI complexes [29, 85, 90]. In the model shown in Scheme 1.6, *re*-chain/Λ isomer A is more stable than *re*-chain/Δ isomer D and isomer A favors propylene coordination at *si*-face, where the chirality of a polymer chain end (α-carbon) is transferred to the coordinating face of a propylene monomer (chain-end control) via migratory insertion. The resulting *si*-chain/Λ isomer B isomerizes into more stable *si*-chain/Δ isomer C, which favors propylene coordination at *re*-face, generating *re*-chain/Δ isomer D, which again isomerizes into more stable *re*-chain/Λ isomer A, coming back right where the whole series of events started. The repetitive cycle of isomerization/insertion certainly affords the syndiotactic sequences of PP. Note that the face selection in this model is made by steric repulsion between the methyl group of the propylene monomer and the R^2 substituents, which is in accord with the R^2-dependent syndioselectivity observed experimentally.

The free energy barriers ($\Delta G^{\ddagger}$) for the Λ/Δ isomerization of related cationic Zr or Hf benzyl complexes were experimentally estimated to be between 10 and 20 kcal/mol [30, 31]. The isomerization of Ti–FI catalysts can be further facilitated for the growing species of FI catalysts that have bulkier secondary alkyls as a growing polymer chain.

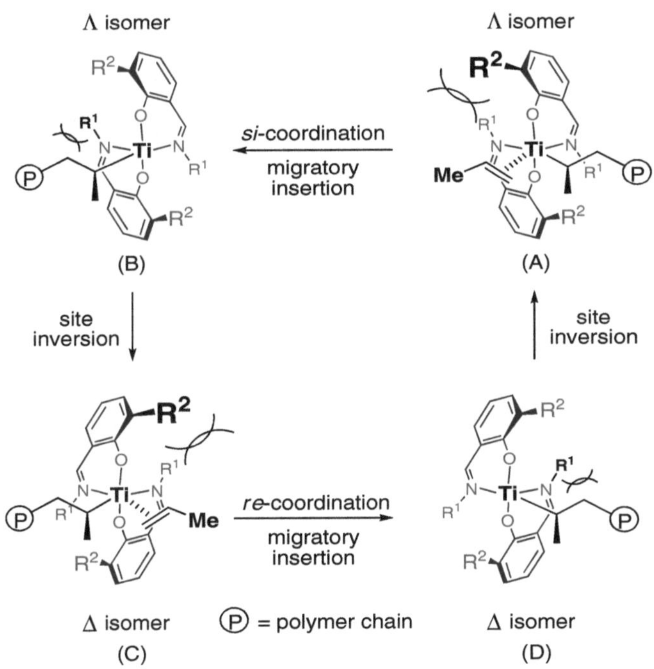

Scheme 1.6 Proposed mechanism for the syndioselective propylene polymerization promoted by Ti–FI catalysts

1.3.2.2 Atactic and Isotactic Polypropylene

When activated with $^{i}Bu_3Al/Ph_3CB(C_6F_5)_4$, complex **1** gave ultra-high-molecular-weight atactic PP (M_w 8,286,000 g/mol, no T_m) with a somewhat broad molecular weight distribution (M_w/M_n 4.15) [91]. On the other hand, corresponding Zr– and Hf–FI catalysts **2** and **40** with $^{i}Bu_3Al/Ph_3CB(C_6F_5)_4$ at 25 °C provide isotactic-rich PPs with high molecular weights (**2**: M_w 209 kg/mol, M_w/M_n 2.42, *mm* 46%, T_m 104 °C; **40**: M_w 412 kg/mol, M_w/M_n 2.15, *mm* 69%, T_m 124 °C). The subsequent optimization of the ligand substituents resulted in the discovery of Zr–FI and Hf complexes **9** and **41** that afforded highly isotactic PPs (**9** *mmmm* 97%, T_m 163 °C; **41** *mmmm* 97%, T_m 165 °C) that are comparable to those obtained with the best heterogeneous Ziegler–Natta catalyst, although atactic PPs were also formed as a byproduct (Table 1.6) [92]. The NMR study on the chain-end structures of the obtained polymer suggests that the propylene monomers were isoselectively enchained by repetitive 1,2-insertions via an enantiomorphic site-control mechanism. This is in sharp contrast to that of highly syndiotactic PPs with Ti–FI catalysts/MAO (active species: phenoxy–imine complexes, chain-end controlled polymerization with 2,1-insertion).

Coates and coworkers demonstrated that certain Ti–FI complexes performed moderately isoselective living propylene polymerization (*mmmm* 73%) [93, 94]. Furthermore, Mazzeo and coworkers reported that Ti–FI complexes bearing I, Br, and Cl, as R^2 (and R^3) substituents afforded prevailingly isotactic PPs via an enantiomorphic site-control mechanism in a non-living fashion (*mm* up to 73% at −20 °C) [89]. Similarly, it was reported that the complex possessing a CF_3 group as R^2 also provided iPP via a site-control mechanism [57].

Table 1.6 Propylene polymerization with FI catalysts using $^{i}Bu_3Al/Ph_3CB(C_6F_5)_4$ as a cocatalyst

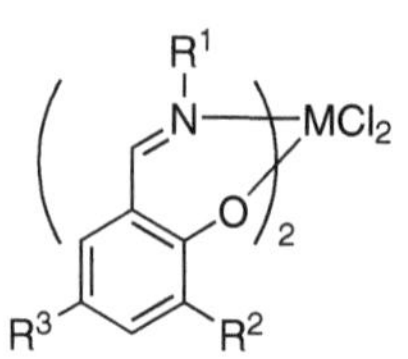

Entry	Complex	M	R^1	R^2	R^3	Activity [g-polymer/(mmol-M h)]	T_m (°C)	M_w[a] (kg/mol)	M_w/M_n[a]
1	**2**	Zr	Ph	^{t}Bu	H	94	103.5	209	2.42
2	**40**	Hf	Ph	^{t}Bu	H	6	123.8	412	2.15
3	**9**	Zr	Cy	1-adamantyl	Me	38	163.3	200	4.72[b]
4	**41**	Hf	Cy	1-adamantyl	Me	62	164.8	530	14.6[b]

Polymerization conditions toluene 250 mL, complex 5 μmol, $^{i}Bu_3Al$ 0.15 mmol, $Ph_3CB(C_6F_5)_4$ 0.01 mmol, propylene 0.1 MPa (100 L/h), 25 °C, 20 min

[a]Determined by GPC (PS calibration)

[b]Bimodal distribution

1.3.3 Higher α-Olefins Polymerization

As described, the combination of an FI complex with $^{i}Bu_3Al/Ph_3CB(C_6F_5)_4$ generates a phenoxy–amine ligated complex as a catalytically active species, which can produce ultra-high-molecular-weight PEs or PPs. As for higher α-olefins (e.g., 1-hexene, 1-octene, 1-decene, and 4-methyl-1-penene), the polymerization conducted by **1** with $^{i}Bu_3Al/Ph_3CB(C_6F_5)_4$ afforded high molecular weight poly(α-olefin)s with narrow polydispersity indexes as summarized in Table 1.7 [95, 96]. Interestingly, the polymerization activities became higher in the order of 4-methyl-1-pentene > 1-decene > 1-octene > 1-hexene in contrast to the completely opposite order for a typical *ansa*-metallocene, *rac*-(C_2H_4)(1-indenyl)$_2ZrCl_2$ (**42**).

The polymerization rate and the polymer molecular weights exhibited a zeroth order dependence on the concentration of 1-hexene and the obtained polymers were completely atactic with almost 50 mol% of regio-irregular units. DFT calculations suggested that the methyl cationic species adopted a trigonal bipyramidal geometry with two oxygen atoms trans to each other. However, upon coordination of 1-hexene, one of the N donors was dissociated from the Ti metal and the species adopted a square pyramidal geometry with the Ti-bound methyl at the apical site, indicating the extremely labile nature of the amine donor derived via activation with $^{i}Bu_3Al/Ph_3CB(C_6F_5)_4$. In this framework, the larger monomer and the resulting sterically encumbered polymer chains are considered to enhance the N donor dissociation, which makes the active species sterically more open and more electrophilic and this accelerates the polymerizations. The zeroth-order dependence on the concentration of the monomer might be attributed either to a stable olefin coordinated species or the N coordinating (olefin non-coordinating) species as a resting state, where the insertion of the coordinated olefin or the dissociation of the amine-N donor are the rate determining step, respectively.

Table 1.7 Higher α-olefin polymerization data for **1** and **42** with $^{i}Bu_3Al/Ph_3CB(C_6F_5)_4$

Entry	Complex	Monomer	TOF (min^{-1})	M_w[a] (kg/mol)	M_w/M_n[a]
1	**1**	1-hexene	257	846	1.65
2	**1**	1-octene	288	906	1.68
3	**1**	1-decene	308	850	1.75
4	**1**	4-methyl-1-pentene	595	1450	1.71
5	**42**	1-hexene	634	60	1.85
6	**42**	1-octene	345	49	1.59
7	**42**	1-decene	295	42	1.77
8	**42**	4-methyl-1-pentene	132	32	1.68

Polymerization conditions *n*-heptane 60 mL, monomer 0.211 mmol, complex 5 μmol, $^{i}Bu_3Al$ 0.15 mmol, $Ph_3CB(C_6F_5)_4$ 6 μmol, 25 °C, 20 min

[a] Determined by GPC (PS calibration)

1.3.4 Styrene Polymerization

Styrene polymerization using Ti–FI complexes **1**, **43**, **21**, and **39** revealed that stereospecificity of the obtained polymers was drastically changed by R^1 substituents (Table 1.8) [97]. At 20 °C, non-fluorinated complexes **1** and **43** (R^1 = Ph) afforded syndiotactic polystyrene (sPS), while isotactic polystyrene (iPS) was obtained from fluorinated complexes **21** and **39** (R^1 = C_6F_5). The obtained sPS and iPS showed high melting transition temperatures consistent with the reported values (T_m ~ 270 °C for sPS, ~ 220 °C for iPS) and had virtually no stereo errors, which precludes the possibility of elucidation of regiochemistry. According to the generally accepted mechanism, syndiospecific styrene polymerization was mediated by a Ti(III) species bearing a benzyl-type polymer chain derived from the 2,1-insertion of styrene [98]. In the polymerization mediated by FI catalysts, such species can be generated via the ligand transfer reaction from an initial cationic Ti(IV) species $[L'_2Ti^{IV}–Me]^+$ to Me_3Al included in MAO, resulting in mono(phenoxy–imine) Ti(III) species (See Sect. 1.2.4). Although the cationic Ti(IV) species derived from complexes **1** and **43** seemed to be inactive toward isospecific styrene polymerization, the fluorinated version of cationic species derived from **21** and **39** were active due to higher electrophilicity from the C_6F_5 groups, furnishing highly isotactic polystyrene. Given the C_2 symmetry of these FI catalysts, and by comparison with the syndioselective propylene polymerizations described above, it is pertinent to postulate that enantiomorphic site control and

Table 1.8 Effects of the substituents of the phenoxy–imine Ti complexes on styrene polymerization behavior

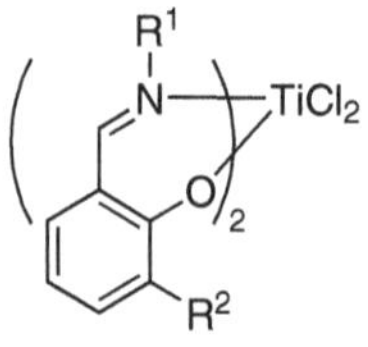

Entry	Complex	R^1	R^2	Activity [g-polymer/(mmol-M h)]	M_n [a] (kg/mol)	M_w/M_n [a]	T_m [b] (°C)	Tacticity [c]
1	**1**	Ph	^{t}Bu	0.56	215	7.5	275	syn
2 [d]	**1**	Ph	^{t}Bu	0.12	779	2.4	275	syn
3	**43**	Ph	TMS	0.64	132	9.0	279	syn
4	**21**	C_6F_5	^{t}Bu	5.21	10	4.4	224	iso
5	**39**	C_6F_5	TMS	10.15	8	2.3	213	iso
6 [d]	**39**	C_6F_5	TMS	1.37	10	2.4	217	iso

Polymerization conditions toluene 30 mL, styrene 100 mL, complex 0.2 mmol, dried MAO 50 mmol, 20 °C, 60 min. Results are based on 2-butanone-insoluble polymers

[a] Determined by GPC (PS calibration)

[b] Determined by DSC

[c] Determined by ^{13}C NMR (syn: syndiotactic, iso: isotactic)

[d] 0 °C, 360 min

slow or no Λ/Δ isomerization may result in iPS. The decomposition to Ti(III) appeared to be suppressed at least below 20 °C for the fluorinated complexes. Consistent with this scenario, raising polymerization temperature or premixing of the fluorinated complexes with MAO gave mixtures of iPS and sPS, presumably because the formation of the assumed Ti(III) species are facilitated under such conditions.

1.4 Copolymerization by FI Catalysts

1.4.1 Ethylene/α-Olefin Copolymers

Copolymerization of olefinic monomers results in a broad range of polyolefin products possessing a wide spectrum of physical and mechanical properties that are tunable according to the types of comonomer, the comonomer composition, and the sequence distribution of the comonomers. The property tuning is a delicate operation that requires production processes and catalyst performance to act in harmony, therefore, a lot of research has gone into developing highly active catalysts capable of producing copolymers with the desired molecular weights and the efficient incorporation of a wide variety of comonomers.

As mentioned in Sect. 1.2, FI catalysts are widely tunable in their steric and electronic properties, and hence systematic studies on the relationship between the polymerization characteristics and the diverse ligand structures have been conducted.

Ethylene/propylene copolymerizations were carried out using Ti–FI complexes **1**, **44–49** bearing a variety of R^2 substituents of various sizes, and the results are summarized in Table 1.9 [99]. Complexes **44–47** bearing secondary alkyl R^2 groups show almost constant propylene uptake (ca. 26 mol%), which was much higher than that for complexes **1** and **48** having tertiary alkyl R^2 groups (6.3 and 4.2 mol%, respectively). These copolymerization results imply that steric congestion near the α-carbon of the R^2 substituents determines propylene uptake rather than the entire volume of the R^2 substituents. On the other hand, the activities among the secondary alkyl R^2 became higher in proportion to the volume of the substituents, that is, cyclododecyl > cyclooctyl > cyclohexyl > ^{i}Pr, although complexes **1** and **48** (tertiary R^2) deviated from the relationship. This activity trend, at least among the secondary R^2 groups, is consistent with the results on Zr–FI catalysts discussed in Sect. 1.3.1.1, that is, the larger R^2 substituents furnish catalysts with higher activity. Interestingly, the trade-off between activity and comonomer uptake induced by the size of the R^2 substituents is not the case for complex **49** having a phenyl group as R^2, which demonstrates much higher activity and propylene uptake than any of the other complexes. The X-ray crystal structures of complexes **1**, **45**, and **49** (Fig. 1.11a–c) suggest that complex **49** is not really less crowded around the reaction sites than complex **45** (R^2 = cyclohexyl), or rather

Table 1.9 Ethylene/propylene copolymerization using complexes **1**, **44–49** with dried MAO

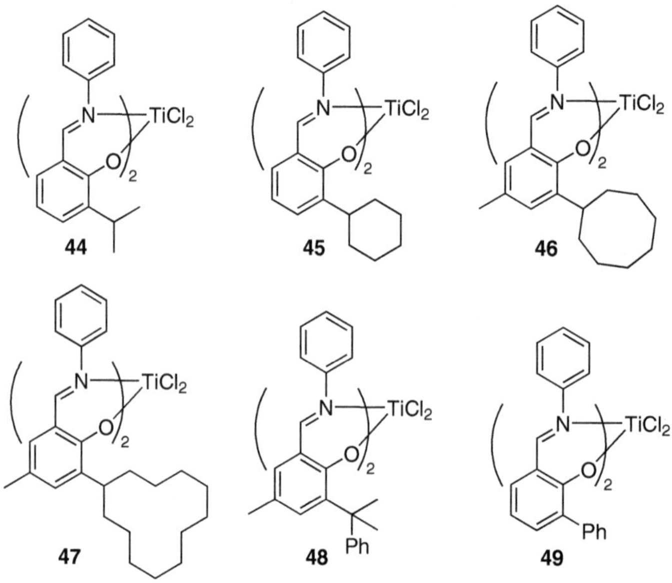

44 **45** **46** **47** **48** **49**

Entry	Complex	Yield (g)	Activity [kg-polymer/(mmol-M h)]	M_w[a] (kg/mol)	M_w/M_n[a]	Propylene content[b] (mol%)	TOF for ethylene (min^{-1})	TOF for propylene (min^{-1})
1	**44**	0.147	0.18	20.4	1.30	26.4	68	24
2	**45**	0.380	0.45	57.3	1.36	26.0	177	62
3	**46**	0.728	0.88	83.2	1.65	26.2	339	120
4	**47**	0.911	1.09	78.2	1.72	25.5	429	147
5	**1**	0.541	0.68	69.1	1.88	6.3	350	24
6	**48**	0.320	0.38	12.7	1.95	4.2	214	9
7	**49**	2.93	3.42	42.1	1.82	38.8	1071	679

Polymerization conditions toluene 250 mL, complex 5.0 μmol, dried MAO 1.25 mmol, ethylene feed 50 L/h, propylene feed 150 L/h, 50 °C, 10 min
[a]Determined by GPC (PE calibration)
[b]Determined by ^{13}C NMR

comparably crowded relative to complex **1** (R^2 = ^{t}Bu), judging from the distances between the chlorine and the nearest carbon in the ligands (**1**: 3.97 Å/4.00 Å, **45**: 4.53 Å/4.64 Å, **49**: 3.76 Å/3.91 Å).

In copolymerization of ethylene with higher α-olefins (Table 1.10), the comonomer uptake shows dependency on the size of the cycloalkyl R^2 substituents: cyclododecyl ≥ Ph > cyclooctyl > cyclohexyl ≫ ^{t}Bu for both 1-hexene and 1-decene. The activities also follow the order of Ph ≫ cyclododecyl > cyclooctyl > cyclohexyl ≫ ^{t}Bu. Again, complex **49** (R^2 = Ph) supports excellent comonomer uptake and copolymerization activities at the same time.

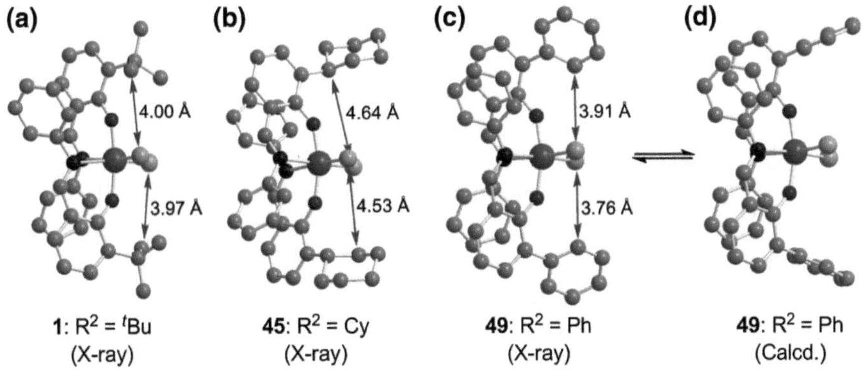

Fig. 1.11 Steric crowdedness of Ti–FI complexes measured by the distances from the Cl to the nearest carbon (**a–c** determined based on the X-ray structures, **d** a hypothetical structure that has the most open coordination sphere)

Table 1.10 Ethylene/1-hexene and ethylene/1-decene copolymerizations using complexes **1**, **45–47** and **49** with dried MAO

Entry	Complex	Comonomer	Activity [kg-polymer/(mmol-M h)]	M_w[a] (kg/mol)	M_w/M_n[a]	Comonomer content[b] (mol%)
1	**45**	1-hexene	0.85	48.3	1.16	5.6
2	**46**	1-hexene	1.05	56.1	1.18	8.4
3	**47**	1-hexene	1.34	42.8	1.27	12.6
4	**1**	1-hexene	0.71	41.7	1.29	1.5
5	**49**	1-hexene	8.42	125.7	1.92	12.0
6	**45**	1-decene	1.60	55.9	1.17	4.0
7	**46**	1-decene	2.41	65.3	1.26	7.2
8	**47**	1-decene	2.96	56.6	1.39	9.6
9	**1**	1-decene	0.79	55.3	1.32	1.0
10	**49**	1-decene	10.90	145.2	1.97	8.7

Polymerization conditions toluene 200 mL, complex 2.5 μmol, dried MAO 1.25 mmol, ethylene feed 100 L/h, 1-hexene or 1-decene 50 mL, 25 °C, 5 min
[a] Determined by GPC (PE calibration)
[b] Determined by ^{1}H NMR

The origin of this unusual behavior for complex **49** is not clear but DFT calculations on a cationic ethylene π-complex having a Ti-bound n-propyl group suggested that the phenyl groups can easily release the steric constraints by rotation and that the species is sterically the most open (Fig. 1.11d).

R^2 effect on comonomer uptake in the copolymerization of ethylene and propylene was also investigated for Zr–FI catalysts, **50–53** [100]. The monomer reactivity ratios in Table 1.11 again demonstrated that the ability of comonomer incorporation is basically attributable to the size of the R^2 substituents. The monomer reactivity ratios (r_1 and r_2, monomer 1 = ethylene, monomer 2 = propylene) represent relative reactivity of an active species to ethylene and propylene.

Table 1.11 Effects of the substituents of the phenoxy–imine Zr complexes on monomer reactivity ratios in ethylene/propylene copolymerizations

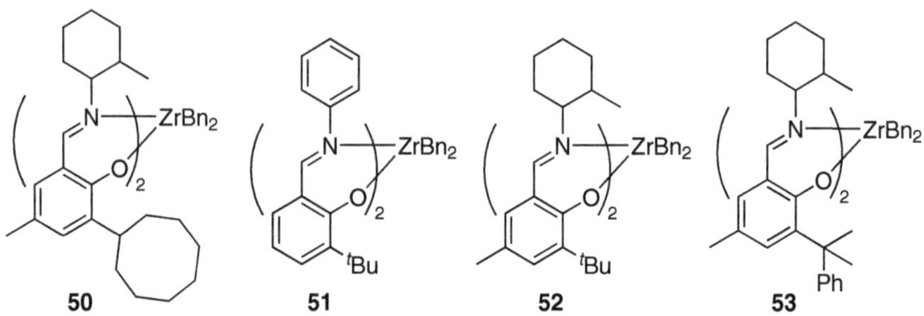

Entry	Complex	Propylene in feed[a] (mol%)	Propylene conversion (%)	Propylene content[b] (mol%)	r_1
1	**50**	32.25	8.6	2.39	16
2	**51**	58.84	7.6	2.59	45
3	**52**	58.84	7.7	1.79	69
4	**53**	58.84	0.42	0.88	152

Polymerization conditions hexane 5.0 mL, dried MAO Al/Zr = 300, 100 °C, 10 min, 7.0 atm gauge pressure maintained by continuous ethylene supply
[a]Calculated initial monomer composition
[b]Determined either by IR or ^{13}C NMR

For example, the r_1 values are defined as k_{11}/k_{12}, where k_{11} is the rate constant of ethylene insertion to the ethylene last inserted species, $L'_2Zr–CH_2CH_2–R$, and k_{12} is the rate constant of propylene insertion to the same species. The r_1 for complex **50** bearing cyclooctyl groups as R^2 is 16, meaning that ethylene is 16 times more reactive than propylene in this copolymerization at 100 °C (ethylene selectivity). The ethylene selectivity of the CMe_2Ph group as R^2 becomes extremely high at $r_1 = 152$, despite the difference of only one carbon in the substrates and a high polymerization temperature of 100 °C. The r_1 values for complexes **51** and **52** having $R^2 = {}^t$Bu fall in between **50** and **53**. DFT calculations on the transition states of ethylene or propylene insertion to the cationic $L'_2Zr–CH_2CH_2CH_3$ species suggested that one of the methyl groups of CMe_2Ph was forced to direct to the metal and immobilized in the conformation due to the higher rotational barrier of the CMe_2Ph group compared to that of the ^{t}Bu group. Therefore, the substrate selection of the Zr–FI catalysts seems to be made based on the size and shape of the reaction sites as if they were a molecular zeolite.

1.4.2 Ethylene/Cyclic Olefin Copolymers

Ethylene/cyclic olefin copolymers (COCs) represented by ethylene/norbornene (NBE) copolymers display, in general, high thermal stability (high T_g) and useful optical properties such as high transparency and low birefringence. Because COCs are heteroatom-free hydrocarbon polymers, they are lightweight and possess high stability to acids and alkalis, and also high moisture barrier properties. Accordingly, COC products are used in food and pharmaceutical packaging, medical appliances, plastic lenses, filters, and optical storage media. Because cyclic olefins used for COCs are mostly bicyclic olefins, they are sterically encumbered and more nucleophilic than ethylene or α-olefins (HOMO energy level, NBE: −7.25 eV, ethylene: −8.15 eV, propylene: −7.66 eV by DFT calculations) due to ring strain [101].

Copolymerization of ethylene and NBE was examined using Ti–FI catalysts as a comparison to typical metallocene catalysts (*rac*-(C_2H_4)(1-indenyl)$_2ZrCl_2$ (**42**) and Me_2C(9-fluorenyl)(Cp)$ZrCl_2$ (**56**)), which are well-known for achieving the high incorporation of NBE [101]. The results compiled in Table 1.12 demonstrate that Ti–FI catalysts can achieve strikingly high NBE incorporation relative to high-performance group 4 metallocene catalysts. While complex **54** (R^2 = Me) combined with MAO was a poor catalyst in terms of activity, complexes **49** and **55** (bearing the phenyl group at the R^2 position) with MAO activation formed copolymers of very high T_g's and NBE contents (**49**: T_g 120 °C, NBE content 45.7 mol%; **55**: T_g 126 °C, NBE content 46.1 mol%) and with high efficiency under the given conditions. These results indicate that the phenyl group as R^2 substituents are again advantageous when incorporating comonomers.

Microstructural analysis by ^{13}C NMR demonstrated that the obtained copolymers were nearly perfectly alternating and consecutive NBE insertions were strictly prohibitive (**49**: –NBE–E–NBE–E–NBE– 96.1%, –E–NBE–$(E)_n$– 3.9%; **55**: –NBE–E–NBE–E–NBE– 96.0%, –E–NBE–$(E)_n$– 4.0%, E stands for an ethylene unit). As implied by the extremely narrow molecular weight distributions, **49**/MAO mediates living ethylene/NBE copolymerization under the conditions examined. Living ethylene/cyclic olefin copolymerization was often observed for several other related catalysts. The living nature observed for these copolymerizations is considered to stem from several factors or a combination of them such as: (1) a certain level of living nature for ethylene homopolymerization; (2) nearly impossible β-H

Table 1.12 Results of ethylene/NBE copolymerization with the phenoxy–imine Ti and typical metallocene-based complexes

Entry	Complex	R^1	R^2	Activity [kg-polymer/(mmol-M h)]	NBE content[a] (mol%)	T_g[b] (°C)	M_w[c] (kg/mol)	M_w/M_n[c]
1	**54**	Ph	Me	0.65	48.7	120	186	1.09
2	**49**	Ph	Ph	3.27	45.7	120	573	1.21
3[d]	**55**	Pyrrolyl	Ph	6.98	46.1	126	215	1.91
4	**42**	–	–	1.75	27.7	49	476	1.74
5	**56**	–	–	1.40	28.0	49	259	1.58

Polymerization conditions toluene 250 mL, complex 1.0 μmol, charged norbornene 5 g, ethylene 0.1 MPa, dried MAO 1.25 mmol, 25 °C, 10 min

[a] Determined by ^{13}C NMR

[b] Measured by DSC

[c] Determined by GPC (PS calibration)

[d] Complex 0.5 μmol

transfer for the NBE-inserted chain end due to ring strain; (3) highly electrophilic NBE to form a stable NBE-coordinated π-complex as a possible dormant species.

1.4.3 Ethylene/Polar Functional Monomer Copolymers

Copolymerization of non-polar olefins with polar functional olefins is one of the long-standing challenges in the field of transition metal catalyzed olefin polymerization. Polyolefins having polar functional groups are believed to add useful material properties to unless otherwise non-polar olefinic materials and copolymerization provides a direct and straightforward way to afford such polymers unlike the energy-demanding, less controllable, post-polymerization radical grafting.

Although late-transition metal complexes have made a lot of progress in this field by taking advantage of their high tolerance to functional groups, examples for early transition metal catalysts have been limited to polymerization in the presence of a large excess of Lewis acids and/or using polar comonomers whose functionalities are masked by innocuous protecting groups to prevent poisoning. Nevertheless, activities are significantly depressed and the comonomer uptake is generally very low. A recent study has revealed that Ti–FI catalysts exhibited significant functional tolerance, much higher than that shown by group 4 metallocene catalysts.

Copolymerization of ethylene with 5-hexen-1-yl acetate (HA) was examined with Ti–FI catalysts activated with MAO [102]. Under the conditions at 4.0 mmol/L of HA, where metallocene compounds, Cp_2MCl_2 (M: Ti, Zr) and $Me_2Si(C_5Me_4)$ (^{t}Bu–N)$TiCl_2$ (CGC), were completely inactive, Ti–FI catalysts **1**, **49**, **57–61** demonstrated appreciable activities and comonomer uptake (Table 1.13).

The R^2 effects on comonomer uptake appear to hold true also for these copolymerizations, that is, the phenyl group as R^2 established higher activities and higher comonomer uptake than the R^2 = ^{t}Bu group as discussed in Sect. 1.4.2 [99]. Significantly, complexes **49** and **57** remained active even at [HA] = 21.0 mmol/L, which is higher than the concentration of MAO (20.0 mmol/L), indicating the remarkable tolerance of the Ti–FI complexes to the polar groups. In order to shed light on the observed tolerance, the calculations on the methyl cationic species coordinated by ethylene and by the carbonyl group of HA were carried out, showing that the energy differences between these two species (ΔE) for the Ti–FI complexes were much smaller (37–61 kJ/mol) than those of the metallocenes and CGC (>100 kJ/mol), indicating that the Ti–FI catalysts were significantly more tolerant to the functional group. These differences in ΔE appear to come from the electronic properties of these two classes of catalyst, probably stemming from the bonding properties between the metal and the ancillary spectator ligands (see Sect. 1.2.3).

As a way of verification that the electronic properties can affect the copolymerization behavior, R^1 substituents were varied in electronic nature. The results revealed that electron withdrawing groups significantly decreased the copolymerization activities (complexes **60** and **61**), which is opposite to the ethylene

Table 1.13 Results of ethylene/HA copolymerization with the Ti–FI complexes

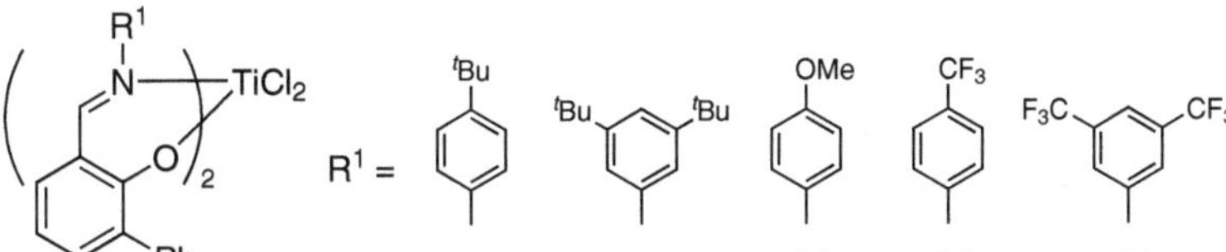

57 58 59 60 61

Entry	Complex	Comonomer (mmol)	Activity [g-polymer/ (mmol-M h)]	Comonomer content[a] (mol%)	M_w[b] (kg/ mol)	M_w/ M_n[b]
1	**1**	1.00	86	0.13	497	2.1
2	**49**	1.00	337	0.81	269	2.2
3	**57**	1.00	341	0.90	273	2.2
4	**58**	1.00	515	0.74	387	2.4
5	**59**	1.00	353	0.66	252	2.2
6	**60**	1.00	178	0.50	190	2.4
7	**61**	1.00	28	–[c]	–[c]	–[c]
8	**49**	2.00	68	1.97	59	1.8
9	**57**	2.00	71	1.90	69	1.7
19	**58**	2.00	61	1.82	67	2.2
11	**59**	2.00	56	1.98	55	1.8
12	**60**	2.00	17	–[c]	–[c]	–[c]
13	**49**	5.25	11	2.45	20	1.8
14	**57**	5.25	15	3.20	23	1.6

Polymerization conditions toluene 250 mL, complex 20 μmol, dried MAO 5.00 mmol, ethylene 0.1 MPa, 25 °C, 10 min

[a] Determined by ^{1}H NMR

[b] Determined by GPC (PS calibration)

[c] Not determined

homopolymerization, probably because of stronger coordination of the carbonyl group to the more electrophilic species. These results suggest that the nature of metals can be electronically tuned by ligands to tolerate the functional groups to some significant extent even with early metals.

1.4.4 Block Copolymers

One of the most promising applications that take advantage of the highly controlled living nature of Ti–FI catalysts possessing *ortho*-fluorinated aryl R^1 is the production of ethylene- and propylene-based block copolymers consisting of multiple segments bearing different properties (crystalline/amorphous, high T_g/low T_g, etc.). The first examples of the synthesis of olefin block copolymers were disclosed by scientists at Mitsui Chemicals in 2000 [50], in which diblock and triblock copolymers such as PE-*b*-sPP, PE-*b*-EPR, PE-*b*-EPR-*b*-sPP, PE-*b*-EPR^1-*b*-EPR^2, PE-*b*-EBR (EBR: poly(ethylene-*co*-but-1-ene)), sPP-*b*-EPR, sPP-*b*-EPR-*b*-PE, EPR^1-*b*-EPR^2-*b*-PE were prepared (EPR^1 and EPR^2 stand for EPR having different compositions) by sequentially adding different monomers or monomer mixtures of different compositions. After that, synthesis and characterization of various block copolymers using living FI catalysts were investigated by many groups [51–53, 93, 94, 103–113] and in some instances evaluation of the block copolymers on physical, thermal, morphological, mechanical, and optical properties were also carried out to assess their properties as novel materials.

Linear block copolymers possessing covalently linked multiple segments having different and contrasting properties (e.g., crystalline and amorphous) are known to show higher order structures through microphase separation [114]. Several groups have reported that block copolymers, including sPP-*b*-EPR [53, 107, 109], sPP-*b*-PE [109], and sPP-*b*-EPR-*b*-sPP [115], can have a variety of ordered or disordered (random) morphologies observed by transmission electron microscopic [53, 109, 110, 115] or small angle neutron scattering analysis [107]. A single molecule of PE-*b*-EPR was characterized by atomic force microscopy, in which the hard PE core surrounded by a soft shell of the EPR segment was clearly observed [106].

Scientists at Mitsui Chemicals successfully prepared a series of diblock copolymers consisting of PE and poly(ethylene-*co*-α-olefins) (α-olefins: 1-hexene, 1-octene, and 1-decene) with well defined structures using complex **62** (Chart 1.2) [108]. Well-controlled living polymerizations allowed for systematic investigations on thermal and mechanical properties of these diblock copolymers. Tensile tests on the diblock copolymer exhibited better extensibility and toughness in comparison to the corresponding random copolymer and the polymer blend.

Scientists at Dow Chemical Co. achieved production of olefin multi-block copolymers using two different catalysts capable of a catalytic chain growth on zinc [116–118]. In catalytic chain growth, a growing polymer chain on a transition metal catalyst is transferred to a main group metal (zinc in this case) via transmetallation, which is transferred back to another transition metal in a reversible manner. When this transmetallation is comparable to chain propagation in rate, all Zn–R species have an equal opportunity, on average, to enter the catalytic chain growth on the catalyst, resulting in a living-like polymerization. When there are two catalysts that are very different in ethylene selectivity in copolymerization, the segments having high and low ethylene contents are covalently connected through the multiple reversible chain transfer reactions, which afford multi-block copolymers. Hf complex **63**, which produces amorphous ethylene/1-octene copolymers (EOR),

Chart 1.2 Structures of complexes **62–64**

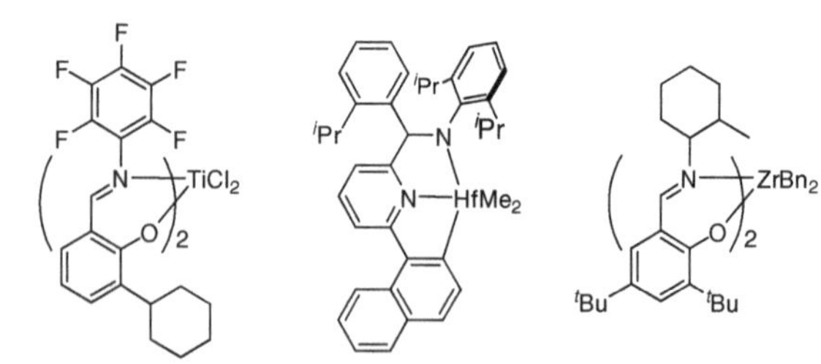

62 **63** **64**

and a Zr–FI catalyst **64** bearing bulky R^2 substituents, which showed very high ethylene selectivity even in the presence of a high concentration of 1-octene, were employed in the copolymerization of ethylene and 1-octene in the presence of diethyl zinc. Since both complexes are capable of the rapid reversible chain transfer with alkyl zinc, the polymerization afforded multi-block copolymers of the type, (PE-*b*-EOR)$_n$. These block copolymers display about 40 °C higher melting temperature (T_m ~ 120 °C) than random copolymers with similar densities, while maintaining excellent elastic properties at a higher temperature range, behaving as typical thermoplastic elastomers [116, 119].

1.5 Concluding Remarks

In order to respond to diverse market demands, polymerization catalysts need to be capable of synthesizing a wide variety of polymers, in other words, polymerization catalysts need to be highly active, very selective, and extremely versatile.

Among the ancillary ligands used for transition metal complexes, phenoxy–imine ligands relative to other ligand motifs possess a larger diversity of structure, which offers more opportunities and a higher degree of freedom for catalyst design. For example, the R^1 and R^2 substituents of FI catalysts are conveniently located at strategic positions, and thus provide us with efficient means to control the polymerization characteristics in a rational manner. Phenoxy–imine ligands also appear to possess an inherent electronic nature, leading to efficient olefin insertion reactions, whatever the combination of transition metals. These unique and key features of FI catalysts are thus responsible for the creation of a wide variety of new polyolefinic value-added materials. In addition to this, research and development of FI catalysts has enhanced our understanding of the causal relationship between catalyst structures, polymerization characteristics, primary and higher order polymer structures, and the chemical and physical properties of the resultant materials, building a bridge across catalysis science and material science.

References

1. Galli P, Vecellio G (2001) Prog Polym Sci 26:1287
2. Sinn H, Kaminsky W, Vollmer HJ, Woldt R (1980) Angew Chem Int Ed Engl 19:390
3. Brintzinger HH, Fischer D, Muelhaupt R, Waymouth RM (1995) Angew Chem Int Ed Engl 34:1143
4. Alt HG, Köppl A (2000) Chem Rev 100:1205
5. Resconi L, Cavallo L, Fait A, Piemontesi F (2000) Chem Rev 100:1253
6. Britovsek GJP, Gibson VC, Wass DF (1999) Angew Chem Int Ed 38:428
7. Ittel SD, Johnson LK, Brookhart M (2000) Chem Rev 100:1169
8. Hou Z, Wakatsuki Y (2002) Coord Chem Rev 231:1
9. Suzuki Y, Terao H, Fujita T (2003) Bull Chem Soc Jpn 76:1493

10. Gibson VC, Spitzmesser SK (2003) Chem Rev 103:283
11. Gambarotta S (2003) Coord Chem Rev 237:229
12. Makio H, Fujita T (2007) Polymerization of alkenes. In: Hiyama T (ed) Comprehensive organometallic chemistry III. Elsevier, Oxford
13. Saitoh J, Mitani M, Matsui S, Sugi M, Tohi Y, Tsutsui T, Fujita T, Nitabaru M, Makio H (1998) EP-0874005, April 26
14. Matsui S, Tohi Y, Mitani M, Saito J, Makio H, Tanaka H, Nitabaru M, Nakano T, Fujita T (1999) Chem Lett 1065
15. Matsui S, Mitani M, Saito J, Tohi Y, Makio H, Tanaka H, Fujita T (1999) Chem Lett 28:1263
16. Matsugi T, Fujita T (2008) Chem Soc Rev 37:1264
17. Mitani M, Saito J, Ishii S, Nakayama Y, Makio H, Matsukawa N, Matsui S, Mohri J, Furuyama R, Terao H, Bando H, Tanaka H, Fujita T (2004) Chem Rec 4:137
18. Strauch J, Warren TH, Erker G, Fröhlich R, Saarenketo P (2000) Inorg Chim Acta 300–302:810
19. Bochmann M (1996) J Chem Soc Dalton Trans 3:255
20. Kaminsky W (1998) J Chem Soc Dalton Trans 9:1413
21. Coates GW, Hustad PD, Reinartz S (2000) Chem Rev 100:1223
22. Mason AF, Tian J, Hustad PD, Lobkovsky EB, Coates GW (2002) Isr J Chem 42:301
23. Tohi Y, Makio H, Matsui S, Onda M, Fujita T (2003) Macromolecules 36:523
24. Makio H, Fujita T (2004) Macromol Symp 213:221
25. Tohi Y, Nakano T, Makio H, Matsui S, Fujita T, Yamaguchi T (2004) Macromol Chem Phys 205:1179
26. Cherian AE, Lobkovsky EB, Coates GW (2005) Macromolecules 38:6259
27. Pärssinen A, Luhtanen T, Klinga M, Pakkanen T, Leskelä M, Repo T (2007) Organometallics 26:3690
28. Johnson AL, Davidson MG, Lunn MD, Mahon MF (2006) Eur J Inorg Chem 15:3088
29. Milano G, Cavallo L, Guerra G (2002) J Am Chem Soc 124:13368
30. Axenov KV, Klinga M, Lehtonen O, Koskela HT, Leskelae M, Repo T (2007) Organometallics 26:1444
31. Bei X, Swenson DC, Fordan RF (1997) Organometallics 16:3282
32. Matsui S, Mitani M, Saito J, Tohi Y, Makio H, Matsukawa N, Takagi Y, Tsuru K, Nitabaru M, Nakano T, Tanaka H, Kashiwa N, Fujita T (2001) J Am Chem Soc 123:6847
33. Makio H, Fujita T (2009) Acc Chem Res 42:1532
34. Bryliakov KP, Kravtsov EA, Lewis BL, Talsi EP, Bochmann M (2007) Organometallics 26:288
35. Bryliakov KP, Kravtsov EA, Pennington DA, Lancaster SJ, Bochmann M, Brintzinger HH, Talsi EP (2005) Organometallics 24:5660
36. Makio H, Oshiki T, Takai K, Fujita T (2005) Chem Lett 34:1382
37. Kravtsov EA, Bryliakov KP, Semikolenova NV, Zakharov VA, Talsi EP (2007) Organometallics 26:4810
38. Saito J, Mitani M, Matsui S, Kashiwa N, Fujita T (2000) Macromol Rapid Commun 21:1333
39. Saito J, Mitani M, Matsui S, Tohi Y, Makio H, Nakano T, Tanaka H, Kashiwa N, Fujita T (2002) Macromol Chem Phys 203:59
40. Chen EY-X, Marks TJ (2000) Chem Rev 100:1391
41. Bochmann M, Lancaster SJ (1994) Angew Chem Int Ed Engl 33:1634
42. Matsui S, Mitani M, Saito J, Matsukawa N, Tanaka H, Nakano T, Fujita T (2000) Chem Lett 554
43. Makio H, Kashiwa N, Fujita T (2002) Adv Synth Catal 344:477
44. Ishii S, Mitani M, Saito J, Matsuura S, Kojoh S, Kashiwa N, Fujita T (2002) Chem Lett 740
45. Matsukawa N, Matsui S, Mitani M, Saito J, Tsuru K, Kashiwa N, Fujita T (2001) J Mol Catal A 169:99

46. Ishii S, Saito J, Mitani M, Mohri J, Matsukawa N, Tohi Y, Matsui S, Kashiwa N, Fujita T (2002) J Mol Catal A 179:11
47. Ishii S, Furuyama R, Matsukawa N, Saito J, Mitani M, Tanaka H, Fujita T (2003) Macromol Rapid Commun 24:452
48. Terao H, Ishii S, Saito J, Matsuura S, Mitani M, Nagai N, Tanaka H, Fujita T (2006) Macromolecules 39:8584
49. Saito J, Tohi Y, Matsukawa N, Mitani M, Fujita T (2005) Macromolecules 38:4955
50. Mitani M, Yoshida Y, Mohri J, Tsuru K, Ishii S, Kojoh S, Matsugi T, Saito J, Matsukawa N, Matsui S, Nakano T, Tanaka H, Kashiwa N, Fujita T (2000) WO Patent 200155231, January 26
51. Saito J, Mitani M, Mohri J, Yoshida Y, Matsui S, Ishii S, Kojoh S, Kashiwa N, Fujita T (2001) Angew Chem Int Ed 40:2918
52. Mitani M, Mohri J, Yoshida Y, Saito J, Ishii S, Tsuru K, Matsui S, Furuyama R, Nakano T, Tanaka H, Kojoh S, Matsugi T, Kashiwa N, Fujita T (2002) J Am Chem Soc 124:3327
53. Kojoh S, Matsugi T, Saito J, Mitani M, Fujita T, Kashiwa N (2001) Chem Lett 822
54. Saito J, Mitani M, Onda M, Mohri J, Ishii S, Yoshida Y, Furuyama R, Nakano T, Kashiwa N, Fujita T (2003) Stud Surf Sci Catal 145:515
55. Mitani M, Nakano T, Fujita T (2003) Chem Eur J 9:2396
56. Furuyama R, Saito J, Ishii S, Makio H, Mitani M, Tanaka H, Fujita T (2005) J Organomet Chem 690:4398
57. Sakuma A, Weiser M-S, Fujita T (2007) Polym J 39:193
58. Kui SCF, Zhu N, Chan MCW (2003) Angew Chem Int Ed 42:1628
59. Chan MCW, Kui SCF, Cole JM, McIntyre GJ, Matsui S, Zhu N, Tam K-H (2006) Chem Eur J 12:2607
60. Chan MCW (2007) Macromol Chem Phys 208:1845
61. Tam K-H, Lo JCY, Guo Z, Chan MCW (2007) J Organomet Chem 692:4750
62. Chan MCW (2008) Chem Asian J 3:18
63. Yu S-M, Mecking S (2008) PMSE Prepr 98:885
64. Yu S-M, Mecking S (2008) J Am Chem Soc 130:13204
65. Suzuki Y, Kinoshita S, Shibahara A, Ishii S, Kawamura K, Inoue Y, Fujita T (2010) Organometallics 29:2394
66. Deckers PJW, Hessen B, Teuben JH (2002) Organometallics 21:5122
67. Nakayama Y, Saito J, Bando H, Fujita T (2006) Chem Eur J 12:7546
68. Nakayama Y, Bando H, Sonobe Y, Fujita T (2004) Bull Chem Soc Jpn 77:617
69. Nakayama Y, Bando H, Sonobe Y, Fujita T (2004) J Mol Catal A 213:141
70. Nakayama Y, Bando H, Sonobe Y, Kaneko H, Kashiwa N, Fujita T (2003) J Catal 215:171
71. Nakayama Y, Bando H, Sonobe Y, Suzuki Y, Fujita T (2003) Chem Lett 32:766
72. Bando H, Nakayama Y, Sonobe Y, Fujita T (2003) Macromol Rapid Commun 24:732
73. Nakayama Y, Bando H, Kaneko H, Sonobe Y, Mitani M, Kojoh S, Kashiwa N, Fujita T (2003) Stud Surf Sci Catal 145:517
74. Nakayama Y, Saito J, Bando H, Fujita T (2005) Macromol Chem Phys 206:1847
75. Nandorf C, Matsui S, Saito J, Fujita T, Klapper M, Müllen K (2006) J Polym Sci A Polym Chem 44:3103
76. Matoishi K, Nakatsuka S, Nakai K, Isokawa M, Nagai N, Fujita T (2010) Chem Lett 39:1028
77. Matoishi K, Nakai K, Nagai N, Terao H, Fujita T (2011) Catal Today 164:2
78. Sainath AVS, Isokawa M, Suzuki M, Ishii S, Matsuura S, Nagai N, Fujita T (2009) Macromolecules 42:4356
79. Makio H, Fujita T (2007) Macromol Rapid Commun 28:698
80. Tian J, Coates GW (2000) Angew Chem Int Ed 39:3626
81. Saito J, Mitani M, Mohri J, Ishii S, Yoshida Y, Matsugi T, Kojoh S, Kashiwa N, Fujita T (2001) Chem Lett 576
82. Mitani M, Furuyama R, Mohri J, Saito J, Ishii S, Terao H, Kashiwa N, Fujita T (2002) J Am Chem Soc 124:7888

83. Mitani M, Furuyama R, Mohri J, Saito J, Ishii S, Terao H, Nakano T, Tanaka H, Fujita T (2003) J Am Chem Soc 125:4293
84. Saito J, Mitani M, Onda M, Mohri J, Ishii S, Yoshida Y, Nakano T, Tanaka H, Matsugi T, Kojoh S, Kashiwa N, Fujita T (2001) Macromol Rapid Commun 22:1072
85. Lamberti M, Milano G, Cavallo L, Guerra G, Pellecchia C (2002) PMSE Prepr 87:44
86. Lamberti M, Pappalardo D, Zambelli A, Pellecchia C (2002) Macromolecules 35:658
87. Hustad PD, Mason AF, Reinartz S, Coates GW (2002) PMSE Prepr 87:54
88. Hustad PD, Tian J, Coates GW (2002) J Am Chem Soc 124:3614
89. Mazzeo M, Strianese M, Lamberti M, Santoriello I, Pellecchia C (2006) Macromolecules 39:7812
90. Corradini P, Guerra G, Cavallo L (2004) Acc Chem Res 37:231
91. Saito J, Onda M, Matsui S, Mitani M, Furuyama R, Tanaka H, Fujita T (2002) Macromol Rapid Commun 23:1118
92. Prasad AV, Makio H, Saito J, Onda M, Fujita T (2004) Chem Lett 33:250
93. Mason AF, Coates GW (2004) J Am Chem Soc 126:16326
94. Edson JB, Wang Z, Kramer EJ, Coates GW (2008) J Am Chem Soc 130:4968
95. Saito J, Suzuki Y, Fujita T (2003) Chem Lett 32:236
96. Saito J, Suzuki Y, Makio H, Tanaka H, Onda M, Fujita T (2006) Macromolecules 39:4023
97. Michiue K, Onda M, Tanaka H, Makio H, Mitani M, Fujita T (2008) Macromolecules 41:6289
98. Schellenberg J, Tomotsu N (2002) Prog Polym Sci 27:1925
99. Terao H, Iwashita A, Matsukawa N, Ishii S, Mitani M, Tanaka H, Nakano T, Fujita T (2011) ACS Catal 1:254
100. Makio H, Ochiai T, Tanaka H, Fujita T (2010) Adv Synth Catal 352:1635
101. Terao H, Iwashita A, Ishii S, Tanaka H, Yoshida Y, Mitani M, Fujita T (2009) Macromolecules 42:4359
102. Terao H, Ishii S, Mitani M, Tanaka H, Fujita T (2008) J Am Chem Soc 130:17636
103. Tian J, Hustad PD, Coates GW (2001) J Am Chem Soc 123:5134
104. Coates GW (2002) PMSE Prepr 86:328
105. Fujita M, Coates GW (2002) Macromolecules 35:9640
106. Ono S, Matsugi T, Matsuoka O, Kojoh S, Fujita T, Kashiwa N, Yamamoto S (2003) Chem Lett 32:1182
107. Radulescu A, Mathers RT, Coates GW, Richter D, Fetters LJ (2004) Macromolecules 37:6962
108. Furuyama R, Mitani M, Mohri J, Mori R, Tanaka H, Fujita T (2005) Macromolecules 38:1546
109. Ruokolainen J, Mezzenga R, Fredrickson GH, Kramer EJ, Hustad PD, Coates GW (2005) Macromolecules 38:851
110. Yoon J, Mathers RT, Coates GW, Thomas EL (2006) Macromolecules 39:1913
111. Weiser M-S, Muelhaupt R (2006) Macromol Symp 236:111
112. Weiser M-S, Thomann Y, Heinz L-C, Pasch H, Muelhaupt R (2006) Polymer 47:4505
113. Weiser M-S, Muelhaupt R (2006) Macromol Rapid Commun 27:1009
114. Park C, Yoon J, Thomas EL (2003) Polymer 44:6725
115. Hotta A, Cochran E, Ruokolainen J, Khanna V, Fredrickson GH, Kramer EJ, Shin Y-W, Shimizu F, Cherian AE, Hustad PD, Rose JM, Coates GW (2006) Proc Natl Acad Sci USA 103:15327
116. Arriola DJ, Carnahan EM, Hustad PD, Kuhlman RL, Wenzel TT (2006) Science 312:714
117. Arriola DJ, Carnahan EM, Hustad PD, Kuhlman RL, Wenzel TT (2006) Polym Prepr 47:513
118. Hustad PD, Kuhlman RL, Carnahan EM, Wenzel TT, Arriola DJ (2008) Macromolecules 41:4081
119. Wang HP, Khariwala DU, Cheung W, Chum SP, Hiltner A, Baer E (2007) Macromolecules 40:2852

Chapter 2
Development of Imine Derivative Ligands for the Exocyclic Activation of Late Transition Metal Polymerization Catalysts

Brian C. Peoples and René S. Rojas

Abstract Transition metal complexes bearing imine and imine derivative ligands represent a growing number of polymerization catalysts in development. The ease of synthesis and large number of structural variations which are readily accessible make these systems of great interest both academically and industrially. One subset of imine-based complexes are those which bear exocyclic functionality which can interact with Lewis acids. These systems are particularly interesting as the activation of the complex occurs remotely, away from the active center, and that the activation can proceed using stoichiometric concentrations of activators. In addition, the presence of the exocyclic functionality may present an effective method to heterogenize polymerization catalysts. In this chapter, the development of such systems and in particular α-iminocarboxamide nickel catalysts and derivative species are discussed.

Abbreviations

Acac	Acetylacetone
Ar	Aryl
ArF	Aryl group containing fluoro-atoms
B	Lewis base
BCF	Tris(pentafluorophenyl)borane

B. C. Peoples
Departamento de Ingeniería Química, Centro para la Investigación Interdisciplinaria Avanzada en Ciencias de los Materiales (CIMAT), Facultad de Ciencias Físicas y Matemáticas, Universidad de Chile, Santiago, Chile
e-mail: bcpeoples@gmail.com

R. S. Rojas (✉)
Departamento de Química Inorgánica, Facultad de Química, Pontificia Universidad Católica de Chile, Casilla 306, Santiago, Chile
e-mail: rrojasg@uc.cl

G. Giambastiani and J. Cámpora (eds.), *Olefin Upgrading Catalysis by Nitrogen-based Metal Complexes II*, Catalysis by Metal Complexes, 36,
DOI: 10.1007/978-94-007-0696-5_2,

bipy	2,2′-Bipyridyl
Bz	Benzyl
C_n	Complex
COD	1,5-Cyclooctadiene
Dipp	Diisopropylphenyl
dme	Dimetoxiethane (ethylene glycol dimethylether)
en	1,2-Diaminoethane
Et	Ethyl
L_n	Ligand
L_s	Ligand salt
Lut	Lutidine (2,6-dimethyl pyridine)
M	Metal
MAO	Methylaluminoxane
Me	Methyl
NacNacH	1,3-Diketimines
P_x	Metal precursor
Ph	Phenyl
Py	Pyridine
r.t.	Room temperature
THF	Tetrahydrofuran
Ts-CN	Para-toluenesulfonylcyanide
tmeda	N,N,N′,N′-tetramethylethylenediamine
X	Halogen or other anionic group
Y	Alkyl or functional groups
Z∩Z′	Bidentate ligand (where Z is the coordinating atom)

2.1 Introduction

Olefins can be upgraded to higher value products through a number of industrial processes, two of the most prevalent are the oligomerization of olefins to higher alkanes and the production of polyolefins via transition metal catalysis. Polyolefin production has grown in step with the development of the global economy and is expected to continue with time [1]. These materials are desirable for many reasons, including their low cost and the ease of which they can be shaped into many forms and sizes. Since the development of the first Ziegler-Natta (1953–1954) and Phillips catalysts (late 1940s) [2] there has been a significant drive to increase catalytic activity, increase the control of polymer structure and comonomer content, and at the same time reduce the cost of the polymerization system [3, 4]. The development of the metallocene catalyst systems has allowed for increased control of polymer structure and molecular weight distribution, resulting in highly stereospecific systems and systems which can produce polymers with novel properties [5–7]. Many of these systems however, require the use of high concentrations of cocatalysts, which has generated considerable interest in the development of catalysts which could produce well defined materials either without the use of cocatalysts, or with only stoichiometric quantities thereof.

It was realized, that such systems could be developed using late transition metals in particular the group 8–10 metals and in particular nickel and iron [8–12]. The relatively low sensitivity of these metals to water and oxygen made the metals less dependent on the use of cocatalysts [13]. This resulted in the development of several generations of late-transition metal catalysts. Initial work with these systems was focused on the modification of the nickel-based shell higher olefins process (SHOP) oligomerization catalysts, it was discovered by Keim et al. [10, 14–15] that the catalysts could be modified via the introduction of steric bulk to the ligand, to produce higher molecular weight materials (Fig. 2.1) [16]. Subsequent variations to the ligand structure resulted in higher molecular weight materials, and eventually very good stereo control of the polymers [8, 10]. Perhaps the most recognized and successful of these variations are the Brookhart diimine catalysts (Fig. 2.1) [17, 18]. These systems produce high molecular weight materials with high activities by controlling the chain termination and transfer rates using the pseudo-axial bulk of the ligand [19, 20]. This understanding of the function of pseudo-axial bulk led to the development of other catalyst systems such as the Grubbs salicylaldimine (Fig. 2.1) [21]. The majority of these systems still require the use of a cocatalyst such as methylaluminoxane (MAO) in order to become active. Therefore, one of the main drives in the design and synthesis of late transition metal catalysts is the development of systems which are either single component, or which can be activated with stoichiometric amounts of activator.

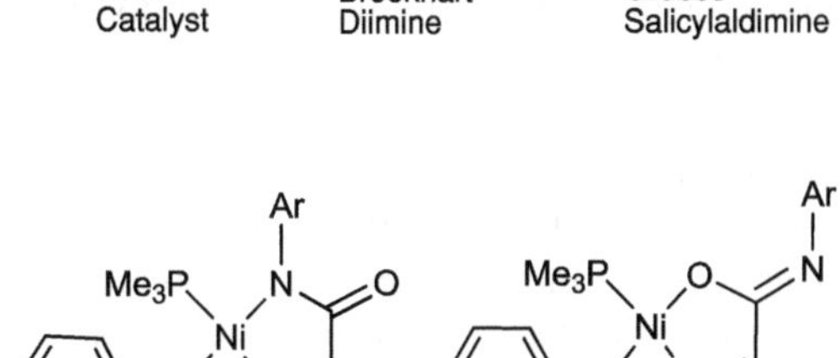

Fig. 2.1 Examples of late transition metal polymerization catalyst structures

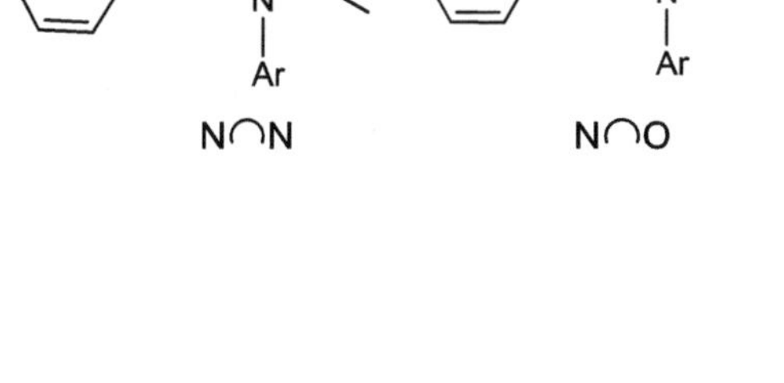

Fig. 2.2 Structure of α-iminocarboxamide catalysts

Nickel α-iminocarboxamide catalysts (Fig. 2.2), have been demonstrated to be active without the use of aluminum-based cocatalysts, and even as single component catalysts [22, 23]. While at first glance the α-iminocarboxamide-based catalysts appear similar to many of the other families of late transmission metal polymerization catalysts, they have many subtle, but critical differences. Similar in structure to the Brookhart diimine catalysts they share some of the basic chemistry, both in the synthesis of the basic ligand set and as well as in the polymerization chemistry. Closer inspection of the ligand structure reveals however, that the systems have properties which are unique. Some of the more notable differences are the anionic nature of the ligand, the presence of the carboxamide group and the presence of a stabilizing base such as PMe_3. Each of these differences imparts to the catalyst certain characteristics which are desirable for either single component activation or stoichiometric activation.

Two of the three substantive differences are linked together. As the metal center of the catalyst is almost always four coordinate, the anionic nature of the ligand structure requires that there be a neutral donor ligand associated with the metal center. This maintains the metal center in the preferred 2+ state, while at the same time allowing the open site to be generated fairly readily. This neutral donor ligand has proven useful, as when removed in the presence of monomer, it has been shown that the catalysts are active for polymerization. When the donor ligand is selected carefully it is even possible to displace the donor ligand using nothing more than increased monomer concentration/pressure [23, 24].

The final difference is the structure and presence of the carboxamide group in the ligand backbone. In α-iminocarboxamide systems the presence of this group allows for the ligand to interact, through the carbonyl, with Lewis acids present in the reaction medium forming a Zwitterion [22, 25]. This withdraws electron density from the metal center resulting in increased polymerization activity, and also provides a method for the catalyst to be supported on solid Lewis acids [26]. Finally, in α-iminocarboxamide systems, the presence of the carboxamide groups allows for the coordination of the ligand to the metal center in two distinct ways, either in an N∩O or a N∩N configuration and is controlled primarily by the steric bulk of the ligands [27].

The α-iminocarboxamide-based ligands have diverse chemical and structural variations which impart to the final polymerization catalysts characteristics which make them particularly well suited to the polymerization of olefins without cocatalysts. Their ability to generate active species via exocyclic activation with Lewis acids, has led to developments in the design and synthesis of related species, in particular novel six-membered ring systems, which have some degree of conjugation as well as an exocyclic functional group through which the electron density at the metal center can be modulated.

In this chapter, several topics related to imine derivative catalysts, including; the synthesis of imine derivative ligands particularly those bearing exocyclic functionality, their metallation, the Lewis acid adduct formation and isolation, and the polymerization behavior of the complexes activated by Lewis acids through the exocyclic functionality, will be discussed.

2.2 Ligand Synthesis

Imines as a class, are important compounds in catalyst ligand design because of the reactivity associated with the free electron pair present in the sp^2 hybridized nitrogen. In addition, the facile synthesis of these compounds allows for the generation of a wide variety of steric and electronic environments (Fig. 2.3). This has led to their widespread application in laboratory and industrial synthetic processes particularly as part of ligand frameworks [8]. In addition, imine ligands are important substrates for the production of chiral amines with high enantioselectivity [28] and the epoxidation of olefins [29]. Traditionally, imines were synthesized from the reaction of ketones or aldehydes with amines in the presence of an acid catalyst (Fig. 2.3). This reaction however, becomes less efficient as the size of the substituents increase, requiring long reaction times (>24 h) at elevated temperatures (>120 °C) and higher concentrations of acid catalysts. Other strategies for the production of imine ligands include the use of $TiCl_4$ [30] and through oxidation of amines by hypervalent iodine [31]. Several methods have been developed to work around the use of these conditions. The use of transition metal complexes (i.e. $TiCl_4$) to form a complex with the ketone/aldehyde to make the group more accessible to nucleophilic attack from the amine has been demonstrated to be an effective method to form imines with large steric impediments [32–35]. A solvent free method has been developed using microwave synthesis which allows for the relatively facile generation of imines derived from amines and ketones [36].

2.2.1 Grafting Functionality onto the Ligand Backbone

The majority of the ligand structures discussed in this chapter, when incorporated into transition metal complexes, are active for the polymerization/oligomerization of olefins. One of the main goals however, in this and other groups, is the remote activation of the precatalyst in the catalytic system. This requires the introduction of functional groups into the ligand backbone which can interact with the activator. In much of the reported literature these activators have been Lewis acids, although other groups could conceivably used. The groups introduced therefore have been Lewis bases, in particular carbonyls (as both ketones and aldehydes),

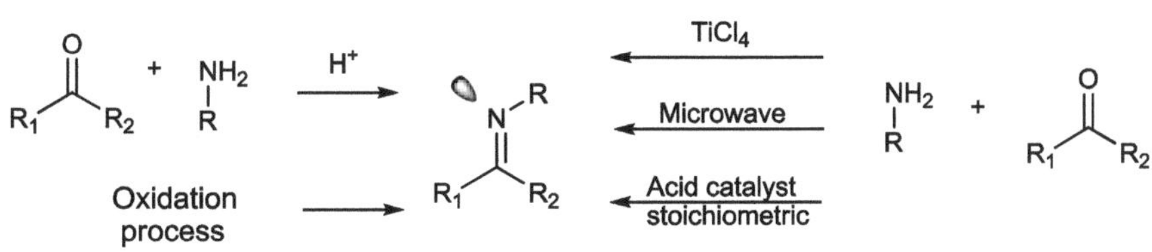

Fig. 2.3 Synthetic methods for the production of imine ligands

nitrile groups, olefins, amines and even nitrous substituents [37]. Some of these groups can be incorporated by selecting the appropriate starting materials, whereas some of them would interfere with the condensation reactions, and as such must be introduced after the ligand backbone has been formed. The introduction of the functional groups into six-membered ring systems is particularly attractive as the protons on the α-carbon (relative to the imine/ketone) can easily be removed (for example by NaH, BuLi, etc.), generating metallated ligands which can be functionalized easily using a variety of reagents.

A ketone can be inserted onto β-ketoimine ligand frameworks at the α-position, via the use of copper acetate in the presence of O_2 at 50 °C (Fig. 2.4) [38, 39]. One of the newest, and most interesting ligands bearing functionalities are those that contain a nitrile group. The nitrile functionalized NacNac–H can be produced via the deprotonation of the NacNac–H ligand with n-butyllithium (Fig. 2.4), followed by the reaction of the resulting anion with the reagent para-toluenesulfonylcyanide (Ts-CN). This selectively introduced the nitrile functional group at the α-position of the ligand framework in good yield [40].

Another facile method to include functionality into the final ligand is through the selection of starting reagents which include either excess functionality, or unreactive functionality. One example of this are ligand systems derived from triacetylmethane. This reagent, which bears three equivalent ketones proved a convenient scaffold onto which 1, 2 or even 3 anilines can be condensed (Fig. 2.5). Regardless of how many condensations reactions take place on the scaffold, only two of the groups can coordinate to a metal center, leaving a third free to interact with other species [41]. Additionally, after deprotonation, the ligand becomes conjugated, allowing for efficient transfer of electron density from the metal center to the exocyclic group.

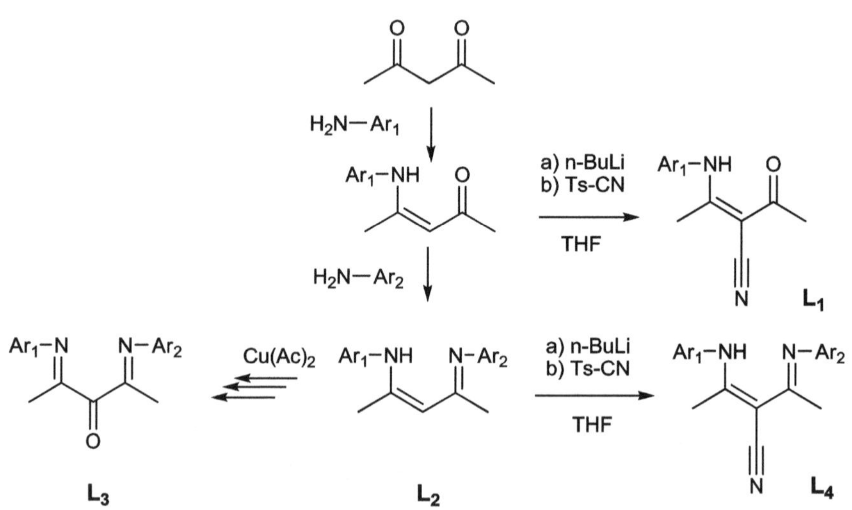

Fig. 2.4 Methods to incorporate functionality into the ligand backbone

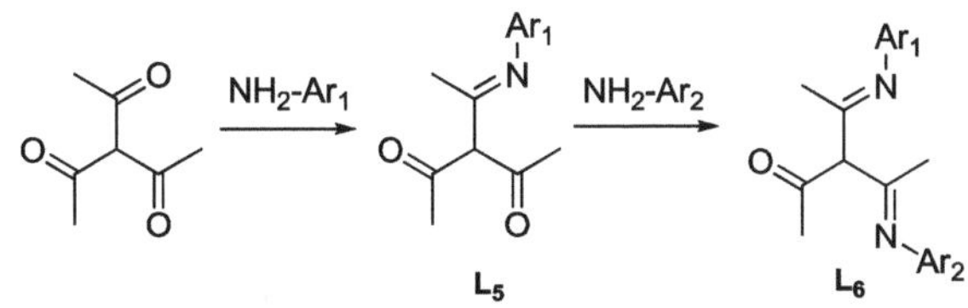

Fig. 2.5 Triacetylmethane as a scaffold for ligand development

Exocyclic activation sites can be introduced in places other than the ligand backbone. The aryl imine substituents themselves can serve as convenient locations to introduce functionality. For instance the use of aminopyridine or para-methoxyaniline can both be used in this manner, allowing for the introduction of functionality while maintaining the ligand core geometry and chemistry [42, 43].

2.2.2 FTIR Characterization of Imine Ligands

While FTIR is not necessarily one of the most commonly used techniques to characterize imine ligands, it can provide information about the structural characteristics of the ligand. The imine functionality has a characteristic band which allows for facile characterization and a limited evaluation of the electronic environment of the group. In many systems there are two nitrogens present in the ligand, the first being an amine (R_1–NHR_2) and the second an imine ($R_1C{=}NR_2$) which are connected via a conjugated structure. This produces a system in which the IR adsorption takes place in two distinct regions. The first is a weak absorption in the range of 3300–3450 cm^{-1}, which is associated with the amine, while the imine (>C=NR) has a signal which occurs in a band centered in the region of 1590–1670 cm^{-1} [28, 44]. These signals are seen in ligands used in both five- and six-membered ring systems. The band shift is mainly dependant on the electronic properties of the imine substituent, with the signal shifted to higher wavenumbers when the substituent is an electron donor, and to lower wavenumbers when the substituent is electron withdrawing [28, 45].

2.2.3 NMR Characterization of Imine Ligands

NMR is the method most often used for the characterization of imine ligands. Generally, ^{13}C NMR is used for the identification of imine complexes as imines lack protons and ^{14}N enriched precursors, which are expensive would be necessary to obtain spectra in reasonable lengths of time.

In order to deal with the large amount of available NMR data on a variety of ligand families, three sets of imine-based ligands were selected to base the NMR section on. In *β*-diimine ligands NMR characterization is the most straightforward. Due to the conjugation of the system, the proton associated with the amine is shifted to lower field in the ^1H NMR, and appears in the range from 11 to 15 ppm, indicating a very acidic nature.[1] In the ^{13}C NMR the signal associated with the imine carbon is located at in the range of 150–165 ppm [32, 48–50]. In the case of ketoimines, the characterization of the ligand is less straightforward due to the tautomerism of the ligands between the keto-amine and alcohol-imine forms. In many instances it is possible to observe both the imine as well as the carbonyl groups. As is expected, the carbonyl is observed at lower field, typically at 190 ppm, whereas the imine is observed at 160 ppm (similar to the range observed in the β-diimines) [41, 51–53]. In the literature, the analysis of the NMR spectra is often not clear on the distinction between the two different tautomers. During the preparation of this chapter the authors came across many instances in the literature where the NMR assignments list situations in which the N was listed in the ^{13}C NMR as a imine, while at same time in the ^{1}H NMR was listed as an amine. Obviously, this cannot be the case, however, it is understandable as the conjugated nature of the ligand makes the location of the proton and the true nature of the carbonyl/imine uncertain.

Finally, the third type of imine system to take into consideration are the salicylamide systems [21]. In these systems the shifts are similar to those observed in ketoimines. It appears from the literature that the hydrogen is shared between the two basic functionalities on the ligand [54, 55]. There is some debate as to the proper assignment in these systems as the evidence for the location of the proton is unclear.

2.2.4 Imine Ligand Properties and Coordination Modes

Much of the design and coordination of polymerization catalysts can be explained using Pearson's hardness/softness theory. This involves matching the relative electron densities of the coordinating ligands to the vacancies on the metal center. While early transition metals are most always hard acids, the late transition metals are normally borderline hard acids, indicating that they prefer to coordinate borderline hard bases. The typical groups used for metal–ligand interactions contain either O, N, or P. Within this group, O is known to be a hard base, while P is known to be a soft base, only nitrogen is a borderline base making it ideal for interactions with late transition metals [56, 57]. This in many ways makes the imine ligand derivatives more suitable for coordination of both early and late transition metals. To a limited extent, this hardness can be modified by variation of the substituents on the imine [58–60] controlling the metal–ligand interaction. In the preparation of coordination complexes, bidentate ligands with at least one imine group in the structure have an outstanding ability to chelate transition metals. Some general examples are shown in Fig. 2.6.

[1] Keep in consideration that solvent effects can greatly change the chemical shifts of these groups and in particular the protons associated with the imine/alcohol, however, the absolute trend should remain consistent.

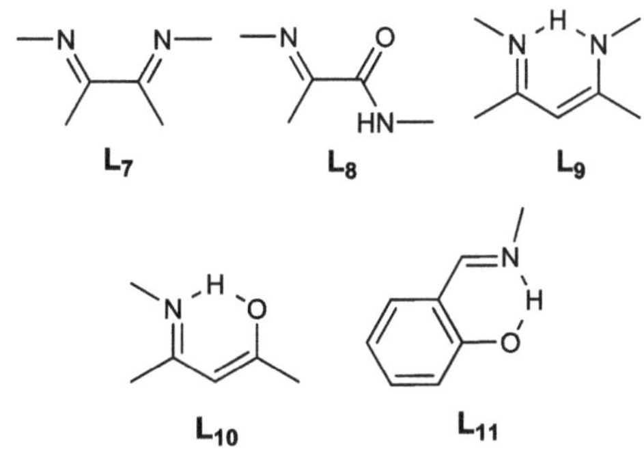

Fig. 2.6 Examples of imine derivative ligand structures

In the ligand types above (L_7–L_{11}), nearly any transition metal can be chelated, however, there are substantial differences if the ligand used is a neutral species or previously deprotonated and metallated (anionic ligand). In addition, the stabilization can vary greatly depending on the size of the chelate ring (Fig. 2.6). In the case of complexes with low coordination numbers, such as the late transition metals (Ni, Pd, Cu) and when using a neutral ligand, a more favorable reaction is observed in (L_7), in the case of (L_9) a higher temperature and longer reaction times are required [61–63]. This is due to the fact that a better electron transference to the metal occurs in ligand systems which are planar (conjugated). In ligands with sp^3 nitrogen or carbon in the ligand framework the coordination is more complicated due to the rotation of the ligand. The bidentate ligands which form five-member rings are easier to synthesize because the bite angle of the ligand is less, which generates a more open steric environment. Complexes with five-membered rings have bite angles of ca. 80°, [64–66], whereas complexes bearing six-membered chelate rings have higher bit angles, ca. 90° [21, 38, 63]. In both cases the metal can assume either a tetrahedral, pseudo-tetrahedral or square planar geometry, depending on the steric environment around the metal center.

The majority of the ligand and derivatives shown in Fig. 2.6, can be deprotonated by reaction with an equivalent of alkaline hydrides, butyl lithium or softer agents such as potassium tert-butoxide [52, 66] and subsequently transmetallated using simple and commercially available transition metal precursors. In late transition metals, the deprotonated ligand allows access to five coordinate (mainly square planar, or distorted square planar) complexes with coordination vacancies occupied by organic monodentate ligands which stabilize the complex. These reactions are rapid in comparison to those conducted using the neutral species, most likely due to the driving force created by the elimination of the salt formed during the reaction.

2.2.5 *Synthesis and Characterization of Specific Imine Derivative Ligands*

Imine-based ligands have played an important role in many aspects of catalysis. As was shown in Sect. 2.2.1, the ligands are, in general, facile to synthesize and they have the ability to be used with a wide range of metals. This has enabled them to be used as supporting ligands in many catalytic processes. In many cases, the synthesis of the ligands can be modified in order to generate complexes which have diverse electronic and steric environments [67–71].

2.2.5.1 α-Iminocarboxamide

α-Iminocarboxamide ligands have been synthesized using a variety of synthetic procedures [22, 24, 72], and more efficient methods have been developed with increased steric bulk in the backbone [73] can be synthesized following standard condensation reaction procedures [22]. An improved synthesis employs the use of oxalyl chloride and pyruvic acid in benzene at room temperature in the presence of triethylamine to generate pyruvic acid chloride. Addition of one equivalent of aniline results in reaction at the acid chloride site. This reaction takes place over a period of 2 h and requires the addition of one equivalent of triethylamine. A different aniline may be added at this stage to generate α-iminocarboxamide ligands with different substituents at the two nitrogen sites. The second step requires overnight heating to 110 °C, as mentioned previously, the more bulky 2,6-substituted phenyl groups require harsher reaction conditions and normally the reaction proceeds with low yield. Two equivalents of aniline are added to pyruvic acid chloride in cases where identical (symmetric) substitution is desired. Purification can be accomplished by chromatography or crystallization, depending on the aryl substituents. Many anilines are commercially available bearing a variety of substituents; other anilines can be readily synthesized using a number of methods, of particular utility is the Suzuki cross-coupling reaction (Fig. 2.7) [27, 74–78].

It is important to note that the α-iminocarboxamide ligand is commonly used as a deprotonated species and in theory, could be considered a tridentate ligand, (N∩N∩O), however, due to the planar sp^2 hybridization of the carboxamidate carbon, only the nitrogen or oxygen can coordinate to a Lewis site together with the imine nitrogen. Instead, the carboxamide group could interact in a single η^3 fashion, however this is not very likely due to the substantial steric restraints.

2.2.5.2 α-Imine-β-ketone Ligand

Ligand frameworks such as 4-(2,6-diisopropylphenylimino)-3,3-dimethylpentan-2-one and 3-(2,6-diisopropylphenylimino)butan-2-one (Fig. 2.8), L_{13} and L_{14}, respectively, can be easily prepared by condensation of the respective 2,3-butadione and 3,3-dimethylpentane-2,4-dione with equimolar amounts of the respective aniline. This ligand framework allows for the mono, bi, or tridentate binding mode to an electrophillic center (N/η^2–C=C or O). This variety of binding modes allows for several structures to be isolated and characterized.

In compounds such as L_{13}, which have a break in the conjugation due to the sp^3 carbon between the imine group and the acetate (–$COCH_3$), the chelation

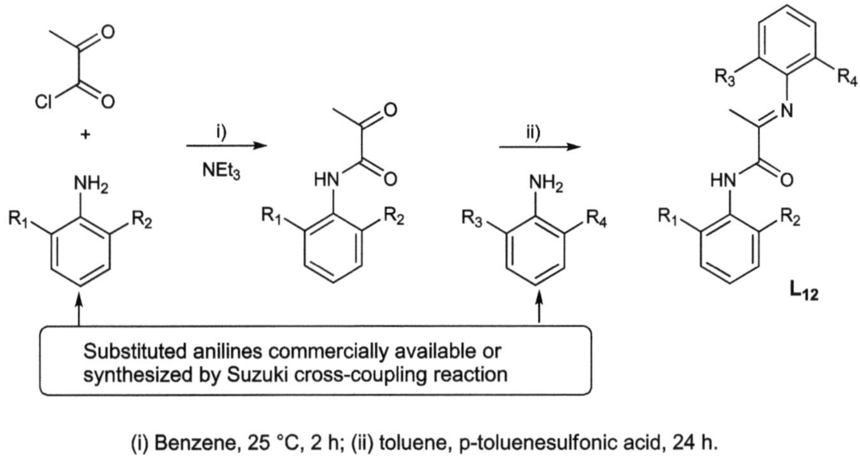

Fig. 2.7 Pathway to α-iminocarboxamide ligand

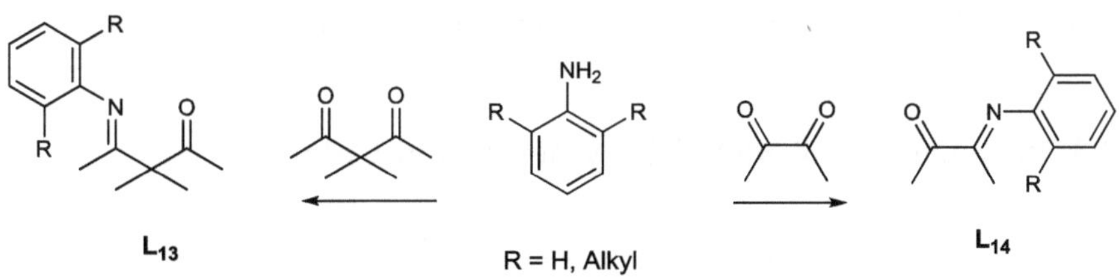

Fig. 2.8 Pathway to α- and β-ketoimines

properties could be depressed and act rather as a monodentate O-binding mode. This is discussed in later sections.

The chelate properties of a similar ligand are increased by the removal of the sp^3 carbon. This allows a better coordination to the metal center and selectively the bidentate system. In order to ensure the desired coordination to and electrophillic site, a conjugated backbone is required, mainly because the delocalization of the electron density helps to stabilize the final compound [67].

L_{13} and L_{14} ligand types or other imine derivatives can be accessed by simple synthetic routes described previously, and enlarge the variety of ligands with conjugated backbones. Some of these ligands are applicable in the preparation of olefin polymerization catalysts. Some examples are shown in Fig. 2.9 [79–85].

2.2.5.3 α-Keto-β-diimine Ligand

α-Keto-β-diimine ligands can be produced starting from 4-(2,6-diisopropyl-phenyl)iminopent-2-en-2-(aryl) amine substrates, whose synthesis is discussed elsewhere in depth [67]. The substrate is added to a solution of copper acetate in

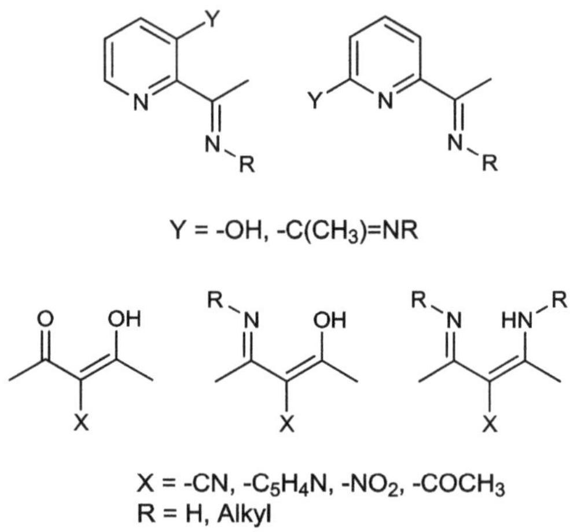

Fig. 2.9 General functionalized bidentate ligand frameworks

methanol/dichloromethane and stirred; an intermediate copper complex bearing the bisimine ligand can be isolated. The complex can be dissolved in methanol, and decomposed in the presence of O_2 forming copper oxide and the oxidized ligand [38].

2.3 Preparation of Nickel Polymerization Catalysts

There are two crucial components when preparing transition metal catalysts. The first is the ligand (deprotonated or neutral) and the second is the metal precursor used. The selection of the metal precursor is critical as it determines the structure of the catalyst (alkylated or halogenated), as well as the reaction conditions to be used (both the ligand and the metal precursor should, ideally, be soluble in the same solvents). It has also been reported in combinatorial synthesis/polymerizations that the same metal complex produced using distinct metal precursors can actually have different polymerization activities [8]. While this result is not explained in detail, it illustrates that some metal precursors work better than others, affecting not only the yield, but also the reactivity of the final complex.

2.3.1 Commonly Used Nickel Precursors

The term metal precursor is commonly applied to a variety of relatively simple organometallic compounds which can readily serve as a metal source in organometallic synthesis. More specifically, in the case of preparing polymerization catalysts, metal precursor is an organometallic compound which contains one or more organic ligands which typically have free lone-pair electrons, common ligands include: ethers, amines, tetrahydrofuran (THF), phosphine groups, and bulky olefins. These groups are covalently and coordinatively bonded to one or more metal atoms. During the synthesis of the desired catalyst, the organic ligand bound to the ‘metal precursor’ rapidly dissociates and is replaced on the metal center in the presence of a “better” ligand.[2] This is often supported by the chelate effect, which is the formation or presence of two or more separate bonds between a multidentate ligand and a single central atom [86]. Some of the more commonly used metal precursors for the preparation of precatalysts and catalysts for olefin oligomerization and polymerization are shown in Fig. 2.10.

2.3.2 Metallation/Transmetallation of Ligands with Metal Precursors

In nickel(II) complexes containing an anionic bidentate ligand (cf. Sect. 2.2), η^1-σ-bound aryl substituents have played an important role in the development of well defined catalyst precursors. In the pioneering work of Keim et al. and others [87–90], the phenyl ligand in Fig. 2.11, originates from the ylid ligand precursor. In subsequent work, related single component catalyst precursors with aryl substituents have been prepared by other routes [91, 92].

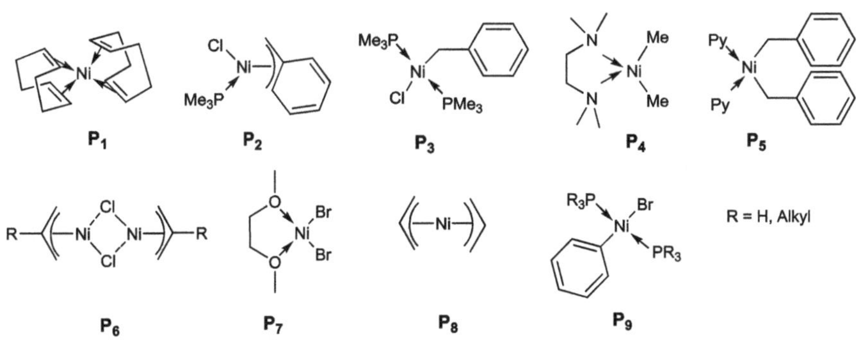

Fig. 2.10 Nickel precursors used in the preparation of polymerization catalysts (Note: For more information on the preparation of nickel precursors see: $Py_2Ni(Bn)_2$ and derivatives [93, 94]$(PMe_3)_2NiMeCl$ [95], [(bipy)$NiMe_2$] [96, 97], [(tmeda)$NiMe_2$] has been obtained from [(tmeda)MgMe] [98, 99] and [(tmeda)Ni(acac)] [100], [$(py)_2Ni(CH_2SiMe_3)_2$], can be prepared from [$(py)_4NiCl_2$] and $Mg(CH_2SiMe_3)Cl$ in good yield, is stable at r.t. under a protective gas atmosphere. In contrast to [(tmeda)$NiMe_2$], the reaction of [$(py)_2Ni(CH_2SiMe_3)_2$] with PMe_3 has been reported to yield the dialkylcomplex, but with PPh_3 [$Ni(PPh_3)_4$] is obtained again. [(allyl)MX_2] can be prepared from allyl halide and [$Ni(COD)_2$] [101, 102] [$Ni(COD)_2$] preparation [103]. Cationic compounds [(allyl)ML_2]Y (e.g. L_2 = COD, L = $NCCH_3$) are usually prepared from the allyl complexes [(allyl)MX_2] (M = Ni, Pd) by halide abstraction with silver or thallium salts in the presence of the appropriate ligand [104–106].)

[2] A “better” ligand is normally one which is more compatible with the metal cation, as described by the hardness/softness of the metal and the ligand. Particularly, those which are multidentate ligands.

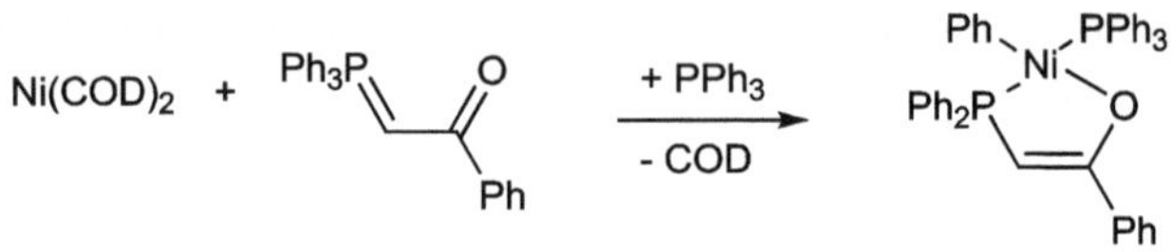

Fig. 2.11 Pathway to Keim oligomerization catalyst

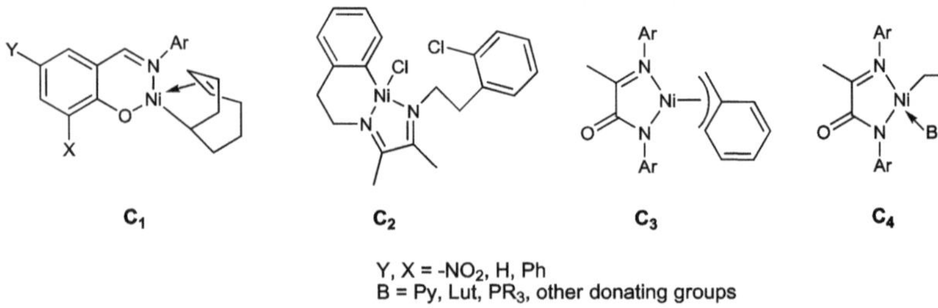

Fig. 2.12 Examples of complexes prepared by oxidative addition to $Ni(COD)_2$

The precursor bis(1,5-cyclooctadiene)nickel, P_1 above, has primarily been used to prepare the precursors such as P_2, P_3 and P_9, through the oxidative addition of BzCl, or PhBr in the presence of a phosphine [107, 108], as well as to produce P_6 and P_8 through the oxidative addition of allylchloride and methyl allyl chloride. Additionally, it has been used as a precursor for the direct preparation of nickel catalysts and precatalysts, some examples based on imine or imine derivative bidentate ligands are shown in Fig. 2.12.

Compound C_1 is prepared by oxidative addition of a free salicylaldimine ligand to $Ni(COD)_2$, and isolated by precipitation in anhydrous n-hexane [37, 109].

Complexes of type C_2 were also obtained from the reaction of $[Ni(COD)_2]$ and the 1,4-diaza-1,3-butadiene ligands. Equimolar amounts of $Ni(COD)_2$ and the ligand were mixed in tetrahydrofuran at −78 °C, the temperature was raised until room temperature, and the stirring was continued for specific periods of time and were obtained in 70–90% yields. In a second step this type of compound are convert in a ionic compounds. The general procedure was by direct abstraction of halide ligand of complexes with $TlBF_4$ in the presence of a neutral ligand such as acetonitrile or collidine [110–112].

The α-iminocarboxamide η^3-benzyl nickel complex (C_3) can be prepared in a one pot reaction by the oxidative addition of benzyl chloride to $Ni(COD)_2$ followed by the addition of the potassium salt of N-(2,6-diisopropylphenyl)-2-(2,6-diisopropylphenylimino) propanamide in THF. This type of reaction is generally applicable with bidentate ligand types (see Sect. 2.2.5.1, L_{12}). Limiting the exposure to light during the synthesis produces better results, most likely due to the light sensitivity of $Ni(COD)_2$, the presence of dimethylaniline has also been shown to increase yields. The role of dimethylaniline is not clear; presumably it improves the product yield by stabilizing reactive intermediates. Successive crystallizations from pentane allow for isolation of C_3 as slighty air sensitive dark orange crystals in 70% yield. While isopropyl substitution on both phenyl rings provides the desired product, similar reactions with ligand frameworks with lesser steric bulk failed. The major product from those reactions is the (bis-α-imino-carboxamide)nickel derivative [$(N{\cap}O)_2Ni$] in a trans conformation and are air and temperature stable complexes [27].

It appears that starting from compound C_3 one could access compound C_4. The strategy here is to force the dissociation of the benzyl η^3-component by pyridine. Previous studies have shown that ligand (L_{12}) (Fig. 2.7), has two coordination modes, N∩N or N∩O, both coordinate the metal center giving complexes, but are differentiated in two remarkable way: first, the nitrogen atoms are a soft Lewis base in comparison to the O atom, this means that it has a higher affinity and better stabilization effect over soft Lewis acids in comparison to oxygen atoms (favorable as the thermodynamic product), second, the N∩O chelate coordination gives a less crowded metal center and as a result will be favorable when L_{12} ligand has bulky substituent or also if the coordinating monodentate base (B) is sufficiently large. Nevertheless, the N∩O coordination mode stabilizes the metal center less and consequently produces the more reactive compound [24].

In addition to the base (B = pyridine) and the η^3-component, the NMR spectroscopy revealed that the product of the reaction contained two isomers in a 1:1 ratio. X-ray crystallography studies show a square-planar arrangement around the nickel center with an N∩N-bound α-iminocarboxamide ligand, and pyridine trans to the imine nitrogen.

The second isomer present in solution was assigned as the N∩N-bound α-iminocarboxamide complex with pyridine trans to the carboxamide nitrogen. On the basis of the previously observed lack of reactivity from N∩N-isomers, a different synthetic method to access the (N∩O)NiBzB complex was developed. The same method used in the synthesis of compound (C_3) was applied to access compound (C_5), thus through the reaction of potassium N-(2,6-diisopropylphenyl)-2-(2,6-diisopropylphenylimino) propanamidate, $Ni(COD)_2$, benzyl chloride, and Py or Lut in THF (Fig. 2.13).

The product is isolated as a single isomer, in the case of B = Lut, in yields of over 60%. The success of this reaction is highly dependent on the size of B. Interestingly, it was also possible to isolate the product from the oxidative addition of benzoyl chloride to $Ni(COD)_2$ in presence of pyridine, producing

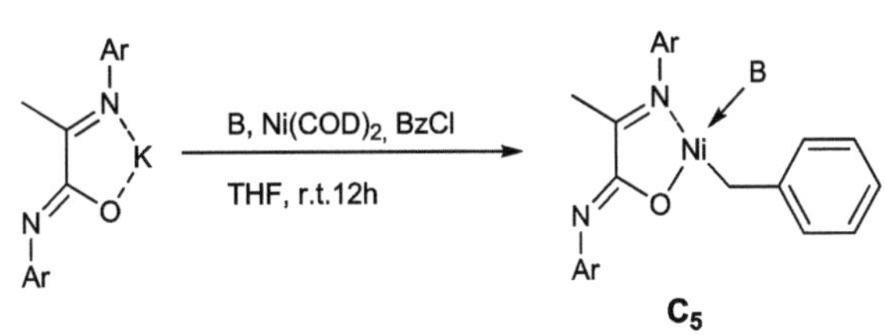

Fig. 2.13 Access to N∩O coordination mode in α-iminocarboxamide nickel complexes

the [N-(2,6-diisopropylphenyl)-2-(2,6-diisopropylphenylimino)-propanamidato-κ^2 N,O](η^1-COPh)nickel(pyridine) complex [24]. The formation of the $(N\cap O)_2Ni$, which bears two α-iminocarboxamide ligands, and is inactive for polymerization, was also observed [27].

Additional complexes with N∩N and N∩O five-member chelate rings have been reported [64, 91, 113–115].

2.3.3 *Synthesis of Nickel Complexes by Using Precursor P_2, P_3 and P_9*

Nickel precursors such as P_2, P_3 and P_9 have been widely used in the preparation of olefin polymerization catalysts and precatalysts. This is because monodentate phosphines (PMe_3 or PPh_3) ligands are easily displaced by bidentate ligands, particularly in transmetallation reactions. Lithium, sodium or potassium salts of ligands such as L_8–L_{11} (Fig. 2.6), and L_{13} and L_{14} (Fig. 2.8) are examples of compounds that react directly with the nickel phosphine precursor giving the respective chelated N∩N, N∩O, and N∩C nickel complexes.

In the case of α-iminocarboxamide ligand derivatives, the reaction of the ligand salt (L_{12S}) with P_2 or P_3 yields the corresponding organometallic product. However, the size of the R substituent defines the coordination mode of the final product. Characterization by NMR spectroscopy indicates that when R is larger than an ethyl group, only a single isomer is formed (see Fig. 2.14). Each complex contains a nickel center ligated by α-iminocarboxamide, benzyl and one phosphine. The ^{31}P NMR spectrum is characteristic and reveals a single peak at $\delta = -7.7$ ppm, which is indicative of an N∩O-coordination of the α-iminocarboxamide ligand to nickel (the N∩N-coordination gives rise to peaks in the −18 to −20 ppm range) [22].

In the case of ligand salts, where the 2,6-dimethylphenyl substituents are present, one obtains an intermediate situation. The N∩O-mode is kinetically preferred, probably because of easier displacement by the sterically less encumbered oxygen. The N∩N isomer is ultimately the thermodynamic product,

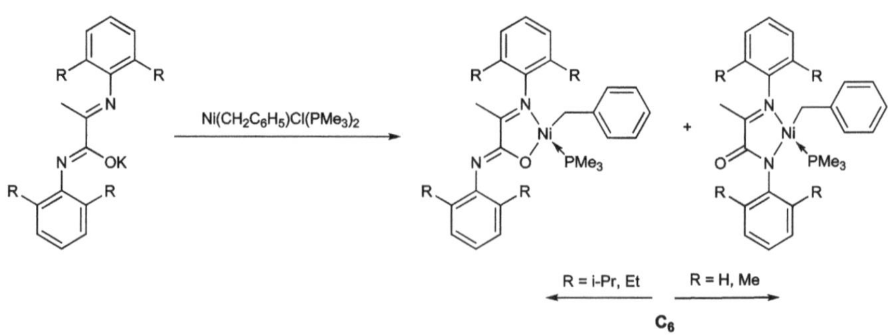

Fig. 2.14 Transmetallation of α-iminocarboxamide ligands

which reflects an electronic preference for nickel to bind to nitrogen over oxygen. The stability and conversion of the N∩O isomer was investigated using ^{1}H NMR spectroscopy starting with a sample of 90% purity. Over a period of 60 min at 40 °C, the methyl peak of PMe_3 ($\delta = 0.56$ ppm) is reduced in intensity by more than 50% with a concomitant increase of two signals at 0.34 and 0.22 ppm, which correspond to the phosphine resonances in the N∩N isomers. Complete conversion from N∩O to N∩N takes place after 2 h at 50 °C and the rearrangement is irreversible. These studies led to fine tuning of the reaction conditions so that each isomer could be produced directly. The N∩O chelated compound (Fig. 2.14) when R = Me, was isolated in greater than 95% purity by running the reaction at −10 °C for 4 h, while the N∩N chelated compound was isolated by running the reaction at 50 °C for 4 h. It should be noted here that there is no change in the binding mode of the α-iminocarboxamide ligands when R is larger than methyl after heating to 60 °C for 1 h [27].

The transmetallation reaction of the compounds shown in Fig. 2.15 was carried out using P_2 or P_3. The ligand is a relatively small framework, which depending on the structure can serve as mono, bi, or tridentate ligand when bound to the nickel center. A variety of binding modes allows for several structures which were isolated and characterized. Synthetic access begins with ligands L_{13} and L_{14} (Fig. 2.8),

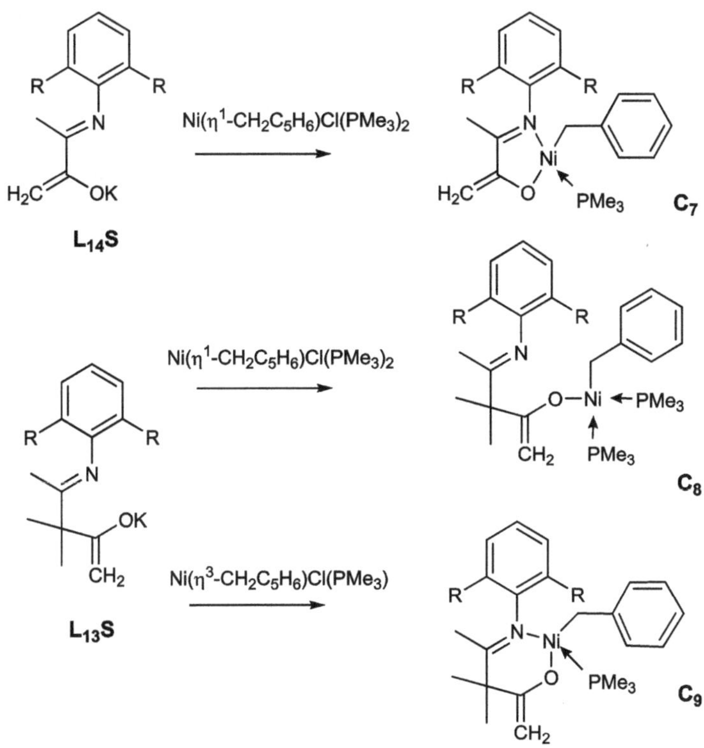

Fig. 2.15 Transmetallation of L_{13S} and L_{14S} to C_7–C_9

deprotonation with NaH or KH in THF provides the salt in over 80% yield. Subsequent reaction of the salt with P_2, yields compounds with different coordination modes as shown in Fig. 2.15.

In the case of L_{14S} (Fig. 2.15), the expected 3-(2,6-diisopropylphenylimino)-but-1-en-2-olato-κ^2N,O](η^1-CH_2Ph)(trimethylphosphine)nickel (C_7) is obtained. While in the case of the reaction of L_{13S} with the same precursor, the formation of a single isomer containing a nickel center ligated by 4-(2,6-diisopropyl-phenylimino)-3,3-dimethylpent-1-en-2-olato, η^1-CH_2Ph and two PMe_3 ligands (as confirmed by X-ray diffraction), C_8 is observed [116]. This difference is likely associated with the sp^3 carbon between the functionalities which depress the chelate effect in comparison with L_{13S} where the functionality is present in a conjugated structure.

In an attempt to isolate an N∩O-bound bidentate complex, with the consideration that the presence of the two phosphines helps to have the ligand bind in a monodentate mode, the monophosphine species [Ni(η^3-CH_2Ph)Cl(PMe_3)] in toluene-pentane was used at r.t. and formed a single isomer containing a nickel center ligated by 4-(2,6-diisopropyl-phenylimino)-3,3-dimethyl-pent-1-en-2-olato, η^1-CH_2Ph and a PMe_3 ligand, C_9. The ^{31}P NMR spectrum reveals a single peak at $\delta = -21.2$ ppm. The structure was also corroborated by X-ray diffraction studies [116].

Similar compounds, prepared by P_2, P_3, P_6 and P_9 are shown in Fig. 2.16 [117, 118].

The compound L_{5S} in Fig. 2.17 was designed with an electronically delocalized conduit in which three atoms that can potentially coordinate to a metal center. As shown in Sect. 2.2.1, when starting from the commercially available triacetylmethane [3-(1-hydroxyethylidene)-2,4-pentanedione], replacement of one of the oxygen atoms by an aryl-substituted nitrogen group provides

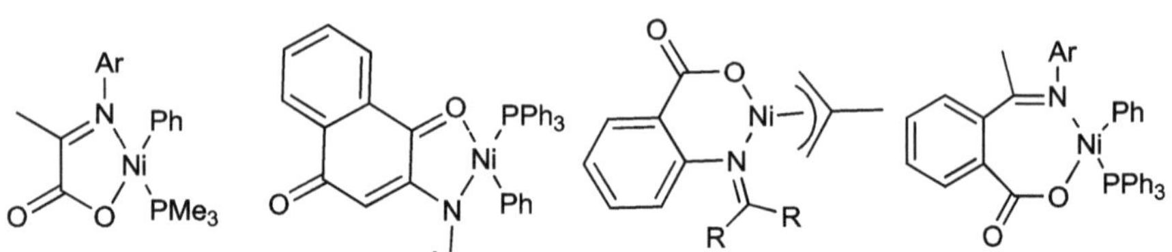

Fig. 2.16 Examples of complexes with exocyclic functionality

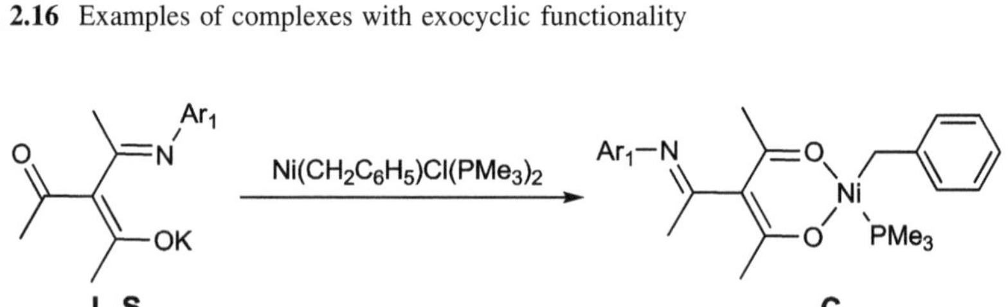

Fig. 2.17 Synthetic pathway to C_{10}

3-[1-(arylamino)ethylidene]pentane-2,4-dione derivatives and anionic structures which can be deprotonated using KH. The reaction of the salt with P_3 provides an organometallic compound (C_{10}) which contains the organic fragment L_5 (anionic), an η^1-benzyl group (^{1}H NMR: $\delta = 2.07, 7.65$ ppm), and a coordinated trimethylphosphine (^{31}P NMR: $\delta = -6.79$ ppm) [43].

These data alone do not allow for the unambiguous determination of the ligand coordination mode. X-ray diffraction studies shows that the nickel center has a distorted square-planar coordination geometry with the 3-[1-(2,6-diisopropylphenylimino)ethyl]acetylacetonate ligand bound through the two oxygen atoms at nearly equal Ni–O bond lengths (1.889 and 1.898 Å). Additionally, the ligand framework is in a coplanar disposition in a delocalized electronic structure [43].

Another "tridentate" imine ligand derivate, shown in Fig. 2.18 (L_1), has been used in the preparation of nickel complexes for ethylene polymerization. As in some of the compounds previously reviewed, this allows access to the investigation of remote activation through an exocyclic CN functionality. This functionality is restricted by geometry to participate in its chelation form to the metal center. The preparation of nickel complex was done by reaction of the potassium salt of the ligand generated by deprotonation with KH in ether. Potassium 3-cyano-4-(2,6-diisopropylphenylimino)pent-2-en-2-olate (L_{1S}) can then be reacted with Ni $(PMe_3)_2(\eta^1\text{-}CH_2C_6H_5)Cl$ to yield 3-cyano-4-(2,6-diisopropylphenylimino)pent-2-en-2-olato-κ^2N,O] $(\eta^1\text{-}CH_2Ph)(PMe_3)Ni$, ($C_{11}$) in diethyl ether at room temperature (Fig. 2.18).

The product is obtained as a mixture of C_{11} and bis[3-cyano-4-(2,6-diisopropylphenylimino)pent-2-en-2-olate-κ2 N,O-]Ni, $(Ni(L_1)_2$. $Ni(L_1)_2$ contamination can be minimized by slow addition of the ligand salt at decreased reaction temperatures. Successive crystallizations from pentane allow isolation of C_{11} as air and thermally sensitive dark orange crystals in 78% yield. The ^{31}P NMR spectrum in C_6D_6 exhibits a single resonance at −12.31 ppm. Single crystals of C_{11} suitable for X-ray diffraction studies were obtained by diffusion of pentane into a toluene solution at room temperature. The result of this study shows a molecular structure with an N∩O-binding mode for the 3-cyano-4-(2,6-diisopropylphenylimino)pent-2-en-2-olate ligand (L_1) and a trans-orientation between PMe_3 and the imine nitrogen as shown in Fig. 2.19.

The square planar geometry around nickel is slightly distorted. However, the CN functionality, in the plane of the chelated ring, is oriented in the opposite

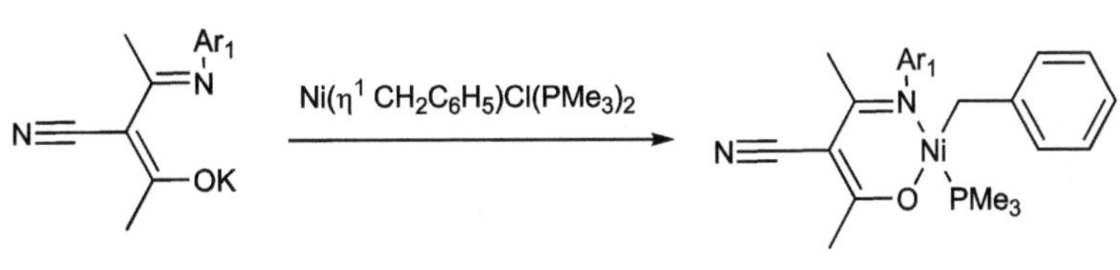

Fig. 2.18 Synthetic pathway to produce C_{11} which contains CN exocyclic functionality

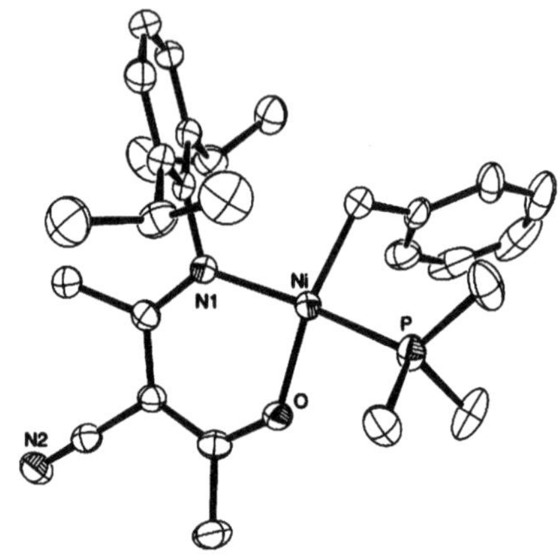

Fig. 2.19 Crystal structure of complex C_{11} (reprinted with permission from [51]. Copyright 2008 American Chemical Society)

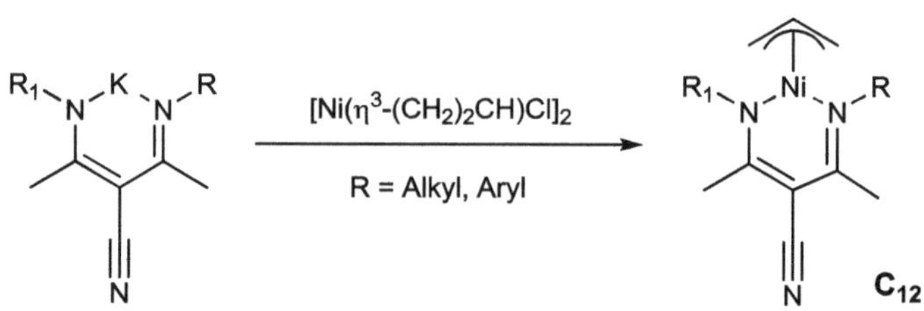

Fig. 2.20 Synthesis of nickel allyl complexes

direction of the N and O atoms and free to coordinate Lewis acids. In the secondary product $[Ni(L_1)_2]$ the CN functionality is not coordinated as well.

Another approach to access catalysts or precatalyst for olefin polymerization was found by using nickel allyl precursors such as P_6 or P_8. These precursors allow the use of the selected ligand in either a deprotonated or neutral form, respectively. An example is the synthesis of C_{12}, which can either be produced by reaction P_6 ($R = H$) with L_{4S} (Fig. 2.20), or through the reaction of P_8 with the neutral ligand L_4 producing the expected $L_4Ni\eta^3$-allyl in yields of over 80% [119]. These complexes crystallize from solution readily compared to other compounds, and the crystal structure of the compound C_{12} is shown in Fig. 2.21. However, due to the high stability of this type of compound, the activation process is not very efficient and the reactivity is lower than the similar η^3-benzyl or η^1-benzyl-(trimethylphosphine or B) nickel complexes.

The nickel precursor P_9, has also been applied in synthesis of nickel compounds via reaction of ligand salts (such as shown in Fig. 2.6). Some remarkable examples are shown in Fig. 2.22.

Another common way to prepare precatalysts for polymerization is by the direct reaction of the neutral ligand with a nickel precursor such as P_7 or similar (e.g. dichloride), through displacement of the previous weakly bound ligand and coordination of the new one to the metal. The tetrahedral geometry of these types

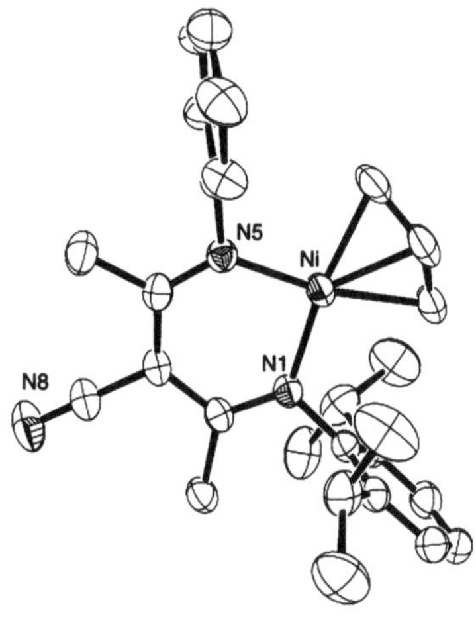

Fig. 2.21 ORTEP structure of C_{12} [119]

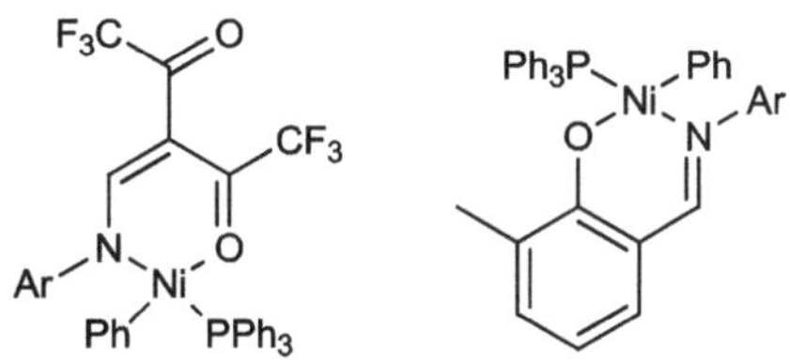

Fig. 2.22 Examples of complexes produced using P_9

of complexes make the characterization by ^1H NMR spectroscopy difficult due to the paramagnetic nature of the compounds [120] however, in the most of the cases the complexes are easily crystallized from common solvents and as a result X-ray crystallography is the spectroscopic analysis of choice for the final determination of the connectivity of these types of complexes in the solid state.

It is useful to describe here the synthesis of nickel complexes with "tridentate" ligands, which, as previously shown, allow only two atoms to be coordinated to the same metal, because of geometric restriction. The ligand 2,4-bis(2,6-diisopropylphenylimino)pentan-3-one (L_3) is a symmetrical compound, which can coordinate through two imine nitrogens or by one nitrogen and the oxygen. Figure 2.23 shows the synthetic pathway into C_{13} nickel dibromide complexes. This compound is obtained by the addition of L_3 to a stirring suspension of P_7 in CH_2Cl_2 at room temperature. The resulting red complex, C_{13}, can be isolated in 49% yield [38].

The IR (and other types of spectroscopy) support the formation of just one isomer. Unfortunately, the ^1H NMR, shows a broad peak associated with a paramagnetic species, which is expected for a nickel compound with a tetrahedral configuration.

Single crystals suitable for X-ray crystallography examination were obtained from a concentrated CH_2Cl_2 solution at -35 °C and the results are shown in Fig. 2.24. The molecular connectivity is consistent with a neutral N∩N-bound 2,4-bis(2,6-diisopropylphenylimino)pentan-3-one ligand. Bond distances within the six-membered chelate ring are consistent with the proposed structure as

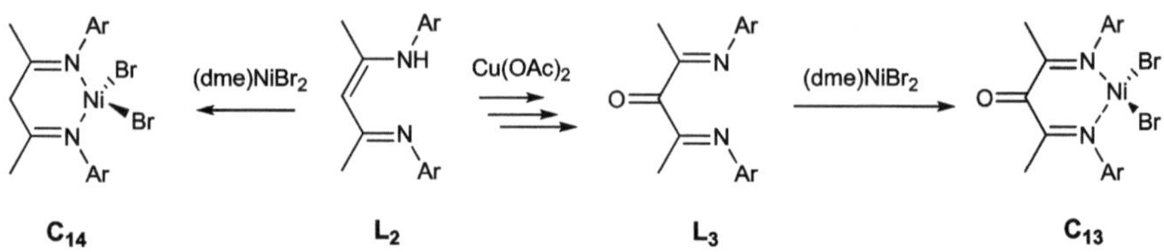

Fig. 2.23 Synthesis of C_{13} and C_{14}

Fig. 2.24 ORTEP drawing of C_{13} shown at the 50% probability level (Reproduced with permission from [38]. Copyright Wiley–VCH Verlag GmbH & Co. KGaA)

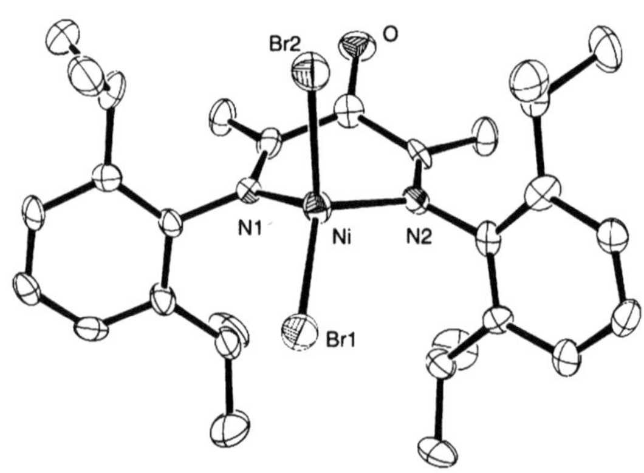

illustrated by the following selected bond lengths (in Å): C(3)–N(2): 1.292(10), C(2)–C(3): 1.518(12), C(1)–N(1): 1.287(11), C(1)–C(2): 1.511(12) and C(2)–O(1): 1.223(10) [13]. The Ni atom adopts a pseudotetrahedral coordination geometry with the chelate ring adopting a boat-like conformation similar to that observed previously in complex C_{14}. Bond distances for complexes with and without the addition of the carbonyl group are in close agreement; the main structural difference stems from the presence of the exocyclic carbonyl functionality on the α-carbon in this compound.

2.4 Activation of Compounds for Olefin Polymerization

2.4.1 Activation with Soluble Lewis Acids

Due to the dependence of metallocene and early transition metal post-metallocene systems on the use of expensive co-catalysts such as methylaluminoxane (MAO), there has been a great deal of interest in developing systems which require less, or ideally no cocatalyst, in order to make the process more profitable [121]. In late transition metals, promising candidates can be activated using alternative reagents, ideally in stoicheometric concentrations. Molecular Lewis acids can be added to those precatalysts which are already alkylated and that contain in the structure

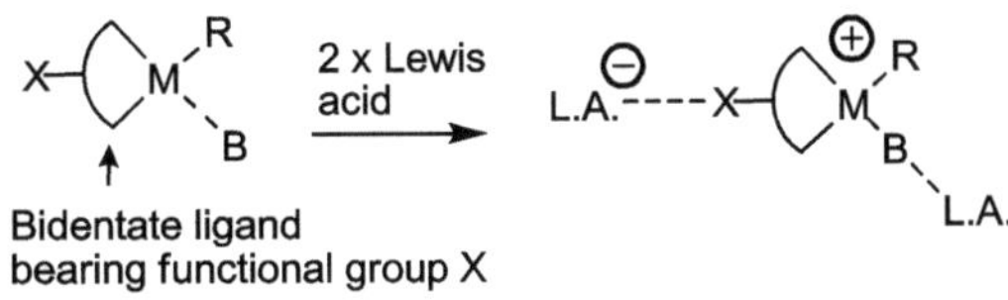

Fig. 2.25 Lewis acid activation of complexes

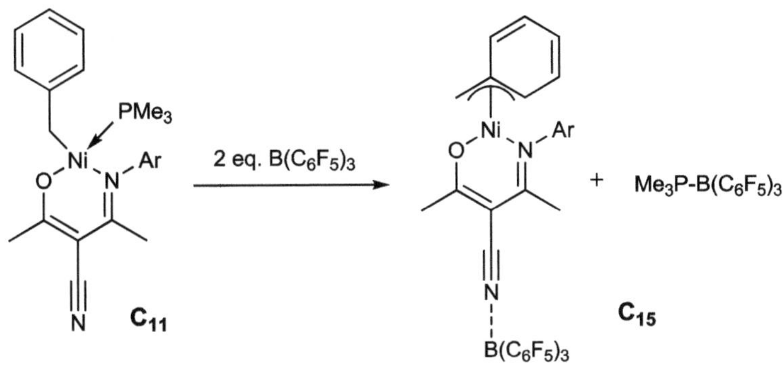

Fig. 2.26 Isolation of BCF adduct C_{15} from C_{11}

Lewis bases such as nitrogen or oxygen moieties which bear a lone pair of electrons which can interact with the Lewis acid (Fig. 2.25). This offers the advantage that the activator is remote, and not directly associated with the metal center. This is beneficial as it opens up the active site for monomer coordination.

From the precatalysts shown in the last section, the first approach to activation different to MAO was done by using molecular Lewis acid $B(C_6F_5)_3$ which is a solid, stable to 200+ °C and is easy to handle [122–124]. In complexes such as C_{11}, which bear a base, B, the addition of two equivalents proceeds quantitatively producing a η^3-benzyl complex, C_{15}, as shown in Fig. 2.26. In this reaction one equivalent of the acid acts to scavenge the base, while the second forms an adduct through the exocyclic functionality. Depending on the basicity of the groups, the order in which the acid reacts with the exocyclic functionality and the base may change.

The crystal structure of C_{15} is shown in Fig. 2.27, and demonstrates the formation of an adduct between the exocyclic functionality and the Lewis acid. This adduct is remote from the metal center, which creates a more open steric environment, for what will become the active site once monomer is present. The hardness of the acid and corresponding basic site on the ligand ensure complete reaction. The resulting adduct crystallizes readily, producing structures of the polymerization species with the activator already present.

In complexes which have secondary groups which can bind to the metal center the addition of the BCF to complexes can produce the isomerization of the ligand. The isomerization from an O∩O to an N∩O binding mode or from an N∩O to an N∩N mode are the most common, two examples are shown in Fig. 2.28.

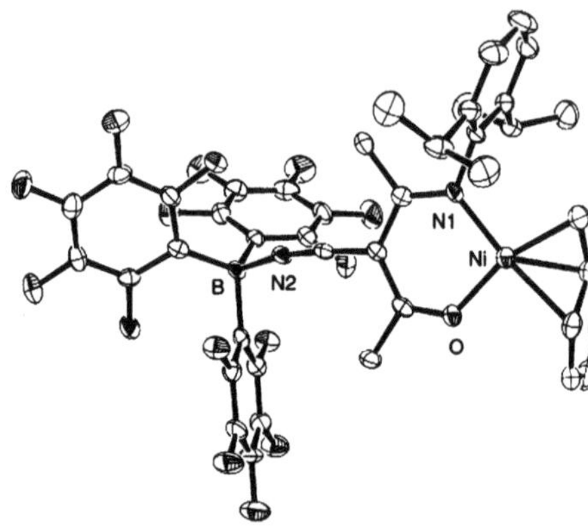

Fig. 2.27 Crystal structure of C_{11}-BCF adduct (C_{15}) (Reprinted with permission from [51]. Copyright 2008 American Chemical Society)

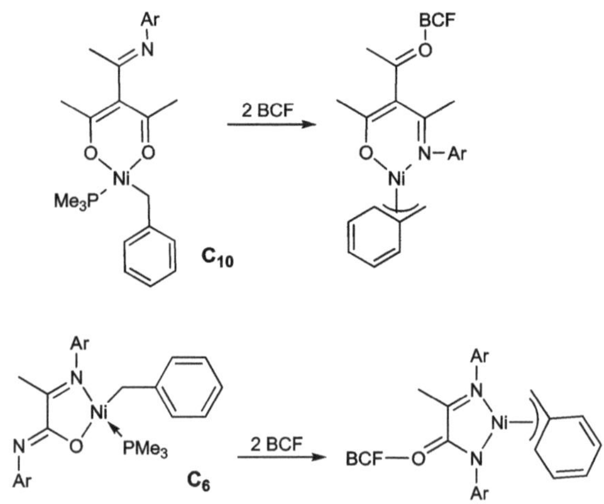

Fig. 2.28 BCF induced isomerization of C_{10} and C_6

This rearrangement is caused by the removal of the base (phosphine) which reduces steric crowding and allows the formation of new, more stable, η^3 nickel complexes.

In systems which do not bear coordinating bases, the reaction with BCF proceeds stoichiometrically and rapidly produces an adduct as shown in Fig. 2.29. The crystal structure of the adduct formed is shown in Fig. 2.30. In this figure it can be observed that the Lewis acid is again remote from the metal center.

2.4.2 Single Component Systems

Single component systems have been known academically and industrially for a long while. These systems do not require the presence of cocatalysts or activators

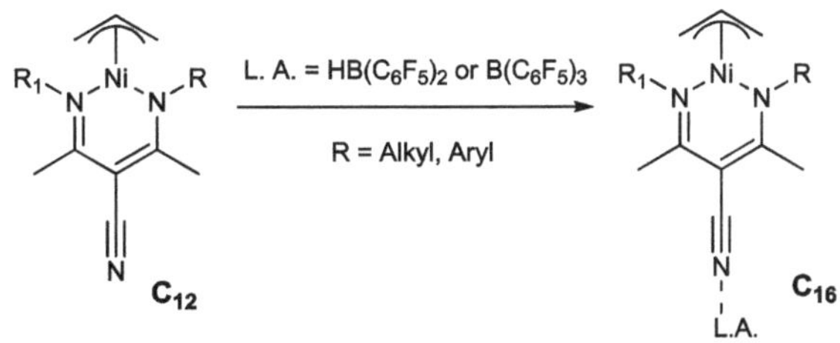

Fig. 2.29 Direct adduct formation complexes without Lewis bases

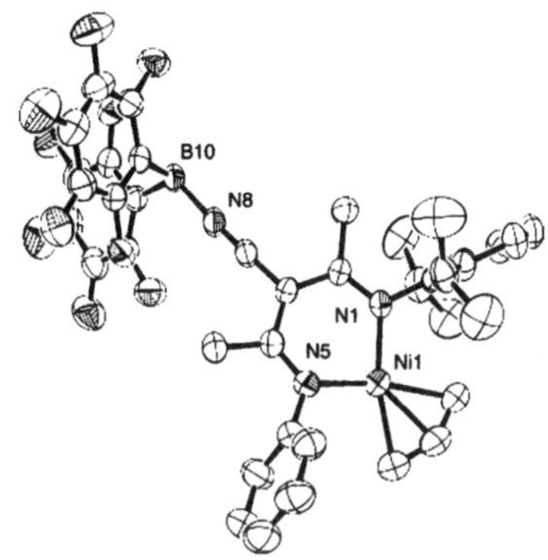

Fig. 2.30 Crystal structure of C_{16} [119]

to function. The SHOP, developed by Keim et al., is a prominent industrial method for the synthesis of linear ethylene oligomers in the C_4–C_{20} range, which are converted to detergents, plasticizers, lubricants and a variety of fine chemicals [15], and are based on neutral nickel catalysts which do not require presence of an activator. Grubbs et al. develop a new family of neutral, single-component catalysts base on salicylaldimine ligand (Fig. 2.1). In this catalyst system, by using sufficient bulk in the ortho-position of the salicylaldimine ligand to aid phosphine dissociation and to prevent disproportionation of the ligand, they were able to produce polyethylene with Mw > 250,000 g/mol and later on functionalized copolymers [125, 126]. In α-iminocarboxamide-based nickel catalysts, single component systems have been developed which rely on the presence of a bulky coordinating base which can be displaced in the presence of monomer (Fig. 2.13) [24]. As this system does not require the presence of a Lewis acid activator, it readily incorporates sterically hindered functional monomers, such as norbornene derivatives.

2.4.3 Polymerization Reactions

As discussed earlier, the structure of the α-iminocarboxamide catalyst, which allows for both a wide variety of steric environments, but also for a number of different activation methods, as such a comparison of the polymerization activities and the resulting polymers is not as straight forward as in other cases. Therefore, in this section general comparisons will be made in order to compare the different activation routes in selected systems, as well as the effect of ligand modification using a single activator.

2.4.3.1 Effect of Activator Choice on Polymerization Activity and Polymer Properties

The presence of the carboxamide group and phosphine ligand on the catalysts allows the catalyst to be activated using a number of different reagents. In the case of the α-iminocarboxamide catalysts, this is normally accomplished using a phosphine scavenger, or in the case of the single component systems, through ethylene pressure [24, 92, 125]. The most reported reagent are borane Lewis acids such as BF_3, BCF and BPh_3. Two equivalents of these reagents are typically added in order to remove the phosphine, as well as to coordinate to the basic site on the ligand backbone. A second reagent used to activate the systems is $Ni(COD)_2$, although the mechanism is less clear it is thought that the second nickel center serves as a phosphine scavenger, allowing the precatalyst to generate an active center and initiate polymerization. While the majority of the activating agents are homogenous, it has also been reported that montmorillonite clays can also serve to activate the catalyst system (Fig. 2.31) [26]. It is thought that the presence of Bronsted acid sites on the surface of the clay act as a phosphine scavengers, while at the same time the presence of the Lewis sites on the surface generate a Zwitterionic site through their coordination to the ligand backbone.

As can be observed in Table 2.1, the choice of activator can significantly affect the polymerization activities as well as the polymer properties. The single component catalyst, which activates via the displacement of lutidine (or pyridine) from the nickel center, has an activity of 43 kg mol^{-1} h^{-1} at 20 °C. Increasing the reaction temperature, which favors the displacement of the lutidine from the metal center, significantly increases the polymerization activity, with the reaction run at 40 °C proceeding with an activity nearly eight times as high (304 kg mol^{-1} h^{-1}). The other activator which produces a neutral system is $Ni(COD)_2$. With this system a polymerization activity of 220 kg mol^{-1} h^{-1} is reported. When the precatalyst is exposed to Lewis acid activators, the polymerization activities increase due to the Zwitterionic charge on the nickel center (Figs. 2.25 and 2.31). The use of BCF to

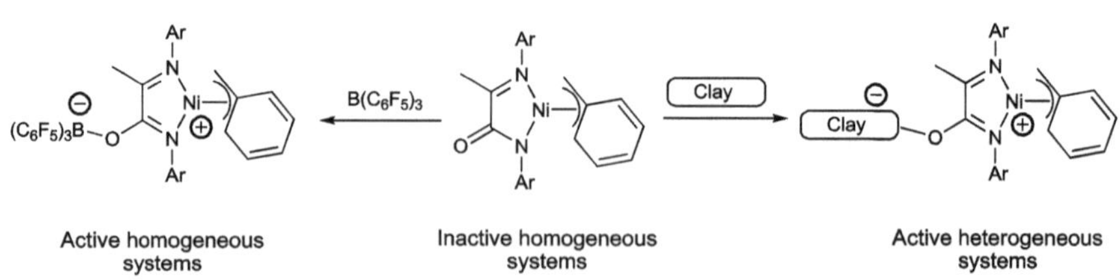

Fig. 2.31 Activation of α-iminocarboxamide complexes for polymerization

Table 2.1 Polymerization activities with different activators

Catalyst	Activator	Temperature/time (°C/min)	A[c]	Tm (°C)	Mw (10^{-3} g/mol)	PDI
C_5[a]	None	20/20	43	–	125	1.8
C_5[a]	None	40/20	304	–	143	2.2
C_4[b]	$Ni(COD)_2$	20/20	220	127	104[d]	1.3
C_4[b]	BCF	25/10	850	122	350	2.3
C_3	MMT[e]	40/40	965	128	500	3.4

Polymerization time = 20 min, pressure = 100 psi, toluene as solvent in a autoclave reactor
[a] B = Lut
[b] B = PMe_3, Ar = 2,6-diisopropylphenyl in all systems
[c] Activity in kg PE mol^{-1} h^{-1}
[d] Derived from Mn to PDI
[e] MMT = acid treated montmorillonite

activate the catalyst results in a polymerization activity of 850 kg mol^{-1} h^{-1}, significantly higher than either the single component or the $Ni(COD)_2$ activated systems. The use of montmorillonite to activate the catalyst results in a system with a polymerization activity of 965 kg mol^{-1} h^{-1} which is in the same range as that observed with BCF activation. Therefore, the trend for polymerization activities is: Lewis Acids > phosphine scavengers ≈ single component systems.

The polymerization activity, however, is not always the deciding factor. The selection of a phosphine scavenger/single component system produces materials which generally have a lower polydispersity, and perhaps more importantly, have the ability to incorporate some polar copolymers. This has allowed for the formation of new materials, particularly gradient copolymers of polar norbornene derivatives [127]. When similar copolymerizations are attempted with the acid activated species, however, the reaction quickly terminates and no product is formed. This suggests that the cationic nature of the metal center prevents the copolymerization of polar monomers either by binding the incoming monomer too tightly or through a back biting mechanism which terminates the polymerization following an insertion by a polar monomer [127–129]. However, the use of milder Lewis acids may permit the limited incorporation of polar comonomers [72, 130].

2.4.3.2 Effect of Ligand Structure on the Polymerization Activities/Properties

There are a number of variations to the ligand structure which provide for interesting comparisons. The ability to vary the substituents on the phenyl rings provides access to complexes with drastically varied pseudo-axial bulk. Variation to the ligand backbone changes the rigidity of the ligand and as a result the steric environment of the metal center. Increases to the bulk of the ligand backbone resulted in increased polymerization activity and higher molecular weight products [73]. The effect of pseudo-axial bulk on the polymerization activity and polymer properties have principally been carried out in systems which are activated with $Ni(COD)_2$. While it would be possible to evaluate the same catalysts using acidic activators which bind to the ligand backbone, it was suspected that the large steric demands imposed by the activators may affect other aspects of the polymerization reaction. A summary of the results are shown in Table 2.2. Several trends were observed in the polymerizations, the first is that the both the polymerization activity and the molecular weight of the materials produced by the systems decrease with reduced pseudo-axial bulk, the second is that the PDI of the materials and the melting temperature increase with decreasing pseudo-axial bulk.

The nature of these trends is well established particularly in systems such as the Brookhart catalysts. The effect of the pseudo-axial bulk on the polymerization properties and activity is twofold. Nickel catalysts have a tendency to under rapid β-hydride transfer followed by elimination of the polymer chain, this is part of what initially made them useful as oligomerization catalysts. However, this property is undesirable in polymerization catalysts for obvious reasons. In order for the catalyst to undergo the elimination step there must be an incoming monomer which can displace the polymer chain. The trajectory of this incoming monomer, however, is important, only those monomers which arrive at the metal center above/below the plain of the ligand/metal center can displace the polymer. Thus, steric bulk above and below the plain of the ligand, prevents this monomer trajectory and in turn reduces the rate of termination/elimination in the system, increased pseudo-axial bulk results in increased molecular weight.

The presence of the pseudo-axial bulk however, has a secondary effect on the polymer properties, in particular the Tm of the polymers which is related to the chain branching. Nickel systems in particular, can undergo a process known as chain walking (or even chain running) [88, 131]. In this process, when the metal center and polymer undergo a β-hydride transfer but lack the incoming monomer to displace the polymer chain, the nickel center can recoordinate to the more stable β-carbon. This process can be repeated, over and over again until either the

Table 2.2 Polymerization data for α-iminocarboxamide catalysts activated by $Ni(COD)_2$

Catalyst	A[a]	Tm (°C)	Mn (10^{-3} g/mol)	PDI
C_6[b]	385	125	110	1.4
C_6[c]	196	132	73	1.8
C_6[d]	210	130	76	1.7
C_6[e]	133	132	61	1.9
C_6[f]	N.A.	–	–	–
C_6[g]	49	–	Oligomers	–

[a] Activity is given in units of kg PE mol^{-1} h^{-1}, N.A. = not active
[b] R = isopropyl
[c] R = ethyl
[d] R_1 = methyl, R_2 = isopropyl
[e] N∩O R = methyl
[f] N∩N R = methyl
[g] N∩N R_1 = phcnyl, R_2 = isopropyl

polymer is terminated, or a monomer is inserted. Once the monomer inserts however, it produces a branch in the polymer. In general it has been found that the larger the steric bulk, the higher the amount of chain branching and the lower the melting temperature.

2.4.3.3 Effect of the Presence of Exocyclic Functionality

In many systems the effect of exocyclic functionality is difficult to evaluate, often times the reactivity of the corresponding unfunctionalized system is not reported, and even at times the synthesis of the system is not even feasible. In many instances the introduction of the exocyclic functionality produces a slight change in the geometry of the ligand, in particular in systems such as C_{13} and C_{14}, in which the beta carbon of the ring is change from an sp^3 hybridization, lacking conjugation to a conjugated sp^2 hybridization (Fig. 2.32). This may introduce secondary effects on the polymerization behavior of the catalysts.

The polymerization activities and polymer properties of the catalysts shown in Fig. 2.32 are shown in Table 2.3. Comparison of the polymerization activity and

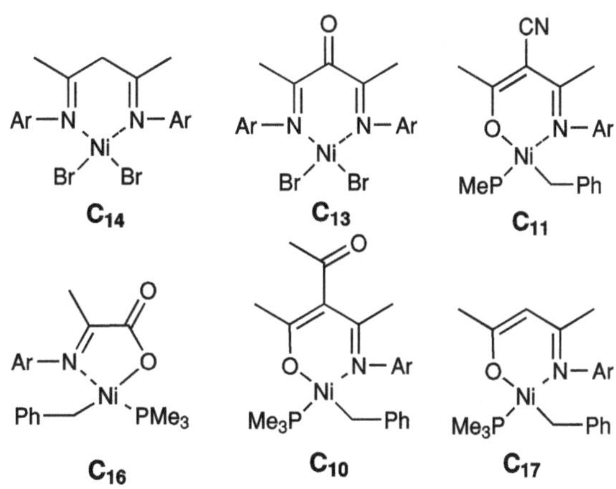

Fig. 2.32 Complexes bearing exocyclic functionality

Table 2.3 Polymerization activities of complexes bearing exocyclic functionalities

Catalyst	Temperature/Time (°C/min)	A[a]	Tm (°C)	Mw (10^{-3} g/mol)	PDI
C_{14}	32/10	8	–	–	–
C_{13}	32/10	5,800	117	282	1.1
C_{11}	20/40	1,070	83	13.9	1.8
C_{16}	25/1,080	2.14	127	249	2.2
C_{10}	30/20	5,450	–	Oligomers	–
C_{17}	30/20	51	–	Oligomers	–

[a] Activity given in units kg PE mol^{-1} h^{-1} . Refs: C_{14}, C_{13} [38], C_{11} [51], C_5 [72], C_{10}, C_{17} [41]

polymer properties of the materials produced by the complexes bearing the basic ligand, one which does not bear any functionality in the backbone, with those that bear some functionality is illustrative of the effect of the functionality, and in particular the remote activation of the complex. Comparing C_{14} with C_{13} for instance reveals that the presence of the carbonyl in the ligand backbone, and its subsequent interaction with a Lewis acid (MMAO in this instance) produces a tremendous increase in the polymerization activity, changing from 8 to 5800 kg mol^{-1} h^{-1} [38]. The effect of the Zwitterionic interaction through the ligand backbone is also clearly demonstrated comparing C_{10}, C_{11} and C_{17}. In these systems the functionality is introduced in such a ways as to not disrupt the ligand geometry. The polymerization activity of the complexes C_{11} and C_{10}, 1070 and 5450 kg mol^{-1} h^{-1}, respectively, is substantially higher than that of complex C_{17}, 51 kg mol^{-1} h^{-1}. This trend is consistent with that observed with C_{13} and C_{14}, and suggests that carbonyl groups are much more effective in producing a cationic metal center when compared with the cyano functionality.

2.5 Conclusions and Future Directions

The field of polymerization catalysis is very large with new ligands or catalysts appearing in the literature on a regular basis. The focus of these investigations is primarily, at least at this moment, focused on the development of either systems with either enhanced insertion control, single component systems, or systems which can tolerate polymer monomers. Many of these systems are based on imine derivative ligands. The ease of synthesis, through condensation reactions, allows for large ligand libraries to be generated. While originally these ligands were focused on late transition metal catalysts, they have been numerous developments for their use in early transition metal catalysts. While α-iminocarboxamide-based catalysts have been shown to be effectively activated exocyclically in the presence of Lewis acids, the modification of the ligand structure is somewhat limited by the nature of the ligand. The expansion of the ligand backbone by a single carbon however, generates what could be considered NacNac derivative ligands. These ligands are already present in a number of catalytic systems including polymerization catalysts. The modification of these ligands such that they too are capable of being activated remotely is the focus of a number of investigations, and the results thus far show increased polymerization activity compared to the unmodified and normally activated systems.

While there is still much to explore in this area, the focus of our lab has expanded to include the use of these systems in early metal, in particular zirconium, polymerization catalysts. The incorporation of NacNac style ligands into zirconium complexes produces catalysts which are active for polymerization of olefins while at the same time generating structurally interesting complexes [128]. In many of the early metal systems the ligand assumes a boat configuration with the β-carbon associating with the metal center, producing a distorted configuration.

While these systems do not show the same activity as their metallocene cousins, they are interesting for reactions in which the use of large quantities of activator is undesirable.

The ability to remotely modify the electronics at a metal center via interactions with Lewis acids presents interesting and potentially useful application. These interactions can not only affect the activity and reactivity of the metal center, but can also serve as catalyst activators. In addition, this interaction can serve as a method for the heterogenization of the complex, in particular with supports which are Lewis acids. Future directions in this area include studying the effect of Lewis acidity on the activity of the resulting adducts, as well as the development of better Lewis acid-complex pairs. This may aid in the development of systems which are capable of polar comonomer insertion.

Acknowledgments The authors wish to acknowledge the kind assistance of M. Valderrama, V. Dougnac and M. Deij in the preparation of this work. New x-ray structures were supplied via collaboration with Professors G. Erker and R. Frölich (Organisch-Chemisches Institut, Universtät Münster). Partial funding was provided through FONDECYT 1100286 and the Alexander von Humboldt Foundation.

References

1. Kaminsky W (2008) Trends in polyolefin chemistry. Macromol Chem Phys 209(5):459–466
2. Moulijn JA, Leeuwen PWNM, Santen RA (1993) Catalysis: an integrated approach to homogeneous, heterogeneous and industrial catalysis. In: Studies in surface science and catalysis, vol 79. Elsevier, Amsterdam, pp 15–16
3. Tuchbreiter A, Mülhaupt R (2001) The polyolefin challenges: catalyst and process design, tailor-made materials, high-throughput development and data mining. Macromol Symp 173(1):1–20
4. Coates GW (2000) Precise control of polyolefin stereochemistry using single-site metal catalysts. Chem Rev 100(4):1223–1252
5. Ribour D, Monteil V, Spitz R (2008) Detitanation of $MgCl_2$ supported Ziegler Natta catalysts for the study of active sites organization. J Polym Sci Part A 46(16):5461–5470
6. Britovsek GJP, Gibson VC, Wass DF (1999) The search for new-generation olefin polymerization catalysts: life beyond metallocenes. Angew Chem Int Ed 38(4):428–447
7. Busico V (2007) Catalytic olefin polymerization is a mature field. Isn't it? Macromol Chem Phys 208(1):26–29
8. Ittel SD, Johnson LK, Brookhart M (2000) Late-metal catalysts for ethylene homo- and copolymerization. Chem Rev 100(4):1169–1204
9. Britovsek GJP, Baugh SPD, Hoarau O, Gibson VC, Wass DF, White AJP, Williams DJ (2003) The role of bulky substituents in the polymerization of ethylene using late transition metal catalysts: a comparative study of nickel and iron catalyst systems. Inorg Chim Acta 345:279–291
10. Mecking S (2001) Olefin polymerization by late transition metal complexes—a root of ziegler catalysts gains new ground. Angew Chem Int Ed 40(3):534–540
11. Britovsek GJP, Gibson VC, McTavish SJ, Solan GA, White AJP, Williams DJ, Kimberley BS, Maddox PJ (1998) Novel olefin polymerization catalysts based on iron and cobalt. Chem Commun 1998(7):849–850
12. Takeuchi D (2010) Recent progress in olefin polymerization catalyzed by transition metal complexes: new catalysts and new reactions. Dalton Trans 39(2):311–328

13. Anselment TMJ, Vagin SI, Rieger B (2008) Activation of late transition metal catalysts for olefin polymerizations and olefin/CO copolymerizations. Dalton Trans (34):4537–4548
14. Keim W, Kowaldt FH, Goddard R, Krüger C (1978) Novel coordination of (benzoylmethylene)triphenylphosphorane in a nickel oligomerization catalyst. Angew Chem Int Ed 17(6):466–467
15. Peuckert M, Keim W (1983) A new nickel complex for the oligomerization of ethylene. Organometallics 2(5):594–597
16. Deng L, Woo TK, Cavallo L, Margl PM, Ziegler T (1997) The role of bulky substituents in brookhart-type Ni(II) diimine catalyzed olefin polymerization: a combined density functional theory and molecular mechanics study. J Am Chem Soc 119(26):6177–6186
17. Johnson LK, Killian CM, Brookhart M (1995) New Pd(II)- and Ni(II)-based catalysts for polymerization of ethylene and α-olefins. J Am Chem Soc 117(23):6414–6415
18. Johnson LK, Mecking S, Brookhart M (1996) Copolymerization of ethylene and propylene with functionalized vinyl monomers by palladium(II) catalysts. J Am Chem Soc 118(1):267–268
19. Camacho DH, Salo EV, Ziller JW, Guan Z (2004) Cyclophane-based highly active late-transition-metal catalysts for ethylene polymerization. Angew Chem Int Ed 43(14):1821–1825
20. Michalak A, Ziegler T (2000) DFT studies on substituent effects in palladium-catalyzed olefin polymerization. Organometallics 19(10):1850–1858
21. Wang C, Friedrich S, Younkin TR, Li RT, Grubbs RH, Bansleben DA, Day MW (1998) Neutral nickel(II)-based catalysts for ethylene polymerization. Organometallics 17(15):3149–3151
22. Lee BY, Bazan GC, Vela J, Komon ZJA, Bu X (2001) α-Iminocarboxamidato nickel(II) ethylene polymerization catalysts. J Am Chem Soc 123(22):5352–5353
23. Bazan GC, Rojas R (2007) Single component, phosphine-free, initiators for ethylene homopolymerization and copolymerization with functionalized co-monomers. USA Patent 11/649,949
24. Rojas RS, Galland GB, Wu G, Bazan GC (2007) Single-component α-iminocarboxamide nickel ethylene polymerization and copolymerization initiators. Organometallics 26(22):5339–5345
25. Kim YH, Kim TH, Lee BY, Woodmansee D, Bu X, Bazan GC (2002) α-Iminoenamido ligands: a novel structure for transition-metal activation. Organometallics 21(15):3082–3084
26. Scott SL, Peoples BC, Yung C, Rojas RS, Khanna V, Sano H, Suzuki T, Shimizu F (2008) Highly dispersed clay-polyolefin nanocomposites free of compatibilizers, via the in situ polymerization of α-olefins by clay-supported catalysts. Chem Commun 35:4186–4188
27. Rojas RS, Wasilke JC, Wu G, Ziller JW, Bazan GC (2005) α-Iminocarboxamide nickel complexes: synthesis and uses in ethylene polymerization. Organometallics 24(23):5644–5653
28. Malkov AV, Vrankovaì K, Stoncius S, Kocovskyì P (2009) Asymmetric reduction of imines with trichlorosilane, catalyzed by sigamide, an amino acid-derived formamide: scope and limitations. J Org Chem 74(16):5839–5849
29. Prieur D, El Kazzi A, Kato T, Gornitzka H, Baceiredo A (2008) Novel *N*-phosphonio imine-catalyzed epoxidation reactions. Org Lett 10(11):2291–2294
30. Tseng CH, Chen YL, Yang SH, Peng SI, Cheng CM, Han CH, Lin SR, Tzeng CC (2010) Synthesis and antiproliferative evaluation of certain iminonaphtho[2, 3-b]furan derivatives. Bioorg Med Chem 18(14):5172–5182
31. Müller P, Gilabert DM (1988) Oxidation of amines to imines with hypervalent iodine. Tetrahedron 44(23):7171–7175
32. Takanashi K, Yamamoto K (2007) Divergent approach for synthesis and terminal modifications of dendritic polyphenylazomethines. Org Lett 9(25):5151–5154
33. Satoh N, Yamamoto K (2009) Self-assembled monolayers of metal-assembling dendron thiolate formed from dendrimers with a disulfide core. Org Lett 11(8):1729–1732
34. Suzuki M, Nakajima R, Tsuruta M, Higuchi M, Einaga Y, Yamamoto K (2006) Synthesis of ferrocene-modified phenylazomethine dendrimers possessing redox switching. Macromolecules 39(1):64–69

35. Higuchi M, Otsuka Y, Shomura R, Kurth DG (2008) Syntheses of novel bis-terpyridine and cyclic phenylazomethine as organic modules in organic-metallic hybrid materials. Thin Solid Films 516(9):2416–2420
36. Gopalakrishnan M, Sureshkumar P, Kanagarajan V, Thanusu J (2007) New environmentally-friendly solvent-free synthesis of imines using calcium oxide under microwave irradiation. Res Chem Intermed 33(6):541–548
37. Carlini C, De Luise V, Martinelli M, Galletti AMR, Sbrana G (2006) Homo- and copolymerization of methyl methacrylate with ethylene by novel Ziegler-Natta-type nickel catalysts based on N, O-nitro-substituted chelate ligands. J Polym Sci Part A 44(1):620–633
38. Azoulay J, Rojas R, Serrano A, Ohtaki H, Galland G, Wu G, Bazan G (2009) Nickel α-keto-β-diimine initiators for olefin polymerization. Angew Chem Int Ed 48(6):1089–1092
39. Yokota S, Tachi Y, Itoh S (2002) Oxidative degradation of β-diketiminate ligand in copper(II) and zinc(II) complexes. Inorg Chem 41(6):1342–1344
40. Allen SD, Moore DR, Lobkovsky EB, Coates GW (2003) Structure and reactivity of mono—and dinuclear diiminate zinc alkyl complexes. J Organomet Chem 683(1):137–148
41. Chen Y, Wu G, Bazan GC (2005) Remote activation of nickel catalysts for ethylene oligomerization. Angew Chem Int Ed 44(7):1108–1112
42. Gong S, Ma H (2008) β-Diketiminate aluminium complexes: synthesis, characterization and ring-opening polymerization of cyclic esters. Dalton Trans (25):3345–3357
43. Gong S, Ma H, Huang J (2008) Zirconium complexes with versatile β-diketiminate ligands: synthesis, structure, and ethylene polymerization. J Organomet Chem 693(23):3509–3518
44. Macho V, Králik M, Hudec J, Cingelova J (2004) One stage preparation of Schiff's bases from nitroarenes, aldehydes and carbon monoxide at presence of water. J Mol Catal A 209(1–2):69–73
45. Rahim M, Taylor NJ, Xin S, Collins S (1998) Synthesis and structure of acyclic bis(ketenimine) complexes of zirconium. Organometallics 17(7):1315–1323
46. Cheng M, Moore DR, Reczek JJ, Chamberlain BM, Lobkovsky EB, Coates GW (2001) Single-site β-diiminate zinc catalysts for the alternating copolymerization of CO_2 and epoxides: catalyst synthesis and unprecedented polymerization activity. J Am Chem Soc 123(36):8738–8749
47. Spencer DJE, Reynolds AM, Holland PL, Jazdzewski BA, Duboc-Toia C, Le Pape L, Yokota S, Tachi Y, Itoh S, Tolman WB (2002) Copper chemistry of β-diketiminate ligands: monomer/dimer equilibria and a new class of bis(β-oxo)dicopper compounds. Inorg Chem 41(24):6307–6321
48. Vitanova DV, Hampel F, Hultzsch KC (2005) Synthesis and structural characterisation of novel linked bis(-diketiminato) rare earth metal complexes. Dalton Trans (9):1565–1566
49. Lee SY, Na SJ, Kwon HY, Lee BY, Kang SO (2004) Syntheses and structures of a macrocyclic β-diketimine and its zinc and copper complexes. Organometallics 23(23): 5382–5385
50. Pilz MF, Limberg C, Ziemer B (2006) Xanthene-based ligand with two adjacent β-diiminato binding sites. J Org Chem 71(12):4559–4564
51. Boardman BM, Valderrama JM, Munoz F, Wu G, Bazan GC, Rojas R (2008) Remote activation of nickel complexes by coordination of $B(C_6F_5)_3$ to an exocyclic carbonitrile functionality. Organometallics 27(8):1671–1674
52. Song D-P, Ye W-P, Wang Y-X, Liu J-Y, Li Y-S (2009) Highly active neutral nickel(II) catalysts for ethylene polymerization bearing modified β-ketoiminato ligands. Organometallics 28(19):5697–5704
53. Lee BY, Kwon HY, Lee SY, Na SJ, Han SI, Yun H, Lee H, Park Y-W (2005) Bimetallic anilido-aldimine zinc complexes for epoxide/CO_2 copolymerization. J Am Chem Soc 127(9):3031–3037
54. Song D-P, Wu J-Q, Ye W-P, Mu H-L, Li Y-S (2009) Accessible, highly active single-component β-ketiminato neutral nickel(II) catalysts for ethylene polymerization. Organometallics 29(10):2306–2314

55. Lippe K, Gerlach D, Kroke E, Wagler J (2008) *N*-(o-Aminophenyl)-2-oxy-4-methoxybenzophenoneimine—Si-chelation by a tridentate ONN ligand system versus benzimidazoline formation. Inorg Chem Commun 11(5):492–496
56. Pearson RG (1963) Hard and soft acids and bases. J Am Chem Soc 85(22):3533–3539
57. Pearson RG, Songstad J (1967) Application of the principle of hard and soft acids and bases to organic chemistry. J Am Chem Soc 89(8):1827–1836
58. Bauers FM, Mecking S (2001) High molecular mass polyethylene aqueous latexes by catalytic polymerization. Angew Chem Int Ed 40(16):3020–3022
59. Zuideveld MA, Wehrmann P, Röhr C, Mecking S (2004) Remote substituents controlling catalytic polymerization by very active and robust neutral nickel(II) complexes. Angew Chem Int Ed 43(7):869–873
60. Schroder DL, Keim W, Zuideveld MA, Mecking S (2002) Ethylene polymerization by novel, easily accessible catalysts based on nickel(II) diazene complexes. Macromolecules 35(16):6071–6073
61. Gibson VC, Spitzmesser SK (2002) Advances in non-metallocene olefin polymerization catalysis. Chem Rev 103(1):283–316
62. Gibson VC, Tomov A, Wass DF, White AJP, Williams DJ (2002) Ethylene polymerisation by a copper catalyst bearing-diimine ligands. J Chem Soc Dalton Trans 11:2261–2262
63. Feldman J, McLain SJ, Parthasarathy A, Marshall WJ, Calabrese JC, Arthur SD (1997) Electrophilic metal precursors and a β-diimine ligand for nickel(II)- and palladium(II)-catalyzed ethylene polymerization. Organometallics 16(8):1514–1516
64. Caris R, Peoples BC, Valderrama M, Wu G, Rojas R (2009) Mono and bimetallic nickel bromide complexes bearing azolate-imine ligands: Synthesis, structural characterization and ethylene polymerization studies. J Organomet Chem 694(12):1795–1801
65. Valdebenito C, Garland MT, Quijada R, Rojas R (2009) Acetamidine complexes as catalysts for ethylene polymerization. J Organomet Chem 694(5):717–725
66. Li J, Song H, Cui C (2010) Synthesis of 2-(*N*-arylimino)pyrrolide nickel complexes and polymerization of methyl methacrylate. Appl Organomet Chem 24(2):82–85
67. Bourget-Merle L, Lappert MF, Severn JR (2002) The chemistry of β-diketiminatometal complexes. Chem Rev 102(9):3031–3066
68. Green SP, Jones C, Stasch A (2007) Stable magnesium(I) compounds with Mg–Mg bonds. Science 318(5857):1754–1757
69. Badiei Y, Dinescu A, Dai X, Palomino R, Heinemann F, Cundari T, Warren T (2008) Copper–nitrene complexes in catalytic C–H amination. Angew Chem Int Ed 47(51):9961–9964
70. Dugan TR, Holland PL (2009) New routes to low-coordinate iron hydride complexes: the binuclear oxidative addition of H_2. J Organomet Chem 694(17):2825–2830
71. Chai J, Zhu H, Stuckl AC, Roesky HW, Magull J, Bencini A, Caneschi A, Gatteschi D (2005) Synthesis and reaction of $[\{HC(CMeNAr)_2\}Mn]_2$ (Ar = 2, 6-$iPr_2C_6H_3$): the complex containing three-coordinate manganese(I) with a Mn–Mn bond exhibiting unusual magnetic properties and electronic structure. J Am Chem Soc 127(25):9201–9206
72. Wang L, Johnson LK, Ionkin AS (2001) Polymerization of olefins. WO 01092348
73. Azoulay JD, Itigaki K, Wu G, Bazan GC (2008) Influence of steric and electronic perturbations on the polymerization activities of α-iminocarboxamide nickel complexes. Organometallics 27(10):2273–2280
74. Hillier AC, Grasa GA, Viciu MS, Lee HM, Yang C, Nolan SP (2002) Catalytic cross-coupling reactions mediated by palladium/nucleophilic carbene systems. J Organomet Chem 653(1–2):69–82
75. Miyaura N, Suzuki A (1995) Palladium-catalyzed cross-coupling reactions of organoboron compounds. Chem Rev 95(7):2457–2483
76. Suzuki A (1982) Organoborates in new synthetic reactions. Acc Chem Res 15(6):178–184
77. Wolfe JP, Wagaw S, Marcoux J-Fo, Buchwald SL (1998) Rational development of practical catalysts for aromatic carbon nitrogen bond formation. Acc Chem Res 31(12):805–818
78. Gassman PG, Drewes HR (1978) The ortho functionalization of aromatic amines. Benzylation, formylation, and vinylation of anilines. J Am Chem Soc 100(24):7600–7610

79. Turner SS, Collison D, Mabbs FE, Halliwell M (1997) Preparation, magnetic properties and crystal structure of bis[3-(4-pyridyl)pentane-2,4-dionato]copper(II). J Chem Soc Dalton Trans (7):1117–1118
80. Chen B, Fronczek FR, Maverick AW (2003) Solvent-dependent 4^4 square grid and $6^4.8^2$ NbO frameworks formed by $Cu(Pyac)_2$ (bis[3-(4-pyridyl)pentane-2,4-dionato]copper(II)). Chem Commun 17:2166–2167
81. Tayyari SF, Reissi H, Milani-Nejad F, Butler IS (2001) Vibrational assignment of α-cyanoacetylacetone. Vib Spectrosc 26(2):187–199
82. Tsiamis C, Tzavellas LC, Stergiou A, Anesti V (1996) Variable coordination and conformation of the 3-cyano-2, 4-pentanedionato anion in a mixed-ligand binuclear copper(II) Chelate. Inorg Chem 35(17):4984–4988
83. Thambidurai S, Jeyasubramanian K, Ramalingam SK (1996) Bis(benzotriazol-1-yl) methylimine as a novel cyanating agent for labile β-diketonates. Polyhedron 15(22): 4011–4014
84. Angelova O, Matsichek I, Atanasov M, Petrov G (1991) Chelating modes of 3-substituted 2, 4-pentanediones. Crystal and electronic structure of bis(3-cyano-2, 4-pentanedionato) cobalt(II). Inorg Chem 30(8):1943–1949
85. Bansleben DA, Friedrich SK, Younkin TR, Grubbs RH, Wang C, Li RT (1998) Catalyst compostitions and processes for olefin polymers and copolymers. PCT/US1998/003165
86. Wiberg E, Wiberg N, Holleman AF (2001) Inorganic chemistry. Academic Press, San Diego
87. Keim W, Appel R, Storeck A, Krüger C, Goddard R (1981) Novel nickel- and palladium-complexes with aminobis(imino)phosphorane ligands for the polymerization of ethylene. Angew Chem Int Ed 20(1):116–117
88. Möhring VM, Fink G (1985) Novel polymerization of α-olefins with the catalyst system nickel/aminobis(imino)phosphorane. Angew Chem Int Ed 24(11):1001–1003
89. Ostoja Starzewski KA, Witte J (1987) Control of the molecular weight of polyethene in syntheses with bis(ylide)nickel catalysts. Angew Chem Int Ed 26(1):63–64
90. Kurtev K, Tomov A (1994) Ethene polymerization by binuclear nickel–ylide complexes. J Mol Catal A 88(2):141–150
91. Desjardins SY, Cavell KJ, Jin H, Skelton BW, White AH (1996) Insertion into the nickel–carbon bond of N–O chelated arylnickel(II) complexes. The development of single component catalysts for the oligomerisation of ethylene. J Organomet Chem 515(1–2): 233–243
92. Desjardins SY, Cavell KJ, Hoare JL, Skelton BW, Sobolev AN, White AH, Keim W (1997) Single component N–O chelated arylnickel(II) complexes as ethene polymerisation and CO/ethene copolymerisation catalysts. Examples of ligand induced changes to the reaction pathway. J Organomet Chem 544(2):163–174
93. Cámpora J, Del Mar Conejo M, Mereiter K, Palma P, Pérez C, Reyes ML, Ruiz C (2003) Synthesis of dialkyl, diaryl and metallacyclic complexes of Ni and Pd containing pyridine, [alpha]-diimines and other nitrogen ligands: crystal structures of the complexes cis-NiR2py2 (R = benzyl, mesityl). J Organomet Chem 683(1):220–239
94. Cámpora J, López JA, Maya C, Palma P, Carmona E, Valerga P (2002) Synthesis and reactivity studies on alkyl-aryloxo complexes of nickel containing chelating diphosphines: cyclometallation and carbonylation reactions. J Organomet Chem 643–644:331–341
95. Klein HF, Karsch HH (1972) Chem Ber 105
96. Wilke G, Herrmann G (1966) Stabilisierte Nickeldialkyle. Angew Chem 78(11):591
97. Wilke G, Herrmann G (1966) Stabilized dialkylnickel compounds. Angew Chem Int Ed 5(6):581–582
98. Greiser T, Kopf J, Thoennes D, Weiss E (1980) Über metallalkyl—und—arylverbindungen: XXV. Die kristallstruktur von dimethyl(N, N, N′, N′-tetramethylethylendiamin)magnesium. J Organomet Chem 191(1):1–6
99. Kaschube W, Poerschke K-R, Angermund K, Kruger C, Wilke G (1988) Chem Ber 121
100. Kaschube W, Pörschke KR, Wilke G (1988) tmeda-Nickel-Komplexe, : III. (N, N, N′, N′-Tetramethylethylendiamin)-(dimethyl)nickel(II). J Organomet Chem 355(1–3):525–532

101. Wilke G, Bogdanovi B, Hardt P, Heimbach P, Keim W, Kröner M, Oberkirch W, Tanaka K, Steinrücke E, Walter D, Zimmermann H (1966) Allyl-transition metal systems. Angew Chem Int Ed 5(2):151–164
102. Wilke G, Bogdanovic B, Hardt P, Heimbach P, Keim W, Kröner M, Oberkirch W, Tanaka K, Steinrücke E, Walter D, Zimmermann H (1966) Allyl-Übergangsmetall-Systeme. Angew Chem 78(3):157–172
103. Schunn RA, Ittel SD, Cushing MA, Baker R, Gilbert RJ, Madden DP (2007) Bis(1,5-Cyclooctadiene)Nickel(0). In: Robert JA (ed) Inorganic syntheses. Wiley, NY, pp 94–98
104. Mabbott DJ, Mann BE, Maitlis PM (1977) Cationic (η_{-3}-allylic)(4-diene)-palladium and—platinum complexes. J Chem Soc Dalton Trans (3):294–299. doi:10.1039/DT9770000294
105. Pardy RBA, Tkatchenko I (1981) Synthesis of (3-allyl)(4-cyclo-octa-1,5-diene) nickel cations: valuable catalysts for the oligomerization of alkenes. J Chem Soc Chem Commun 49–50
106. White DA, Doyle JR, Lewis H (2007) Cationic diene Complexes of palladium (II) and platinum (II). In: Cotton FA (ed) Inorganic syntheses. Wiley, NY, pp 55–65
107. Carmona E, Paneque M, Poveda ML (1989) Synthesis and characterization of some new organometallic complexes of nickel(II) containing trimethylphosphine. Polyhedron 8(3):285–291
108. Carmona E, Marin JM, Paneque M, Poveda ML (1987) New nickel o-methylbenzyl complexes. Crystal and molecular structures of $Ni(\mu^3\text{-}CH_2C_6H_4\text{-o-Me})Cl(PMe_3)$ and $Ni(\eta^1\text{-}CH_2C_6H_4\text{-o-Me})_4(PMe_3)_2(\mu^3\text{-}OH)_2$. Organometallics 6(8):1757–1765
109. Carlini C, Martinelli M, Raspolli Galletti AM, Sbrana G (2003) Homopolymerization of methyl methacrylate by novel salicylaldiminate-nickel/methylaluminoxane catalysts obtained by oxidative addition of the chelate ligand to a nickel(0) precursor. J Polym Sci Part A 41(11):1716–1724
110. Carlini C, Martinelli M, Passaglia E, Raspolli Galletti A, Sbrana G (2001) Homopolymerization of methyl methacrylate by novel ziegler-natta-type catalysts based on bis(chelate)–nickel(II) complexes and methylaluminoxane. Macromol Rapid Commun 22(9):664–668
111. Carlini C, Martinelli M, Raspolli Galletti AM, Sbrana G (2002) Copolymerization of ethylene with methyl methacrylate by Ziegler-Natta-type catalysts based on nickel salicylaldiminate/methylalumoxane systems. Macromol Chem Phys 203(10–11): 1606–1613. doi:10.1002/15213935(200207)
112. Ceder RM, Muller G, Ordinas M, Font-Bardia M, Solans X (2003) Nickelacycles with anionic C-N-N terdentate α-diimine based ligands. Reaction with ethylene. Dalton Trans (15):3052–3059
113. Hu WQ, Sun XL, Wang C, Gao Y, Tang Y, Shi LP, Xia W, Sun J, Dai HL, Li XQ, Yao XL, Wang XR (2004) Synthesis and characterization of novel tridentate [NOP] titanium complexes and their application to copolymerization and polymerization of ethylene. Organometallics 23:1684–1688
114. Bellabarba RM, Gomes PT, Pascu SI (2003) Synthesis of allyl- and aryl-iminopyrrolyl complexes of nickel. Dalton Trans 23:4431–4436
115. Lee BY, Bu X, Bazan GC (2001) Pyridinecarboxamidato nickel(II) complexes. Organometallics 20(25):5425–5431
116. Boardman BM, Wu G, Rojas R, Bazan GC (2009) Binding modes of a dimethyliminopentanone ligand on nickel pre-catalysts toward olefin polymerization. J Organomet Chem 694 (9–10):1380–1384
117. Okada M, Nakayama Y, Ikeda T, Shiono T (2006) Synthesis of uniquely branched polyethylene by anilinonaphthoquinone ligated nickel complex activated with tris(pentafluorophenyl)borane. Macromol Rapid Commun 27(17):1418–1423
118. Shim CB, Kim YH, Lee BY, Dong Y, Yun H (2003) [2-(Alkylideneamino)benzoato]nickel(II) complexes:active catalysts for ethylene polymerization. Organometallics 22(21):4272–4280

119. Rojas R, Frohlich R, Erker G, Kehr G (2010) Exocyclic activation of allyl nickel complexes based on Nacnac derivatives
120. Sacconi L, Ciampolini M (1963) The occurrence of paramagnetic tetrahedral forms of *N*-arylsalicylaldiminonickel (II) complexes in solution at elevated temperatures. J Am Chem Soc 85(12):1750–1753
121. Chen EY-X, Marks TJ (2000) Cocatalysts for metal-catalyzed olefin polymerization: activators, activation processes, and structure activity relationships. Chem Rev 100(4): 1391–1434
122. Focante F, Mercandelli P, Sironi A, Resconi L (2006) Complexes of tris(pentafluorophenyl)boron with nitrogen-containing compounds: synthesis, reactivity and metallocene activation. Coord Chem Rev 250(1–2):170–188
123. Jacobsen H, Berke H, Doring S, Kehr G, Erker G, Frohlich R, Meyer O (1999) Lewis acid properties of tris(pentafluorophenyl)borane. structure and bonding in L–$B(C_6F_5)_3$ complexes. Organometallics 18(9):1724–1735
124. Piers WE, Chivers T (1997) Pentafluorophenylboranes: from obscurity to applications. Chem Soc Rev 26(5):345–354
125. Younkin TR, Connor EF, Henderson JI, Friedrich SK, Grubbs RH, Bansleben DA (2000) Neutral, single-component nickel (II) polyolefin catalysts that tolerate heteroatoms. Science 287(5452):460–462
126. Boffa LS, Novak BM (2000) Copolymerization of polar monomers with olefins using transition–metal complexes. Chem Rev 100(4):1479–1494
127. Diamanti SJ, Khanna V, Hotta A, Coffin RC, Yamakawa D, Kramer EJ, Fredrickson GH, Bazan GC (2006) Tapered block copolymers containing ethylene and a functionalized comonomer. Macromolecules 39(9):3270–3274
128. Diamanti SJ, Ghosh P, Shimizu F, Bazan GC (2003) Ethylene homopolymerization and copolymerization with functionalized 5-norbornen-2-yl monomers by a novel nickel catalyst system. Macromolecules 36:9731–9735
129. Diamanti SJ, Khanna V, Hotta A, Yamakawa D, Shimizu F, Kramer EJ, Fredrickson GH, Bazan GC (2004) Synthesis of block copolymer segments containing different ratios of ethylene and 5-norbornen-2-yl acetate. J Am Chem Soc 126(34):10528–10529
130. Carone CLP, Bisatto R, Galland GB, Rojas R, Bazan G (2008) Ethylene polymerization and copolymerization with a polar monomer using an α-iminocarboxamide nickel complex activated by trimethylaluminum. J Polym Sci Part A 46(1):54–59
131. Guan Z, Cotts PM, McCord EF, McLain SJ (1999) Chain walking: a new strategy to control polymer topology. Science 283(5410):2059–2062

Chapter 3
Oligomerization and Polymerization of Olefins with Iron and Cobalt Catalysts Containing 2,6-Bis(imino)pyridine and Related Ligands

Lidong Li and Pedro T. Gomes

Abstract The key advances in oligo- or polymerization of olefins catalyzed by the 2,6-bis(imino)pyridine-based iron and cobalt complexes were presented. Particular attention was paid to trace the structural evolution of the 2,6-bis(imino)pyridine and related ligands, recent understandings of the relationship between catalyst structure and catalytic activity, and applications of iron and cobalt catalysts in oligo- or polymerization of olefins. Additionally, the mechanistic aspects of the oligo- or polymerization reactions, involving catalyst activation, oligo- or polymerization initiation, chain propagation and chain transfer, activators or cocatalysts, monomers and comonomers, and immobilization of iron and cobalt catalysts, have also been reviewed and discussed.

List of Abbreviations

AIBN	Azobis-*iso*-butyronitrile
ATRP	Atom transfer radical polymerization
BIP	2,6-Bis(imino)pyridine
BTCA	Benzene-1,3,5-tricarboxylic acid
Bz	Benzyl
^{13}C NMR	Carbon-13 nuclear magnetic resonance spectroscopy
CGC	Constrained geometry catalysts
CRYSTAF	Crystallization analysis fractionation
DFT	Density functional theory

L. Li (✉) · P. T. Gomes (✉)
Centro de Química Estrutural, Departamento de Engenharia Química e Biológica, Instituto Superior Técnico, Universidade Técnica de Lisboa, Av. Rovisco Pais, 1049-001 Lisbon, Portugal
e-mail: pedro.t.gomes@ist.utl.pt

L. Li
e-mail: dick_lld@hotmail.com

G. Giambastiani and J. Cámpora (eds.), *Olefin Upgrading Catalysis by Nitrogen-based Metal Complexes II*, Catalysis by Metal Complexes, 36,
DOI: 10.1007/978-94-007-0696-5_3, © Springer Science+Business Media B.V. 2011

DSC	Differential scanning calorimetry
EPR	Electron paramagnetic resonance spectroscopy
ESI-MS	Electrospray ionization mass spectrometry
EVE	Ethyl vinyl ether
GPC/SEC	Gel permeation chromatography/size-exclusion chromatography
^{1}H NMR	Proton nuclear magnetic resonance spectroscopy
i-BVE	*iso*-Butyl vinyl ether
LDA	Lithium diisopropylamide
LLDPE	Linear low-density polyethylene
MA	Methyl acrylate
MAO	Methylaluminoxane
MM	Molecular mechanics
MMA	Methyl methacrylate
MMAO	Modified methylaluminoxane
M_n	Number-average molecular weight
M_w	Weight-average molecular weight
M_w/M_n	Polydispersity index
n-BMA	*n*-Butyl methacrylate
n-BVE	*n*-Butyl vinyl ether
NMR	Nuclear magnetic resonance spectroscopy
OTf	Triflate group (trifluoromethanesulfonate, $CF_3SO_3^-$)
OTs	Tosylate group (*p*-toluenesulfonate, *p*-$CH_3C_6H_4SO_3^-$)
PE	Polyethylene
PES	Potential energy surface
PMAO	Polymeric methylaluminoxane
PP	Polypropylene
PS	Polystyrene
QM	Quantum mechanics
SAXS	Small-angle X-ray scattering
SGEF	Solvent gradient elution fractionation
SHOP	Shell higher olefin process
t-BA	*tert*-Butyl acrylate
t-BMA	*tert*-Butyl methacrylate
TEAO	Tetraethylaluminoxane
WAXD	Wide-angle X-ray diffraction
XPS	X-ray photoelectron spectroscopy

3.1 Introduction and Scope

The discovery of methylaluminoxane (MAO), by Kaminsky and coworkers in 1980 [1, 2], boosted the development of homogeneous olefin polymerization catalysts in the past three decades, both at academic and industrial levels.

This Lewis acidic alkylating species, which results from the partial hydrolysis of trimethylaluminum ($AlMe_3$), can activate very efficiently a great variety of pre-catalyst metal complexes (e.g., derivatives of metal halides or alkyls) to form single-site homogeneous catalysts for the homo- and copolymerization of olefins. Compared to the traditional heterogeneous Ziegler–Natta catalysts, the homogeneous catalysts possess active species with well-defined structures, which can be closely related to those of their catalyst precursors, making possible the establishment of structure–activity relationships. In addition, their architectures and symmetries can be varied through the modification of their ligands, leading to a vast number of different homogeneous catalysts that produce many kinds of polymers with new microstructures, such as those of the hemiisotactic polypropylene (PP) [3] or the isotactic-atactic stereoblock PP [4, 5].

From the metallocenes to post-metallocenes and then to late transition metal catalysts, a series of landmarks can be defined during the chronological development of the homogeneous olefin polymerization catalysts. These include: the bis(cyclopentadienyl) titanium or zirconium dichloride [1, 2], the C_2 symmetrical *rac*-ethylenebis(indenyl)titanium dichloride (isotactic PP) [6–8], the C_s symmetrical isopropyl(cyclopentadienyl-1-fluorenyl)zirconium dichloride (syndiotactic PP) [9], the *ansa*-cyclopentadienyl-amide titanium or zirconium dichlorides (constrained geometry catalysts, CGC) (ethylene/α-olefin copolymers) [10–12], the α-diimine nickel(II) and palladium(II) dihalides (branched polyethylenes (PE)) [13], the 2,6-bis(imino)pyridine iron(II) and cobalt(II) dihalides (linear PE) [14–16], the phenoxy-imine-based Group 4 transition metal catalysts (FI catalysts) (living polymerization of ethylene and propylene) [17] and the neutral single-component salicylaldimine-based nickel(II) alkyl or aryl catalysts [18]. From the comparison of these different homogeneous catalysts, one can notice that they are based on a limited number of metals (Ti, Zr, Hf, Fe, Co, Ni, Pd), the main differences lying on the diversity of their ligands. Additionally, it can be observed that only when a particular ligand set combines with a certain metal a desirable catalytic activity is promoted, whereas many other ligands fail. The reason why it happens still remains a mystery, which is in fact one of the driving forces for further research in the development of new and high performance catalysts. Coordination of a ligand to a metal not only alters its basic physical and chemical properties, such as color, solubility, stability, symmetry, etc., but affects its electronic distribution and coordination environment leading to varied electronic, magnetic and catalytic properties. It is worth to note that, sometimes, the ligands can also influence the catalytic performances of the catalysts in olefin polymerization through the establishment of non-bonded interactions of the ligands with the growing polymer chains, which are coordinated to the metal centers. For instance, FI catalysts containing fluorinated aryl phenoxy-imine chelate ligands demonstrated to induce unprecedented living polymerization effects with both ethylene and propylene, through an attractive interaction between one of the fluorine atoms in the ligand and a β-hydrogen atom on the growing polymer chain [19–21].

Due to the significance and versatility of the ligands in the homogeneous olefin polymerization catalysts, the design of the precatalysts is mainly focused on the

Fig. 3.1 BIP iron and cobalt precatalysts used by Brookhart and Gibson in the oligomerization and polymerization of ethylene

design and modification of the ligand itself. The 2,6-bis(imino)pyridine (BIP) is one of the most well-known ligands in olefin polymerization since, during the late 1990s it was employed simultaneously by Brookhart and Gibson groups in the synthesis of metal complex precatalysts [14–16] (Fig. 3.1). These authors found that sterically demanding BIP ligands can impart high ethylene polymerization catalytic activities to late transition metals, especially iron and cobalt. A large amount of work has been devoted to the modification of this ligand and to the understanding of the chemistry of its metal derivatives, these subjects being addressed in reviews published recently [22–24]. BIP-based iron and cobalt complexes, when activated with MAO, are efficient catalysts for the conversion of ethylene either to high molecular weight linear polyethylenes or to α-olefins with Schulz–Flory distribution, which can be readily controlled by tuning the electronic and steric nature of the BIP ligands. Besides iron and cobalt, the BIP ligands can also coordinate to a variety of other transition metals, such as titanium [25], zirconium [25], hafnium [25], vanadium [26–33], chromium [31, 32, 34–38], molybdenum [39], nickel [40–42], etc., to generate the precatalysts that are active for olefin polymerization, when activated with MAO or MMAO (modified methylaluminoxane, $AlMeO{:}Al^{i}BuO = 3{:}1$). Furthermore, when reacted with representative or transition metal alkyls, the BIP ligands exhibit an unexpected and amazing chemistry, in which its several positions can be attacked by the alkylating agents, thereby leading to several different reactions, such as alkylation, deprotonation, dimerization, etc. [43].

In this chapter, we present a comprehensive review of the developments on the iron and cobalt catalyst systems based on the BIP and related ligands, and discuss their catalytic performances in olefin polymerization, with a particular emphasis on the ethylene monomer.

3.2 Bis(arylimino)pyridine Iron and Cobalt Complexes

3.2.1 Syntheses of 2,6-Bis(arylimino)pyridine Ligands

It is generally accepted that the polymerization of ethylene or α-olefins promoted by homogeneous single-site catalyst systems commonly involves, among others, two basic mechanistic steps: chain growth (propagation) by migratory insertion of the coordinated olefin into the metal-alkyl chain bond, which occurs in cationic alkyl-olefin metal intermediates, and chain transfer, taking place by various mechanisms, which is responsible for the decrease of the polymer molecular weight and broadening of the corresponding molecular weight distribution. In the case of late transition metal catalysts such as those based on nickel, it is well known that they usually yield olefin dimers or oligomers since their alkyl complexes readily undergo chain transfer by β-hydrogen elimination [44]. A typical example is the Shell higher olefin process (SHOP), developed by Keim and coworkers, in which linear α-olefins are obtained by nickel-catalyzed oligomerization of ethylene [45, 46]. In fact, although the tridentate BIP ligands and their metal derivatives have been known earlier [47–50], the use of BIP-based cobalt and iron precatalysts in the polymerization of ethylene to high molecular weight polyethylenes was only reported in 1998, when sterically demanding BIP ligands were employed by Brookhart's and Gibson's groups [14–16]. After this breakthrough, a large number of BIP ligands with different substitution patterns have been designed and synthesized to date.

Table 3.1 summarizes most of the symmetrical 2,6-bis(arylimino)pyridine ligands reported so far. According to the variation of the iminic carbon substituents (R) and the aryl groups employed, these ligands can be classified into several different types.

In the case of aldimine or acetimine ligands (e.g., **L1–L13** and **L14–L131**, Table 3.1), where the substituent R is H or methyl, respectively, the ligands are commonly prepared by the condensation reactions of either pyridine-2,6-dicarboxaldehyde or 2,6-diacetylpyridine with two equivalents of the appropriate substituted anilines. These reactions can be performed either in strongly polar (e.g., methanol [14, 51] and ethanol [16, 52, 53]) or nonpolar/weakly polar solvents (e.g., benzene [54, 55] and toluene [28, 56, 57]), at room or elevated temperature, being often promoted by the addition of catalytic amounts of acids (e.g., formic [15, 51, 58], acetic [16, 53] and *p*-toluenesulfonic [28, 56, 57, 59] acids), or by alternative or simultaneous removal of the water formed using the Dean–Stark apparatus [28, 57, 59] or dehydrating agents (e.g., molecular sieves [55, 60], sodium sulfate [14, 61], tetraethyl orthosilicate [62]), or a combination of the two (Scheme 3.1).

Generally, the preparations of aldimine ligands are much more straightforward than those of ketimine ligands, the latter requiring always a catalytic amount of acid and harsher reaction conditions. The yields obtained in these condensations often depend upon the reaction conditions and the nature of the aryl substituents. For instance, in the syntheses of the *o*-methoxy-substituted ligands **L74** and **L119** (Table 3.1) [34], when the reaction is carried out in refluxing ethanol, in the presence of acetic acid as catalyst and molecular sieves as water absorbents, even after a prolonged reaction period (up to 5 days), the yields are very low (ca. 5%), whereas the use of toluene as solvent, *p*-toluenesulfonic acid as catalyst, and a Dean–Stark trap to remove the water formed in the reaction, leads to high yields (ca. 80%) after a few hours. Strongly electron-withdrawing substituents on the aryl rings (e.g., F [60, 63], CF_3 [55], Cl [56, 60] and NO_2 [62]) lead to a relatively low yield, possibly due to the reduced nucleophilicity of the corresponding anilines,

Table 3.1 Symmetrical 2,6-bis(arylimino)pyridine ligands, 2,6-($ArN{=}CR)_2C_5H_3N$

	Ar	R	Refs.[a]
Mono-o-substituted phenyl			
L1	Phenyl	H	[291]
L2	2-methylphenyl	H	[52]
L3	4-methylphenyl	H	[292]
L4	2-biphenyl	H	[52]
L5	4-methoxyphenyl	H	[291, 292]
L6	2-(methylthio)phenyl	H	[293]
Di-o-substituted phenyl			
L7	2,6-dimethylphenyl	H	[16, 53]
L8	2,4,6-trimethylphenyl	H	[53, 294]
L9	2,6-diethylphenyl	H	[53, 56]
L10	2-methyl-6-*iso*-propylphenyl	H	[51, 56]
L11	2,6-di-*iso*-propylphenyl	H	[53, 293]
L12	2,6-diphenylphenyl	H	[52]
Others			
L13	1-naphthyl	H	[52, 116]
Mono-o-substituted phenyl			
L14	Phenyl	Me	[28, 58]
L15	2-methylphenyl	Me	[15, 28, 52, 54, 295]
L16	4-methylphenyl	Me	[54, 296]
L17	2,3-dimethylphenyl	Me	[52, 297]
L18	2,4-dimethylphenyl	Me	[16, 52, 59, 296]
L19	2,5-dimethylphenyl	Me	[28, 102]
L20	3,5-dimethylphenyl	Me	[296]
L21	2-ethylphenyl	Me	[15, 28]
L22	2-*n*-propylphenyl	Me	[103]
L23	2-*iso*-propylphenyl	Me	[15, 28, 54]
L24	2-*iso*-propyl-4-methylphenyl	Me	[59]
L25	2-*n*-butylphenyl	Me	[103]
L26	2-*tert*-butylphenyl	Me	[14, 28, 53, 54, 296]
L27	4-*tert*-butylphenyl	Me	[54, 296]
L28	2,5-di-*tert*-butylphenyl	Me	[37, 91]
L29	2-cyclopentylphenyl	Me	[94, 298]
L30	2-cyclohexylphenyl	Me	[94, 298]
L31	4-heptylphenyl	Me	[91]
L32	2-cyclooctylphenyl	Me	[94, 298]
L33	2-cyclododecylphenyl	Me	[94, 298]
L34	2-benzylphenyl	Me	[58]
L35	2-methyl-4-(3-(1-(trimethylsilyl) -1*H*-inden-1-yl)prop-1-ynyl)phenyl	Me	[57]
L36	4-((9*H*-fluoren-2-yl)ethynyl) -2-methylphenyl	Me	[57]
L37	4-(5-(cyclopenta-1,3-dienyl)-5 -(9*H*-fluoren-9-yl)hex-1-ynyl) -2-methylphenyl	Me	[57]

(continued)

Table 3.1 (continued)

	Ar	R	Refs.[a]
L38	2-biphenyl	Me	[78]
L39	2-flourophenyl	Me	[60, 72]
L40	2,4-difluorophenyl	Me	[63, 104]
L41	2,5-difluorophenyl	Me	[63]
L42	2-fluoro-4-methylphenyl	Me	[72]
L43	2-fluoro-5-methylphenyl	Me	[104]
L44	3-(4-fluorophenyl)-2-methylphenyl	Me	[69]
L45	2,5-di-(4-fluorophenyl)-4-methylphenyl	Me	[69]
L46	2-trifluoromethylphenyl	Me	[34, 74]
L47	3-trifluoromethylphenyl	Me	[55]
L48	4-trifluoromethylphenyl	Me	[55]
L49	4-fluoro-2-trifluoromethylphenyl	Me	[74]
L50	2-methyl-3-(3,5-bis(trifluoromethyl) phenyl)phenyl	Me	[69]
L51	2-methyl-5-(3,5-bis(trifluoromethyl) phenyl)phenyl	Me	[69]
L52	2-chlorophenyl	Me	[60, 282]
L53	2-chloro-4-methylphenyl	Me	[104]
L54	2-chloro-5-methylphenyl	Me	[104]
L55	3-chloro-2-methylphenyl	Me	[103]
L56	4-chloro-2-methylphenyl	Me	[28]
L57	5-chloro-2-methylphenyl	Me	[297]
L58	2-chloro-4-fluorophenyl	Me	[104]
L59	4-chloro-2-trifluoromethylphenyl	Me	[91]
L60	2-bromophenyl	Me	[60]
L61	2-bromo-4-methylphenyl	Me	[104]
L62	4-bromo-2-methylphenyl	Me	[28]
L63	3,5-dibromo-4-methylphenyl	Me	[69]
L64	4-bromo-2-trifluoromethylphenyl	Me	[91]
L65	2-iodophenyl	Me	[60]
L66	4-iodophenyl	Me	[57]
L67	4-iodo-2-methylphenyl	Me	[57]
L68	4-(diethylamino)-2-methylphenyl	Me	[76]
L69	4-hydroxyphenyl	Me	[136]
L70	4-hydroxy-2-methylphenyl	Me	[136]
L71	3-methoxyphenyl	Me	[55]
L72	4-methoxyphenyl	Me	[296]
L73	4-methoxy-2-methylphenyl	Me	[42, 254]
L74	2,4-dimethoxyphenyl	Me	[34]
L75	2-cyclopentyl-4-ethoxyphenyl	Me	[94, 298]
L76	2-nitrilophenyl	Me	[75]
L77	2-methyl-4-nitrilophenyl	Me	[75]
L78	4-nitrophenyl	Me	[76]
L79	2-methyl-3-nitrophenyl	Me	[76]
L80	2-methyl-4-nitrophenyl	Me	[76]

(continued)

Table 3.1 (continued)

	Ar	R	Refs.[a]
L81	2-(4,4,5,5-tetramethyl-[1,2,3] -dioxaborolan-2-yl)phenyl	Me	[299]
L82	2-methyl-3-(5-methyl-2-thienyl)phenyl	Me	[69]
L83	4-ferrocenyl-2-methylphenyl	Me	[70]
Di-o-substituted phenyl			
L84	2,6-dimethylphenyl	Me	[14, 16, 34, 53, 296]
L85	2,4,6-trimethylphenyl	Me	[16, 34, 53]
L86	2,6-diethylphenyl	Me	[36, 59]
L87	2,6-di-*iso*-propylphenyl	Me	[14, 16, 34, 36, 53]
L88	2,4,6-tri-*iso*-propylphenyl	Me	[92]
L89	2-methyl-6-*iso*-propylphenyl	Me	[34, 51, 54]
L90	2,4-di-*tert*-butyl-6-methylphenyl	Me	[56]
L91	2,6-dicyclopentylphenyl	Me	[94, 298]
L92	2-cyclopentyl-6-methylphenyl	Me	[94, 298]
L93	2-cyclopentyl-4,6-dimethylphenyl	Me	[94, 298]
L94	2,6-dihexylphenyl	Me	[228, 229]
L95	2,6-dicyclohexylphenyl	Me	[94, 298]
L96	2-cyclohexyl-6-methylphenyl	Me	[94, 298]
L97	2-cyclohexyl-4,6-dimethylphenyl	Me	[94, 298]
L98	2-cyclooctyl-6-methylphenyl	Me	[94, 298]
L99	2-cyclooctyl-4,6-dimethylphenyl	Me	[94, 298]
L100	2-cyclododecyl-6-methylphenyl	Me	[94, 298]
L101	2-cylcododecyl-4,6-dimethylphenyl	Me	[94, 298]
L102	4-(5-(cyclopenta-1,3-dienyl)-5 -(9*H*-fluoren-9-yl)hex-1-ynyl) -2,6-dimethylphenyl	Me	[57]
L103	2,4,6-triphenylphenyl	Me	[234]
L104	2,6-difluorophenyl	Me	[36, 63, 72]
L105	2-fluoro-6-methylphenyl	Me	[77]
L106	2-fluoro-6-trifluoromethylphenyl	Me	[74]
L107	pentafluorophenyl	Me	[36]
L108	2,6-dichlorophenyl	Me	[60]
L109	2-chloro-4,6-dimethylphenyl	Me	[84, 91]
L110	4-chloro-2,6-dimethylphenyl	Me	[300]
L111	2-chloro-6-cyclopentylphenyl	Me	[94, 298]
L112	2,6-dibromophenyl	Me	[36, 60]
L113	2-bromo-4,6-dimethylphenyl	Me	[300]
L114	3-bromo-2,4,6-trimethylphenyl	Me	[59]
L115	4-bromo-2,6-dimethylphenyl	Me	[84, 91, 301]
L116	2-bromo-4,6-di-*iso*-propylphenyl	Me	[92]
L117	4-bromo-2,6-di-*iso*-propylphenyl	Me	[92]
L118	4-iodo-2,6-dimethylphenyl	Me	[57]
L119	2-methoxy-6-methylphenyl	Me	[34]
L120	4-methoxy-2,6-dimethylphenyl	Me	[229]
L121	2-cyclopentyl-6-methoxyphenyl	Me	[94, 298]

(continued)

Table 3.1 (continued)

	Ar	R	Refs.[a]
L122	2,6-dicyclopentyl-4-ethoxyphenyl	Me	[94, 298]
L123	2,6-dimethyl-4-nitrophenyl	Me	[62]
L124	4-nitro-2,6-di-*iso*-propylphenyl	Me	[62]
L125	4-ferrocenyl-2,6-dimethylphenyl	Me	[70]
L126	4-ferrocenyl-2,6-di-*iso*-propylphenyl	Me	[70]
Others			
L127	1-naphthyl	Me	[98]
L128	2-methyl-1-naphthyl	Me	[56, 99]
L129	1-(5,6,7,8-tetrahydro)naphthyl	Me	[85]
L130	1-anthracenyl	Me	[78]
L131	4-pyrenyl	Me	[58]
L132	2,6-dimethylphenyl	Et	[59]
L133	2,4,6-trimethylphenyl	Et	[66]
L134	2,6-di-*iso*-propylphenyl	Et	[66]
L135	2,4,6-trimethylphenyl	^{i}Pr	[66]
L136	2,6-di-*iso*-propylphenyl	^{i}Pr	[66]
L137	2-methyl-6-*iso*-propylphenyl	^{n}Bu	[56]
L138	2,6-dimethylphenyl	^{t}Bu	[67]
L139	2,6-dimethylphenyl	$(CH_2)_2Ph$	[59]
L140	2,4,6-trimethylphenyl	$(CH_2)_2Ph$	[66]
L141	2,4,6-trimethylphenyl	$CH(Bz)_2$[b]	[66]
L142	2,6-dimethylphenyl	Ph	[59]
L143	2,4,6-trimethylphenyl	Ph	[34]
L144	2,6-di-*iso*-propylphenyl	Ph	[64]
L145	2,5-di-*tert*-butylphenyl	Ph	[65]
L146	2,4,6-trimethylphenyl	MeO	[68]
L147	2,6-di-*iso*-propylphenyl	MeO	[64]
L148	2,4,6-trimethylphenyl	$2,6\text{-}(Me)_2C_6H_3O$	[68]
L149	2,4,6-trimethylphenyl	MeS	[68]
L150	2,4,6-trimethylphenyl	$2,6\text{-}(Me)_2C_6H_3S$	[68]

[a] Only references reporting compounds for the first time are given; references are also given when different reaction conditions were used
[b] Bz = benzyl

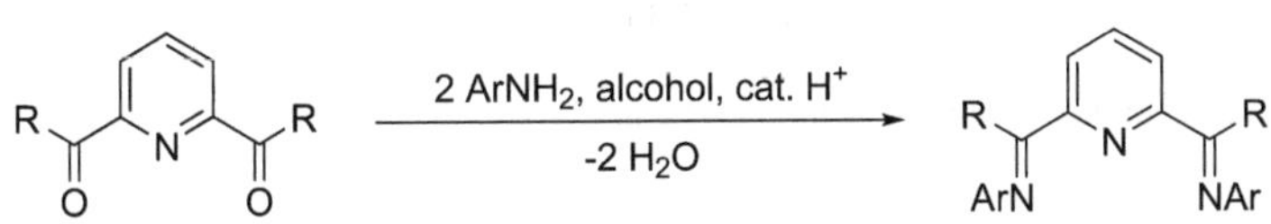

Scheme 3.1 General syntheses of symmetrical 2,6-bis(arylimino)pyridine ligands

and forcing conditions are always required. For example, the fluoro-containing ligand **L104** (Table 3.1) [63] was obtained in low yield (13%) when the corresponding reagents were refluxed in toluene for 3 days, with removal of water using a Dean–Stark trap and in the presence of *p*-toluenesulfonic acid as catalyst, whereas the use of silica–alumina catalyst support and molecular sieves as water absorbents, in toluene, at 30–40 °C, for 24 h, leads to a moderate yield (65%). Su et al. found out that microwave irradiation improves the conversion for this kind of condensation reactions [42].

For the ketimine ligands other than acetimine (e.g., **L132**, **L137**, **L139**, **L142**, **L144**, **L145**, Table 3.1), where R is ethyl [59], *n*-butyl [56], 2-phenylethyl [59] or phenyl [59, 64, 65], the abovementioned direct condensation reactions give also satisfactory conversions. However, in the case of ligand **L143** [34], attempts to obtain it by direct condensation of 2,6-dibenzoylpyridine with 2,4,6-trimethylaniline were unsuccessful. Esteruelas et al. [34] found that this condensation reaction can be promoted by a template reaction with anhydrous nickel dichloride, as shown in Scheme 3.2. The treatment of 2,6-dibenzoylpyridine with one equivalent of anhydrous nickel dichloride and two equivalents of 2,4,6-trimethylaniline in acetic acid, under refluxing conditions, afforded the nickel complex [$NiCl_2$(**L143**)] in high yield (89%). Demetalation of the latter with aminopropyl silica gel, in dichloromethane, gave the ligand **L143** in moderate yield (51%).

Gibson and coworkers [66] reported a straightforward protocol for the syntheses of ketimine ligands (e.g., **L133**, **L135**, **L140**, **L141**, Table 3.1) from the corresponding 2,6-diacetylpyridinebis(imine) ligands, in which the methyl substituents of the α-imino carbon moiety are deprotonated using lithium diisopropylamide (LDA) at –78 °C, followed by alkylation using a primary alkylhalide at 0 °C (Scheme 3.3). This procedure can be repeated to build up the desired substitution pattern at the α-imino carbon position.

The ketimine ligands containing the more sterically encumbered *tert*-butyl substituents at the α-imino carbon arms (e.g., **L138**, Table 3.1) cannot be prepared by the direct condensation reaction of 2,6-diketopyridine with the appropriate

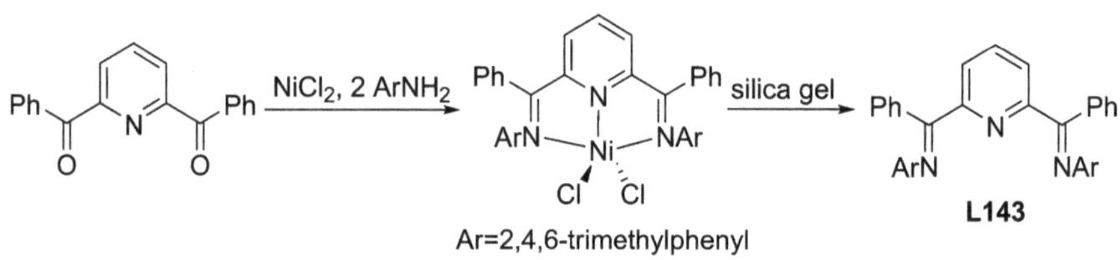

Scheme 3.2 Template synthesis of the BIP ligand **L143**

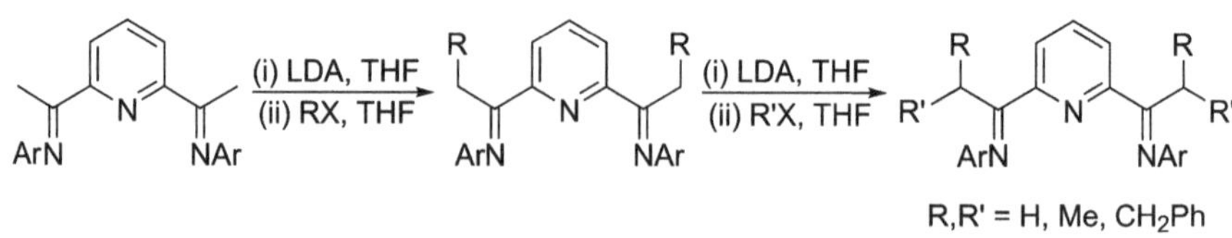

Scheme 3.3 Syntheses of the ketimine ligands with different substituents at the α-imino carbon position, as proposed by Gibson et al.

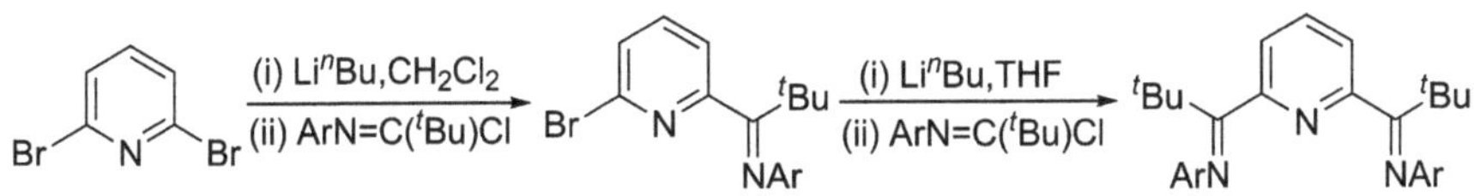

Scheme 3.4 Syntheses of ketimine ligands containing *tert*-butyl substituents at the α-imino carbon

Scheme 3.5 Syntheses of ketimine ligands with ether or thioether groups at the α-imino carbon

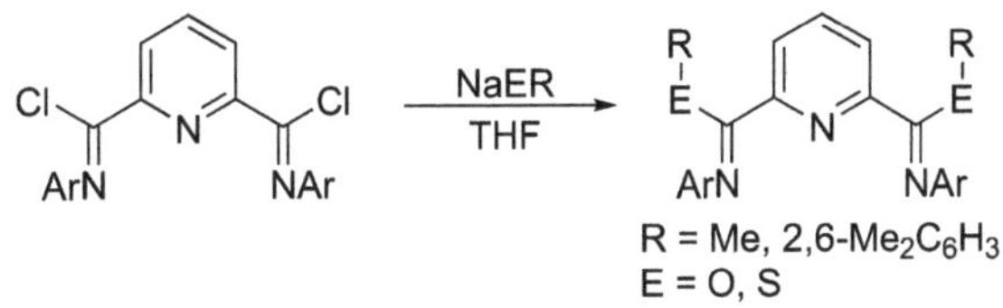

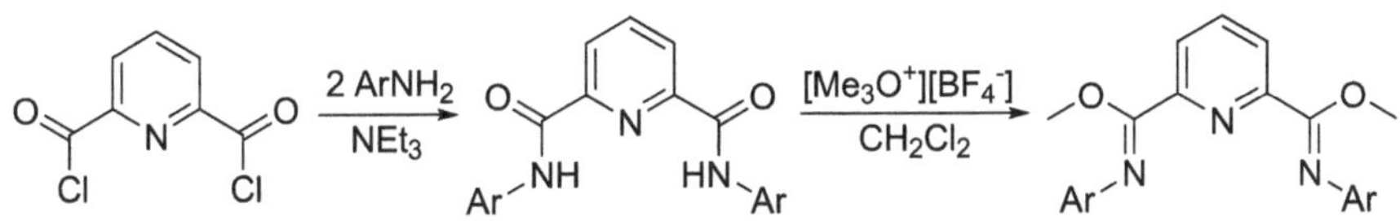

Scheme 3.6 Syntheses of ketimine ligands with methoxy substituents at the α-imino carbon

anilines even if under forcing conditions [67]. However, they are readily accessible using Peterson's method, in which the mono-lithiation of 2,6-dibromopyridine, in dichloromethane at –78 °C, followed by addition of the imidoyl chloride ArN=C(^{t}Bu)Cl, generates the mono(imino)pyridine compound 2-Br-6-(ArN=C(^{t}Bu))C_5H_3N. Repetition of the same reaction, in tetrahydrofuran, gives the BIP ligand 2,6-(ArN=C(^{t}Bu))$_2C_5H_3N$ (Scheme 3.4) [67].

The ligands containing ether or thioether substituents at the α-imino carbon (e.g., **L146**, **L148–L150** in Table 3.1) can also be readily prepared, as shown in Scheme 3.5. Treatment of the pyridine-2,6-dicarboxyimidoyl dichloride with the corresponding NaER salt (R = Me, 2,6-$Me_2C_6H_3$; E = O, S) gives rise to the BIP ligand 2,6-(ArN=C(ER))$_2C_5H_3N$ [68]. Alternatively, the methoxy groups can also be attached to the ligand by treatment of the pyridine-2,6-dicarbonyl dichloride with the appropriate anilines, in triethylamine, to afford the pyridine-2,6-dicarboxamide, which is then further reacted with the Meerwein's reagent [Me_3O^+][BF_4^-] to generate the corresponding BIP ligand (e.g., **L147** in Table 3.1) (Scheme 3.6) [64].

The introduction of aryl groups containing bulky substituents, such as **L35–L37**, **L44**, **L45**, **L50** and **L51** (Table 3.1), can also be carried out by performing coupling reactions in halogen-containing aryl groups of BIP ligands. For instance, Alt and Görl [57] utilized the palladium catalyzed Sonogashira coupling of iodinated BIP ligands with alkynyl-substituted indene or fluorene derivatives, to form a new family of aryl-BIP ligands (Scheme 3.7). Ionkin et al. [69] employed the palladium catalyzed Suzuki cross-coupling reactions of brominated aryl-BIP

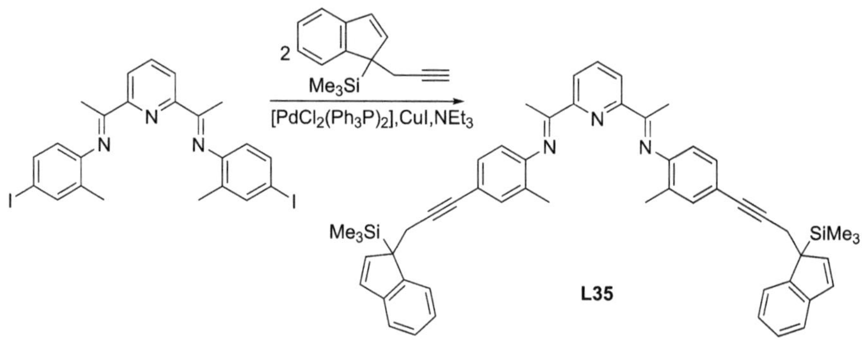

Scheme 3.7 Synthesis of the BIP ligand **L35**

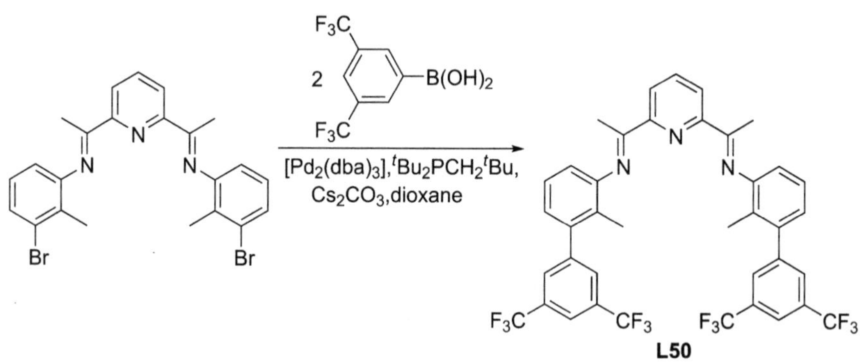

Scheme 3.8 Synthesis of the BIP ligand **L50**

ligands with fluorine-containing aryl boronic acids, to afford a series of bulky CF_3-containing aryl-BIP ligands (Scheme 3.8).

It is worth to note that Gibson et al. [70] synthesized a series of ferrocene-substituted BIP ligands (**L83**, **L125** and **L126**, Table 3.1) by direct condensation of the respective ferrocenyl anilines with 2,6-diacetylpyridine, in moderate yields (34–62%) (Scheme 3.9). The reactions were conducted in refluxing ethanol, using acetic acid as catalyst, and the new ligands are air stable.

Table 3.2 lists the unsymmetrical 2,6-bis(arylimino)pyridine ligands published to date. These type of ligands are generally prepared by consecutive condensation reactions of 2,6-diacetylpyridine with two different anilines, as shown in Scheme 3.10, in which treatment of 2,6-diacetylpyridine with one equivalent of the aniline $ArNH_2$ gives the mono(imino)pyridine 2-(O=C(Me))-6-(ArN=C(Me))C_5H_3N, followed by addition of a further equivalent of the other different aniline $Ar'NH_2$ to afford BIP ligands of the type 2-(ArN=C(Me))-6-(Ar'N=C(Me))C_5H_3N. Conversely, Ionkin et al. [69] used the same reaction employed for the synthesis of the BIP ligands represented in Scheme 3.8 to introduce sterically hindered CF_3-containing substituents in unsymmetrical BIP ligands. For instance,

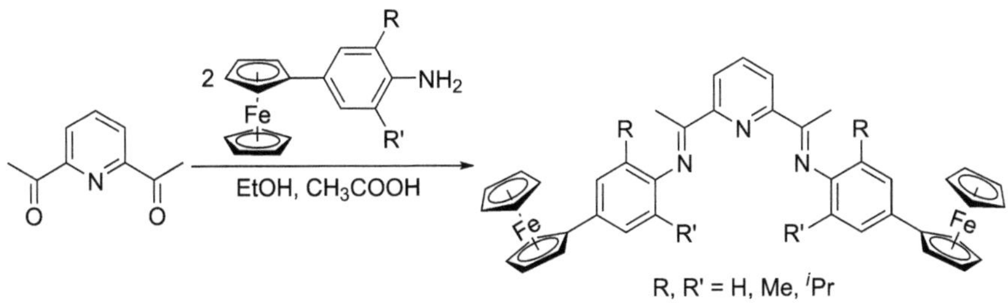

Scheme 3.9 Syntheses of ferrocene-substituted BIP ligands

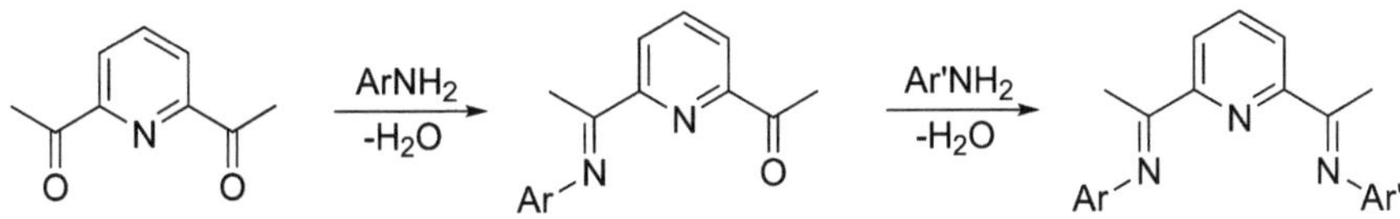

Scheme 3.10 Syntheses of unsymmetrical 2,6-bis(arylimino)pyridine ligands

the palladium catalyzed Suzuki cross-coupling reaction of the unsymmetrical BIP ligand 2-(ArN=C(Me))-6-(Ar′N=C(Me))C_5H_3N (Ar = 4-Br-2,6-$Me_2C_6H_2$, Ar′ = 3,5-Br_2-4-MeC_6H_2) (**L180**) with 3,5-bis(trifluoromethyl)phenylbonoric acid leads to the ligand 2-(ArN=C(Me))-6-(Ar′N=C(Me))C_5H_3N (Ar = 4-(3,5-$(CF_3)_2C_6H_3$)-2,6-$Me_2C_6H_2$, Ar′ = 3,5-(3,5-$(CF_3)_2C_6H_3)_2$-4-MeC_6H_2) (**L178**).

Due to the existence of hindered aryl ring rotation, variable substitution patterns can lead to the formation of BIP ligands with diverse configurations. As illustrated in Fig. 3.2, the arylimino groups and the aryl groups can rotate around the $C_{pyridyl}$–C_{imino} and N_{imino}–C_{aryl} single bonds, respectively. Generally, the arylimino groups are stretched away from the central pyridyl ring to release the steric constraints, being coplanar with the pyridyl ring, and the aryl groups are nearly perpendicular to the pyridyl ring, adopting a (*E*,*E*) configuration with respect to the C=N double bonds, as depicted in Fig. 3.3a for **L123** [62]. The substituents R, R′ and R″ (Fig. 3.2) have crucial impacts on the configurations of BIP ligands. As the bulkiness of the substituent R″ increases, the imino groups noticeably deviate from the pyridyl rings. For instance, in the case of the ligand **L138** [67], the two imino C=N bonds are nearly orthogonal to the pyridyl ring and oriented in a *syn* fashion with respect to the pyridyl ring, leading to a reduced ability in the coordination to the metal center (Fig. 3.3b). For the ligand **L144** [64], the two imino C=N bonds are also perpendicular to the pyridyl ring, but exist in an *anti* configuration in relation to the pyridyl ring, being their iron and cobalt complexes readily accessible (Fig. 3.3c). The rotation of the aryl groups about the N_{imino}–C_{aryl} single bonds depends on both aryl substituents R and R′, and on the substituent R″ at the α-imino carbon. Reduction of the steric bulk at R″, e.g., from an alkyl group to a hydrogen atom, would result in much lower rotational barriers of the aryl groups. Conversely, the rotation barriers for *o*-disubstituted aryl groups are much higher

Table 3.2 Unsymmetrical 2,6-bis(arylimino)pyridine ligands, 2-(ArN=C(R))-6-(Ar′N=C(R))-C_5H_3N

	Ar	Ar′	R	Refs.[a]
L151	Phenyl	2,6-dimethylphenyl	Me	[75]
L152	2-methyl-6-*iso*-propylphenyl	2,6-dimethylphenyl	Me	[100]
L153	2-*tert*-butylphenyl	2,6-dimethylphenyl	Me	[51]
L154	4-nitrilophenyl	2,6-dimethylphenyl	Me	[75]
L155	Phenyl	2,4,6-trimethylphenyl	Me	[75]
L156	2-methylphenyl	2,4,6-trimethylphenyl	Me	[35]
L157	2-ethylphenyl	2,4,6-trimethylphenyl	Me	[35]
L158	2-*iso*-propylphenyl	2,4,6-trimethylphenyl	Me	[56]
L159	2-*tert*-butylphenyl	2,4,6-trimethylphenyl	Me	[34]
L160	4-*tert*-butylphenyl	2,4,6-trimethylphenyl	Me	[35]
L161	4-nitrilophenyl	2,4,6-trimethylphenyl	Me	[75]
L162	4-nitrophenyl	2,4,6-trimethylphenyl	Me	[76]
L163	2-(4,4,5,5-tetramethyl-[1,3,2]-dioxaborolan-2-yl)phenyl	2,4,6-trimethylphenyl	Me	[299]
L164	3-(4,4,5,5-tetramethyl-[1,3,2]-dioxaborolan-2-yl)phenyl	2,4,6-trimethylphenyl	Me	[299]
L165	4-(4,4,5,5-tetramethyl-[1,3,2]-dioxaborolan-2-yl)phenyl	2,4,6-trimethylphenyl	Me	[299]
L166	2-trifluoromethylphenyl	2,4,6-trimethylphenyl	Me	[34]
L167	Phenyl	2,6-di-*iso*-propylphenyl	Me	[101]
L168	2-methylphenyl	2,6-di-*iso*-propylphenyl	Me	[54, 101]
L169	2-*iso*-propylphenyl	2,6-di-*iso*-propylphenyl	Me	[78, 100]
L170	2-*tert*-butylphenyl	2,6-di-*iso*-propylphenyl	Me	[100]
L171	2,6-dimethylphenyl	2,6-di-*iso*-propylphenyl	Me	[100, 101]
L172	2,4,6-trimethylphenyl	2,6-di-*iso*-propylphenyl	Me	[34]
L173	2-methyl-6-*iso*-propylphenyl	2,6-di-*iso*-propylphenyl	Me	[51]
L174	2-cyclohexyl-6-*iso*-propylphenyl	2,6-di-*iso*-propylphenyl	Me	[54]
L175	1-anthracenyl	2,6-di-*iso*-propylphenyl	Me	[78]
L176	2-methylphenyl	2-fluoro-6-methylphenyl	Me	[77]
L177	2,4-dimethylphenyl	2-fluoro-6-methylphenyl	Me	[77]
L178	4-methyl-3,5-di-(3,5-bis(trifluoromethyl)phenyl)phenyl	2,6-dimethyl-4-(3,5-bis(trifluoromethyl)phenyl)phenyl	Me	[69]
L179	Phenyl	4-bromo-2,6-dimethylphenyl	Me	[75]
L180	3,5-dibromo-4-methylphenyl	4-bromo-2,6-dimethylphenyl	Me	[69]
L181	4-nitrilophenyl	4-bromo-2,6-dimethylphenyl	Me	[75]
L182	2-(4,4,5,5-tetramethyl-[1,3,2]-dioxaborolan-2-yl)phenyl	4-bromo-2,6-dimethylphenyl	Me	[299]
L183	4-(4,4,5,5-tetramethyl-[1,3,2]-dioxaborolan-2-yl)phenyl	4-bromo-2,6-dimethylphenyl	Me	[299]

[a] Only references reporting compounds for the first time are given; references are also given when different reaction conditions were used

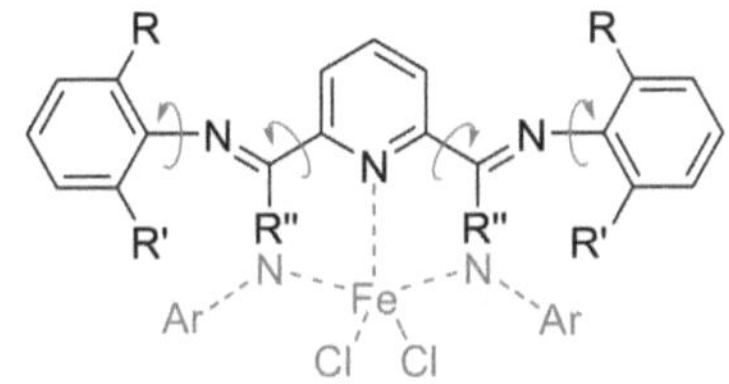

Fig. 3.2 Possible changes in the configurations of the free BIP ligand resulting from group rotations about single bonds

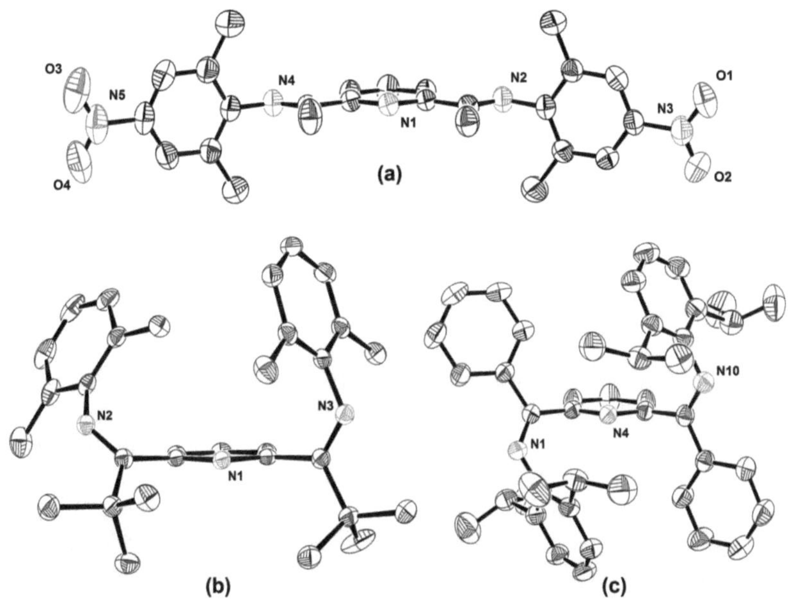

Fig. 3.3 Molecular structures of the free BIP ligands **L123** (**a**), **L138** (**b**) and **L144** (**c**) [62, 64, 67]

than those containing exclusively a single *o*-substituent. Small and Brookhart [51] found that the ^{13}C NMR alkyl resonances of R and R′ are sensitive to the aryl group rotation, and Cámpora et al. [56] showed that the *iso*-propyl ^{1}H NMR resonances of *o*-disubstituted aryl rings are also sensitive to the rotation of these rings. Despite all the possible existing configurations, the formal symmetry assignments in free BIP ligands are based on the U-shaped ligand configurations, similar to those displayed in the corresponding complexes, with the aryl groups orthogonal to the N–N–N plane (gray part of Fig. 3.2). For instance, the symmetrical BIP ligands **L10** (2,6-(ArN=CH)$_2$C$_5$H$_3$N, Ar = 2-Me-6-^{i}PrC$_6$H$_3$) and **L87** (2,6-(ArN=C(Me))$_2$C$_5$H$_3$N, Ar = 2,6-^{i}Pr$_2$C$_6$H$_3$) have C_s/C_2 and C_{2v} symmetries, respectively, and the unsymmetrical BIP ligands **L153** (2-(ArN=C(Me))-6-(Ar′N=C(Me))C$_5$H$_3$N, Ar = 2-^{t}BuC$_6$H$_4$, Ar′ = 2,6-Me$_2$C$_6$H$_3$) and **L171** (2-(ArN=C(Me))-6-(Ar′N=C(Me))C$_5$H$_3$N, Ar = 2,6-Me$_2$C$_6$H$_3$, Ar = 2,6-^{i}Pr$_2$C$_6$H$_3$) possess C_1 and C_s symmetries, respectively.

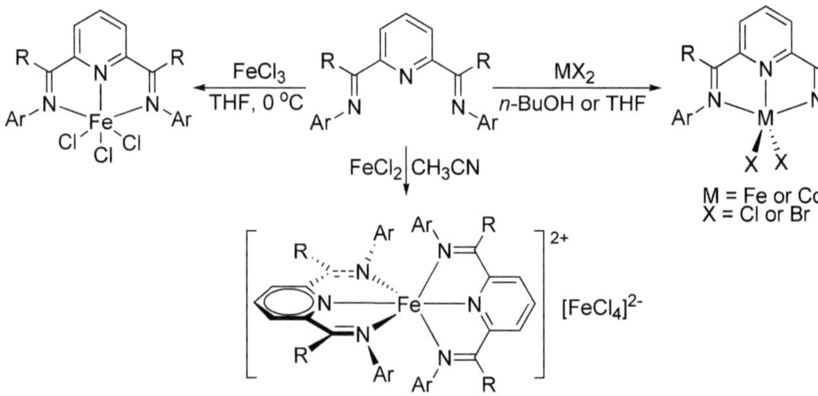

Scheme 3.11 Syntheses of the iron and cobalt complexes based on the BIP ligands

3.2.2 Syntheses of Iron and Cobalt Complexes with 2,6-Bis(arylimino)pyridine Ligands

In general, the iron and cobalt complexes of BIP are readily accessible by treatment of the anhydrous or hydrated iron(II) or cobalt(II) dihalides (MX_2 or $MX_2{\cdot}xH_2O$, where M = Fe or Co; X = Cl or Br) with the BIP ligands 2,6-(ArN=C(R))$_2C_5H_3N$, in tetrahydrofuran or *n*-butanol, giving rise to adduct complexes of the type [MX_2L], where L denotes [2,6-(ArN=C(R))$_2C_5H_3N$] (Scheme 3.11). Reaction of the BIP ligands with iron(III) trichloride, in tetrahydrofuran, leads to complexes [$FeCl_3$L] (Scheme 3.11). The trivalent iron complexes [$FeCl_3$L] can be reduced, in tetrahydrofuran, to the divalent iron complexes [$FeCl_2$L], the reducing agent possibly being the solvent. For instance, the trivalent iron complex [$FeCl_3$(**L50**)] (**L50** = 2,6-(ArN=C(Me))$_2C_5H_3N$, Ar = 2-methyl-3-(3,5-bis(trifluoromethyl)phenyl)phenyl), in tetrahydrofuran, was gradually reduced to the divalent [$FeCl_2$(**L50**)] within a week [69]. In order to suppress the reduction of iron(III) to iron(II), it is necessary to lower the reaction temperature during the formation of the trivalent iron complexes [71].

The formation of bisligand iron(II) complexes $[FeL_2]^{2+}[FeCl_4]^{2-}$ depends on the polarity of the solvent and on the BIP aryl substituents. For instance, the complex [Fe(**L39**)$_2]^{2+}[FeCl_4]^{2-}$ (**L39** = 2,6-(ArN=C(Me))$_2C_5H_3N$, Ar = 2-fluorophenyl) is easily prepared by the treatment of iron(II) dichloride tetrahydrate with the ligand **L39**, in the weakly polar tetrahydrofuran [60, 72]. However, the reactions of the difluoro-substituted BIP ligands 2,6-(ArN=C(Me))$_2C_5H_3N$, where Ar is 2,4-$F_2C_6H_3$ (**L40**), 2,5-$F_2C_6H_3$ (**L41**) or 2,6-$F_2C_6H_3$ (**104**), with iron(II) dichloride tetrahydrate give various products. In the case of the ligand **L41**, the monoligand complex [$FeCl_2$(**L41**)] is the sole product obtained, either in tetrahydrofuran or in the strongly polar acetonitrile, while the ligands **L40** and **L104** afford the monoligand complexes [$FeCl_2$(**L40**)] and [$FeCl_2$(**L104**)], in tetrahydrofuran, and the bisligand complexes [Fe(**L104**)$_2]^{2+}[FeCl_4]^{2-}$ and

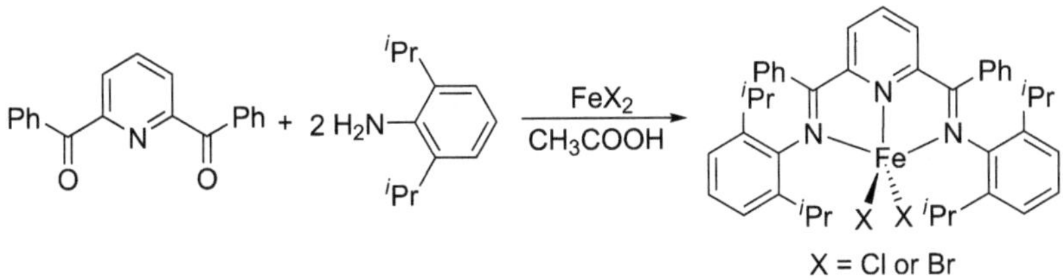

Scheme 3.12 One-pot syntheses of BIP iron(II) complexes

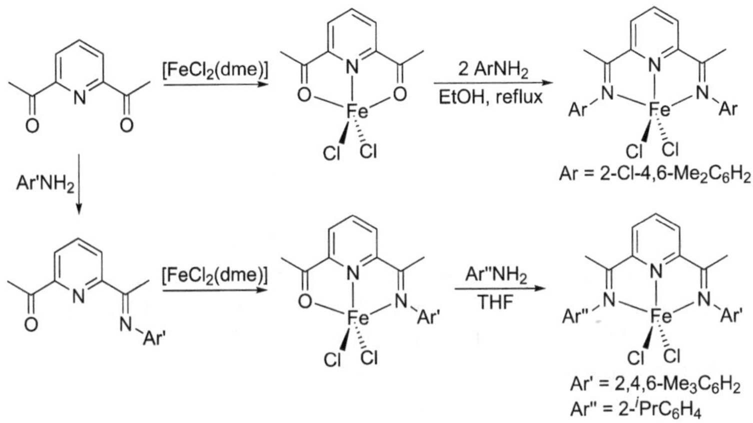

Scheme 3.13 Syntheses of the BIP iron(II) complexes reported by Cámpora et al. [56]

$[Fe(\mathbf{L40})_2]^{2+}[Cl_3FeOCl_3]^{2-}$, in acetonitrile, respectively [63]. However, the ligands containing *o*-methyl or bulkier aryl groups lead exclusively to the monoligand complexes $[FeCl_2L]$, regardless of the solvent. Therefore, it seems that the bisligand complexes $[FeL_2]^{2+}[FeCl_4]^{2-}$ are formed only when the BIP ligands have very small *o*-substituents on the aryl rings, such as F and H [69, 72]. The only exceptional case is that of ligand **L14** (2,6-(ArN=C(Me))$_2$C$_5$H$_3$N, Ar = phenyl), having no aryl substituents other than H, which gives exclusively the monoligand complex [FeCl$_2$(**L14**)] [28, 55, 58, 59].

Archer et al. [73] described a one-pot protocol to prepare BIP iron(II) complexes, in which the condensation of 2,6-dibenzoylpyridine and 2,6-diisopropylaniline in the presence of iron(II) dihalide, in refluxing acetic acid, affords the iron(II) complex [FeX$_2$(**L144**)] (**L144** = 2,6-(ArN=C(Ph))$_2$C$_5$H$_3$N, Ar = 2,6-^{i}Pr$_2$-C$_6$H$_3$, X = Cl or Br) (Scheme 3.12).

Alternatively, to avoid the time-consuming purification process of the free BIP ligands, and the corresponding unsatisfactory yields, Cámpora et al. [56] reported a new synthetic procedure for BIP iron(II) complexes. As shown in Scheme 3.13, treatment of 2,6-diacetylpyridine with [FeCl$_2$(dme)] (dme = dimethoxyethane) gives [FeCl$_2$(2,6-diacetylpyridine)], which, after subsequent addition of two equivalents of 2-chloro-4,6-dimethylanilines, in refluxing ethanol, affords iron(II)

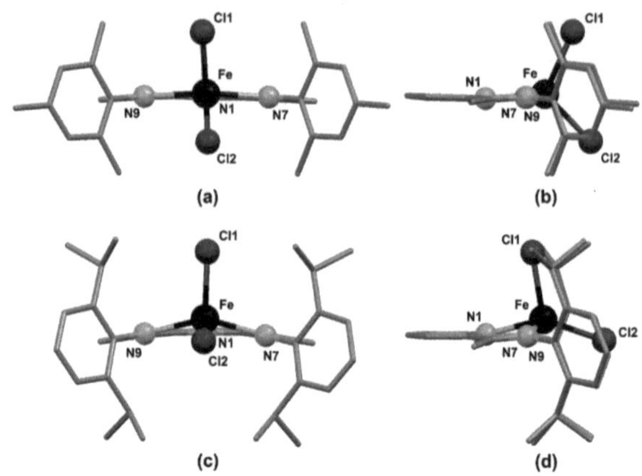

Fig. 3.4 Molecular structures of the complexes [$FeCl_2$(**L85**)] (**a** *front view*; **b** *side view*) and [$FeCl_2$(**L87**)] (**c** *front view*; **d** *side view*) [53]

complexes [$FeCl_2$(**L109**)] (**L109** = 2,6-(ArN=C(Me))$_2C_5H_3N$, Ar = 2-Cl-4,6-$Me_2C_6H_2$). This procedure can be repeated to prepare the unsymmetrical BIP iron(II) complex [$FeCl_2$(**L158**)] (**L158** = 2-(Ar′N=C(Me))-6-(Ar″N=C(Me)) C_5H_3N, Ar = 2-^{i}PrC_6H_4, Ar″ = 2,4,6-$Me_3C_6H_2$) (Scheme 3.13).

The molecular structures of several BIP iron(II) and cobalt(II) complexes have been determined by single crystal X-ray diffraction. The cobalt(II) complexes are commonly isomorphous with those of iron(II) possessing the same BIP ligands. Their coordination geometries around the metal centers are quite flexible, varying from trigonal bipyramidal to square pyramidal, with various degrees of distortion from the ideal geometries, depending on the substitution patterns on the aryl groups. Generally, the complexes containing less bulky *o*-substituents on the aryl rings (e.g., H, F and Me) adopt distorted trigonal bipyramidal geometries, with the pyridyl nitrogen atom and the two chlorine atoms forming the equatorial plane, the two imino nitrogen atoms occupying the axial positions (views (a) and (b), Fig. 3.4), such as in the cases of complexes [$FeCl_2$(**L17**)] [52], [$FeCl_2$(**L41**)] [63], [$FeCl_2$(**L85**)] [53], and [$CoCl_2$(**L40**)] [63]. Conversely, the complexes bearing much bulkier *o*-substituents (e.g., CF_3, ^{i}Pr and ^{t}Bu) exhibit distorted square pyramidal geometries, with the three nitrogen atoms and one of the chlorine atoms forming the base, and the remaining chlorine atom occupying the apical position (views (c) and (d), Fig. 3.4), which are the cases of [$FeCl_2$(**L87**)] [14, 16, 53], [$CoCl_2$(**L87**)] [14, 53], [$CoCl_2$(**L26**)] [14], [$CoBr_2$(**L106**)] [74] and [$FeBr_2$(**L49**)] [74]. The complexes based on the unsymmetrical BIP ligands such as [$FeCl_2$(**L154**)] [75], [$FeCl_2$(**L180**)] [75], [$FeCl_2$(**L162**)] [76] and [$FeCl_2$(**L176**)], follow this rule as well [77].

Regardless of the bulkiness of the BIP ligand aryl substituents, the aryl rings are oriented essentially orthogonal to the backbone, the latter being essentially planar. The substituents at the α-imino carbons also have prominent influences on the geometries of the complexes. These geometries shift from trigonal bipyramidal to square pyramidal with increasing bulkiness of the substituents. For example, in the case of complexes [$FeCl_2${(ArN=C(R))$_2C_5H_3N$}], where Ar = 2,4,6-$Me_3C_6H_2$,

R = Me (**L85**) [53], OMe (**L146**) [68], Et (**L133**) [66], ^{i}Pr (**L135**) [66], or 2,6-$Me_2C_6H_3S$ (**L150**) [68], the first three adopt trigonal bipyramidal geometries. As the bulkiness of the aryl substituents increases, the complex [$FeCl_2$(**L135**)] reveals two independent molecules, one being trigonal bipyramidal and the other square pyramidal, whereas the complex [$FeCl_2$(**L150**)] only show molecules with square pyramidal geometry. Likewise, for the complexes of the type [$FeCl_2${(ArN=C(R))$_2C_5H_3N$}], where Ar = 2-PhC_6H_4, R = H (**L4**) [52] or Me (**L38**) [78], the complex [$FeCl_2$(**L4**)] exhibits trigonal bipyramidal geometry, while complex [$FeCl_2$(**L38**)] shows a square pyramidal one.

In the bisligand complexes [Fe(**L39**)$_2$]$^{2+}$[$FeCl_4$]$^{2-}$ [60] and [Fe(**L45**)$_2$]$^{2+}$ [$FeCl_4$]$^{2-}$ [69], the iron atom of the cationic part is coordinated by the six nitrogen atoms of the two ligands, its geometry at the metal center being described as distorted octahedral, whereas the [$FeCl_4$]$^{2-}$ counteranion geometry can be described as distorted tetrahedral. The geometries of the trivalent iron complexes [$FeCl_3$(**L14**)] [71], [$FeCl_3$(**L155**)] [75], [$FeCl_3$(**L161**)] [75] and [$FeCl_3$(**L50**)] [69] can be generally described as distorted octahedral, where the three nitrogens and one of the chlorine atoms form the equatorial plane and the other two chlorine atoms occupy the axial positions.

The magnetic properties of iron and cobalt complexes have been determined using Evans (NMR) [53], Gouy (balance) [52, 53, 74] or Faraday (balance) [58] methods. All of the monoligand iron and cobalt complexes [MX_2L] (M = Fe or Co, X = Cl or Br) are paramagnetic, with magnetic moments typically in the ranges of 4.8–5.8 BM, for the iron(II) complexes, and 3.9–4.6 BM, for the cobalt(II) ones, in agreement with high spin, four and three unpaired electrons species, respectively [52, 53, 56, 58, 74].

Despite the paramagnetic nature of these complexes, ^{1}H NMR spectroscopy can be informative. The ^{1}H resonances, which are paramagnetically shifted and broadened, can be assigned on the basis of integration and proximity of the nuclei to the paramagnetic centers [53]. Cámpora et al. [56] found that ^{1}H NMR spectroscopy is a powerful tool to investigate the atropisomerism phenomenon in the monoligand iron and cobalt complexes.

3.2.3 Oligo- and Polymerization of Ethylene

Because the metal center at the cationic part of the bisligand iron complexes [FeL_2]$^{2+}$[$FeCl_4$]$^{2-}$ is coordinatively saturated and sterically hindered, these compounds showed to be inactive toward ethylene polymerization, when activated with MAO [60, 63, 69, 72, 76]. The trivalent iron complexes [$FeCl_3L$] showed rather similar catalytic performances when compared with the analogous divalent ones [$FeCl_2L$], when activated with MAO, suggesting that the same active species are generated from both precatalysts [53, 75]. In this context, both reduction and oxidation transformations between Fe(III) and Fe(II) species in the presence of MAO have been observed [79–83], and the "real" active species is unclear to date.

Treatment of the monoligand iron(II) and cobalt(II) complexes [MX_2L] (M = Fe or Co, X = Cl or Br) with MAO or MMAO generates highly active ethylene oligo- or polymerization catalysts, which lead to hydrocarbons ranging from α-olefin oligomers with Schulz–Flory distributions to highly linear polyethylenes with high molecular weights. These iron and cobalt catalytic systems show substantially low productivities for the polymerization of propylene or higher α-olefins, and limited capabilities to incorporate higher α-olefins into the growing polymer chains for the production of branched polymers [51, 74, 78, 84–90]. Therefore, herein, we will lay stress on the catalytic performance of BIP iron and cobalt complexes toward ethylene polymerization, whereas their applications in the polymerization of propylene and higher α-olefins will be discussed in Sect. 3.8.

3.2.3.1 Polymerization of Ethylene

The monoligand iron and cobalt complexes, when activated with MAO, can convert ethylene to oligomers or polymers, substantially depending on the bulkiness of the aryl ring *o*-substituents. In general, mono-*o*-substitution leads to ethylene oligomerization catalysts, e.g., **L2**–**L6** and **L15**–**L83** (Table 3.1), except the *o*-*tert*-butyl (**L26** and **L28**) [37, 53, 91], *o*-benzyl (**L34**) [58], *o*-trifluoromethyl (**L46**, **L59** and **L64**) [74, 91] and *o*-methyl (**L37**, with a very bulky substituent on the *p*-position) [57] cases. Conversely, the di-*o*-substitution gives rise to ethylene polymerization catalysts, e.g., **L7**–**L11** and **L84**–**L126** (Table 3.1), except for the 2,6-difluoro (**L104**) [63, 72] and 2-fluoro-6-methyl (**L105**) [77] substituents.

The electronic and steric environments of the complexes have substantial influences on ethylene polymerization, which can be in principle divided into three aspects: the metal center, the halide group, and the substituents and substitution pattern on the ligand backbone.

The nature of the metal center has a marked influence on the catalytic performances of the complexes. Generally, iron complexes are much more productive and afford polyethylenes with higher molecular weights than those of the corresponding cobalt analogues, under the same polymerization conditions [53, 58, 60, 62, 91, 92]. For instance, iron complexes [$FeCl_2$(**L87**)], [$FeCl_2$(**L85**)] and [$FeCl_2$(**L26**)] show activities of 5.34, 20.6 and 3.75 $\times$ 10^6 g/(mol h bar) and afford polyethylenes with molecular weights (M_w) of 6.11, 1.48 and 3.13 $\times$ 10^5 g/mol, respectively, while the corresponding cobalt complexes [$CoCl_2$(**L87**)], [$CoCl_2$(**L85**)] and [$CoCl_2$(**L26**)] give activities of 0.45, 1.70 and 1.74 $\times$ 10^6 g/(mol h bar) and yield polyethylenes with molecular weights of 0.14, 2.57 and 2.34 $\times$ 10^5 g/mol, respectively [53].

It is noteworthy that the ^{13}C NMR end group analysis of the polyethylenes obtained with iron precatalysts reveals an excess of saturated chain ends in relation to the vinyl unsaturated ones, indicating that chain transfer at the iron catalytic systems involves both β-H elimination and chain transfer to aluminum centers. In contrast, all the polyethylenes formed by cobalt complexes exhibit a 1:1 ratio of saturated to unsaturated chain ends, regardless of the polymerization conditions, which is characteristic of chain transfer exclusively by β-H elimination [53, 60, 62].

The halide group has a slight influence on the catalytic performance of iron and cobalt precatalysts because complexes based on the same BIP ligand, being structurally isomorphous, will form the same active species upon alkylation and activation by MAO. Generally, the BIP iron and cobalt dichlorides are more productive than the corresponding dibromides, and they afford polyethylenes with similar properties. For instance, the iron complex [$FeCl_2$(**L85**)] (20.6×10^6 g/(mol h bar)) is only slightly more active than [$FeBr_2$(**L85**)] (17.6×10^6 g/(mol h bar)) [53, 93]. Through the correlation of the net charge at the metal center with the catalytic activity, Zhang et al. [93] demonstrated that the difference in activity can be ascribed to the variation on the net charge of the iron center. For instance, [$FeCl_2$(**L85**)] (0.6935) has a very slightly higher net charge on the iron center than [$FeBr_2$(**L85**)] (0.6932), leading to the observed small difference in activity.

The substituents and substitution pattern on the BIP ligand play crucial roles on determining the catalytic performance of iron and cobalt complexes. Replacement of the hydrogen atom at the α-imino carbon (aldimine ligands) by alkyl groups (ketimine ligands) results in an increase both in the productivity and in the molecular weight. For instance, the ketimine-based iron complexes [$FeCl_2$(**L84**)], [$FeCl_2$(**L85**)] and [$FeCl_2$(**L87**)] (9.34, 20.6 and 5.34×10^6 g/(mol h bar), respectively), containing a methyl group at the α-imino carbon, are significantly more active than the corresponding aldimine-based ones [$FeCl_2$(**L7**)], [$FeCl_2$(**L8**)] and [$FeCl_2$(**L11**)] (0.56, 0.55 and 0.31×10^6 g/(mol h bar), respectively), and afford polyethylenes with significantly higher molecular weights (M_w) (2.42 vs. 1.08×10^5 g/mol, 1.48 vs. 1.52×10^5 g/mol, 6.11 vs. 1.32×10^5 g/mol, respectively) [53]. Substituents such as the ethyl, *iso*-propyl, 2-phenylethyl and 1,3-diphenylprop-2-yl groups at the α-imino carbon display a similar effect. For example, iron complexes [$FeCl_2$(**L133**)], [$FeCl_2$(**L135**)], [$FeCl_2$(**L140**)] and [$FeCl_2$(**L141**)], bearing those substituents, show activities of 17.1, 17.8, 16.3 and 16.8×10^6 g/(mol h bar), respectively, being much higher than that of [$FeCl_2$(**L8**)] (0.55×10^6 g/(mol h bar)), and yield polyethylenes with molecular weights (M_w) of 1.65, 4.56, 1.98 and 4.68×10^5 g/mol, which are values higher than that obtained with the latter precatalyst (1.52×10^5 g/mol) [66]. Introduction of a phenyl group at the α-imino carbon (e.g., [CoMe(**L144**)], 1.5×10^4 g/(mol h bar)) gives a comparable activity to that obtained with the complex containing a methyl group at the α-imino carbon (e.g., [CoMe(**L87**)], 1.6×10^4 g/(mol h bar)) [64]. Substitution of a methoxy group at the α-imino carbon gives rise to deactivation of the iron precatalyst, e.g., [$FeCl_2$(**L146**)], possibly due to irreversible bonding of the activator to the heteroatom, leading to destabilization of the catalyst via ligand dissociation or decomposition of the active iron-alkyl propagating species [68]. To prevent the establishment of this bonding to the ether oxygen atom, substituents bearing a soft atom such as sulfur, or large steric hindrances such as 2,6-dimethylphenoxy and 2,6-dimethylphenylthio, e.g., **L148**–**L150**, were introduced. As a result, the corresponding iron complexes

[$FeCl_2$(**L148**)], [$FeCl_2$(**L150**)] and [$FeCl_2$(**L151**)] exhibited comparable catalytic activities to that of [$FeCl_2$(**L85**)] [68].

The effects of the presence of alkyl substituents on the aryl rings on the polymerization of ethylene are dependent upon their nature and substitution pattern, leading to various electronic and steric effects at the metal center. Generally, an increase in the steric bulk at the aryl ring *ortho* positions results in a decrease in the activity and an increase in the molecular weight, whereas a replacement at the *para* position leads to an increase in the activity and a decrease in the molecular weight. For instance, with the increase of the steric bulk (2,6-Me_2 < 2-^{t}Bu < 2,6-$^i Pr_2$), the catalytic activities of the corresponding iron complexes [$FeCl_2$(**L84**)], [$FeCl_2$(**L26**)] and [$FeCl_2$(**L87**)] tend to decrease ($9.34 > 3.75 < 5.34 \times 10^6$ g/(mol h bar)) and the molecular weights (M_w) markedly increase ($2.42 < 3.13 < 6.11 \times 10^5$ g/mol) [53]. The replacement of the *para*-aryl proton of [$FeCl_2$(**L84**)] by a methyl group ([$FeCl_2$(**L85**)]) results in an increase in the activity from 9.34 to 20.6×10^6 g/(mol h bar) and a decrease in the molecular weight (M_w) from 2.42 to 1.48×10^5 g/mol [53]. Substitution with cycloalkyl groups has no identical effect on ethylene polymerization. For a cyclopentyl group, the increase of the steric bulk from a mono-*o*-cyclopentyl group ([$FeCl_2$(**L92**)]) to a di-*o*-cyclopentyl ([$FeCl_2$(**L91**)]) one leads to the increase in both activity and molecular weight, whereas, for a cyclohexyl group, the opposite trends are observed when the steric bulk is increased from mono-*o*-cyclohexyl ([$FeCl_2$(**L96**)]) to di-*o*-cyclohexyl ([$FeCl_2$(**L95**)]) [94]. A striking feature of the cycloalkyl effect is that these groups may dramatically improve the temperature stability of the corresponding iron active species, the activity maximum shifting toward higher temperatures with the increasing size of the cycloalkyl group [94]. For example, the activity maxima shift to 50 °C, for cyclopentyl, and to 70 °C, for cyclohexyl and cyclooctyl groups, while for the cyclododecyl group the catalytic system still has a good activity even over 80 °C.

Incorporating phenyl groups into the aryl *ortho* positions substantially depresses the activity. For instance, the aldimine-based iron complex [$FeCl_2$-(**L12**)], bearing 2,6-diphenyl substituents on the aryl rings, displays an unusual low activity (4.2×10^4 g/(mol h bar)) and affords low molecular weight polyethylene ($M_w = 1.2 \times 10^3$ g/mol), while the ketimine-based complexes, bearing 2,4,6-triphenyl substituted aryl rings, present no activity for the cobalt complex [$CoCl_2$(**L103**)] and very low activity for the iron one [$FeCl_2$(**L103**)] [91].

In the case of the di-*o*-substitution with halogen atoms at the aryl rings, the fluorine substitution gives rise to an ethylene oligomerization catalyst (e.g., [$FeCl_2$(**L104**)]), and chlorine and bromine substituents lead to ethylene polymerization catalysts (e.g., [$FeCl_2$(**L108**)] and [$FeCl_2$(**L112**)], respectively). On increasing the size from chlorine to bromine, a reduction in the activity, e.g., 12.8×10^6 g/(mol h bar) for [$FeCl_2$(**L108**)] vs. 8.6×10^6 g/(mol h bar) for [$FeCl_2$(**L112**)], and an increase in the molecular weight (M_w), e.g., 1.33×10^4 g/mol for [$FeCl_2$(**L108**)] vs. 10.1×10^4 g/mol for [$FeCl_2$(**L112**)], are observed [60, 95]. Precatalysts [$FeCl_2$(**L108**)] and [$FeCl_2$(**L112**)] are more active than [$FeCl_2$(**L87**)], which bears di-*o*-*iso*-propyl substituents on the aryl ring, but the molecular weights obtained are substantially lower than that obtained by the latter. The cobalt complexes show the same trends as those of iron complexes. Chen et al. [60] thought that the electron-withdrawing nature of chlorine and bromine substituents, which results in more electrophilic metal centers in the complexes, would be responsible for their enhanced activities in ethylene polymerization. In fact, semiempirical calculations indicate that the atomic (Mulliken) charges on the iron centers of the complexes bearing *iso*-propyl or methyl substituents are very close, –0.2859 ([$FeCl_2$(**L87**)]) and –0.2854 ([$FeCl_2$(**L84**)]), respectively, whereas the chlorine- or bromine-substituted iron complexes have lower charges at the iron centers, –0.2354 ([$FeCl_2$(**L108**)]) and –0.2502 ([$FeCl_2$(**L112**)]), respectively [60]. Although the size of the trifluoromethyl group lies in between the methyl and *iso*-propyl groups, the *o*-trifluoromethyl-aryl monosubstituted iron complexes (e.g., [$FeCl_2$(**L46**)], [$FeCl_2$(**L49**)], [$FeCl_2$(**L59**)] and [$FeCl_2$(**L64**)]) exclusively afford polymers, while the corresponding cobalt complexes (e.g., [$CoCl_2$(**L46**)], [$CoCl_2$(**L49**)], [$CoCl_2$(**L59**)] and [$CoCl_2$(**L64**)]) yield a mixture of oligomers along with some polymeric fraction [74, 91]. These observations are in sharp contrast to those reported for the aryl *o*-alkyl-substituted iron and cobalt complexes, which exclusively lead to oligomers. Pelascini et al. [91] attribute this result to the interaction of one of the fluorine atoms with the β-hydrogens of the polymer chain, as in the case of the fluorine-containing phenoxy-imine titanium complexes [19, 96, 97]. It is noteworthy that the presence of a *o*-trifluoromethyl substituent in the aryl rings of the BIP ligand boosts the activities of cobalt and iron complexes by up to 2 and 6 orders of magnitude, respectively [74].

The replacement of the phenyl ring of the BIP ligand by groups containing extended π-conjugation, such as 1-naphthyl (**L127**), 2-methyl-1-naphthyl (**L128**), 1-anthracenyl (**L130**) or 4-pyrenyl (**L131**), significantly improves the catalytic activity of the corresponding complexes, along with varied effects on the molecular weight [58, 78, 98, 99].

The catalytic performance of the complexes based on unsymmetrical BIP ligand is largely dependent on the combined bulkiness of the *o*-substituents on both aryl rings [100, 101]. For instance, the precatalyst [$FeCl_2$(**L167**)], containing simultaneously phenyl and 2,6-di-*iso*-propylphenyl groups, affords oligomers when exposed to ethylene and MAO, while [$FeCl_2$(**L171**)], containing 2,6-dimethylphenyl and 2,6-di-*iso*-propylphenyl groups, yields polymers [101]. A mixture of oligomer and polymer is obtained by [$FeCl_2$(**L168**)], which contains 2-methylphenyl and 2,6-di-*iso*-propylphenyl rings. Bianchini et al. [101] ascribed this simultaneous oligomerization and polymerization processes to the C_1-symmetry of the complex, in which, due to the hindered rotation of the tolyl group, two atropisomeric propagating alkyl active species are formed, one of them being responsible for the ethylene polymerization and the other for the ethylene oligomerization. A similar phenomenon is observed with the iron complex [$FeCl_2$(**L262**)], which contains (*R*)-1-phenylethyl and 2,6-di-*iso*-propylphenyl rings (see below in Table 3.4) [101].

3.2.3.2 Oligomerization of Ethylene

Treatment of iron and cobalt complexes, exclusively bearing mono-*o*-alkyl substituents on the aryl rings, with MAO generates highly active ethylene oligomerization catalysts, such as in the case of [$FeCl_2$(**L15**)], with activities reaching 1.21 × 10^8 g/(mol h bar), at 90 °C and 600 psig [15]. In general, iron complexes are more active than the corresponding cobalt analogues, and the ketimine-based complexes give much higher activities than the aldimine-based ones [52, 60, 63, 102]. For instance, under the same oligomerization conditions, the acetimine-based iron complex [$FeCl_2$(**L15**)] is substantially more active than the corresponding cobalt one [$CoCl_2$(**L15**)], 1.30 vs. 0.02 × 10^6 g/(mol h bar), and more active than the aldimine-based iron complex [$FeCl_2$(**L2**)], 1.30 vs. 0.48 × 10^6 g/(mol h bar) [52]. These trends parallel those observed for the ethylene polymerization with iron and cobalt complexes containing *o*-disubstituted aryl rings, although the difference is more pronounced in the case of the oligomerization systems. In all cases, the oligomers obtained follow a Schulz–Flory distribution, which is characterized by the constant *K*, representing the probability of chain propagation ($K = \text{rate}_{\text{propagation}}/(\text{rate}_{\text{propagation}} + \text{rate}_{\text{chain transfer}})$ = (moles of C_{n+2})/(moles of C_n)). The *K* values are experimentally determined by the molar ratio of two oligomer fractions, generally the C_{14} and C_{12} ones. A high *K* value means that a catalyst produces higher molecular weight oligomers. In the case of industrial-grade α-olefin oligomers, the *K* value is commonly preferred to be in the range of 0.6–0.8 [75, 76]. If this value exceeds 0.8, unwanted polymer fractions will be present. Generally, the ketimine-based iron complexes give higher *K* values than those of the aldimine-based ones, whereas the *K* values for both ketimine- and aldimine-based cobalt complexes lie in between those obtained for the ketimine- and aldimine-based iron systems. For instance, the ketimine-based iron complex [$FeCl_2$(**L15**)] gives a higher *K* value of 0.79 when compared with 0.50 for the corresponding aldimine-based one [$FeCl_2$(**L2**)], while the ketimine- and aldimine-based cobalt complexes [$CoCl_2$(**L15**)] and [$CoCl_2$(**L2**)] give moderate *K* values of 0.67 and 0.74, respectively [52].

Similar to the ethylene polymerization precatalysts, the influences of alkyl substituents on the ligand aryl rings toward ethylene oligomerization depend on their nature and substitution pattern. Generally, the more sterically congested complexes result in somewhat lower activities and higher *K* values, this effect being not as evident as in the case of ethylene polymerization precatalysts. For instance, when the size of the *o*-substituents on the aryl rings increases in the order Me < Et < ^{i}Pr, the corresponding iron complexes [$FeCl_2$(**L15**)], [$FeCl_2$(**L21**)] and [$FeCl_2$(**L23**)] show decreasing activities (2.80, 2.24 and 2.24 × 10^6 g/(mol h bar), respectively) and increasing *K* values (0.81, 0.81 and 0.87, respectively) [15]. The same trends are also observed in iron complexes bearing *o*-ethyl, *o*-*n*-propyl and *o*-*n*-butyl substituents on the aryl rings [103]. Britovsek [52] and Kim [102] groups investigated the effect of the alkyl group substitution patterns on the oligomerization of ethylene, and found that additional alkyl substitution at the *meta* and *para* positions of the aryl rings substantially enhances the catalytic activities, but has a little effect on the *K* value. For example, iron complexes [$FeCl_2$(**L15**)], [$FeCl_2$(**L17**)] and [$FeCl_2$(**L18**)], bearing 2-methyl, 2,3-dimethyl and 2,4-dimethyl substituents on the aryl rings, respectively, exhibit activities of 2.01, 4.17 and 2.66 $\times$ 10^6 g/(mol h bar) and give *K* values of 0.73, 0.77 and 0.70, respectively [52]. Zhang et al. [93] correlated the variation in the activity with the net charge at the metal center, demonstrating that the activity decreases when the net charge is higher than 0.7000, and increases when it is lower than 0.7000. In fact, for the latter series of iron complexes, the net charge at the iron center increases in the order [$FeCl_2$(**L17**)] (0.7014) < [$FeCl_2$(**L18**)] (0.7044) < [$FeCl_2$(**L15**)] (0.7053), while the activity decreases in the reverse order [$FeCl_2$(**L17**)] > [$FeCl_2$(**L18**)] > [$FeCl_2$(**L15**)]. Ivanchev et al. [94] found that mono-*o*-cycloalkyl substitution at the aryl ring, e.g., cyclopentyl, cyclohexyl, cyclooctyl and cyclododecyl, exclusively leads to oligomers, but the substituent effect was not discussed in detail. Görl and Alt [57] revealed that sterically congested substituents at the *para* positions of the aryl rings also exert a marked influence on the polymer molecular weights. Introduction of the relatively less bulky group (9*H*-fluoren-2-yl)ethynyl at the *para* positions of the mono-*o*-methyl substituted aryl rings of BIP ligand **L15** leads to ligand **L36**, whose iron complex [$FeCl_2$(**L36**)] exclusively produces oligomers. Conversely, the incorporation of the relatively much bulkier 5-(cyclopenta-1,3-dienyl)-5-(9*H*-fluoren-9-yl)hex-1-ynyl group results in the ethylene polymerization precatalyst [$FeCl_2$(**L37**)], although ligand **L37** is mono-*o*-methyl substituted at the aryl rings [57].

The influence of halogen atom substitution at the aryl ring on the oligomerization of ethylene is also dependent on their nature and substitution pattern, as in the case of alkyl substituents. Due to the reduced size of the fluorine atom, the mono-*o*-fluorine substitution at the aryl rings gives the bisligand iron complex $[Fe(\mathbf{L39})_2]^{2+}[FeCl_4]^{2-}$, which is inactive toward ethylene oligomerization [60, 72]. In opposition, the corresponding di-*o*-fluorine substitution leads to the highly active ethylene oligomerization iron catalyst [$FeCl_2$(**L104**)] and to the inactive analogous cobalt complex [$CoCl_2$(**L104**)] [63, 72]. The corresponding chloro-, bromo- and iodo-mono-*o*-substituted BIP ligand derivatives give rise to the highly active ethylene oligomerization catalysts [$FeCl_2$(**L52**)], [$FeCl_2$(**L60**)] and [$FeCl_2$(**L65**)], respectively, and to the inactive analogous cobalt complexes [$CoCl_2$(**L52**)], [$CoCl_2$(**L60**)] and [$CoCl_2$(**L65**)]. An increase in the size of the halogen atoms (Cl < Br < I) leads to a decrease in the catalytic activity [$FeCl_2$(**L52**)] (5.19 $\times$ 10^6 g/(mol h bar)) > [$FeCl_2$(**L60**)] (2.95 $\times$ 10^6 g/(mol h bar)) > [$FeCl_2$(**L65**)] (0.93 $\times$ 10^6 g/(mol h bar)), but the corresponding *K* values vary in the reverse order: [$FeCl_2$(**L52**)] (0.59) < [$FeCl_2$(**L60**)] (0.63) < [$FeCl_2$(**L65**)] (0.67) [60]. The same trends are observed for the series [$FeCl_2$(**L42**)], [$FeCl_2$(**L53**)] and [$FeCl_2$(**L61**)], bearing 2-fluoro-4-methyl, 2-chloro-4-methyl and 2-bromo-4-methyl groups on the aryl rings, respectively [104]. In the case of [$FeCl_2$(**L104**)], [$FeCl_2$(**L41**)] and [$FeCl_2$(**L40**)], bearing 2,6-, 2,5- and 2,4-difluoro groups on the aryl rings, respectively, they show activities of 4.07, 9.33 and 11.1 $\times$ 10^6 g/(mol h bar), at 60 °C, respectively [63]. It is obvious that the substitution of the fluorine atom at the *meta* and *para* positions leads to higher catalytic activities than that at the *ortho* position. Their cobalt analogues are inactive for ethylene oligomerization. It is worth to note that the *K* values obtained by [$FeCl_2$(**L104**)], [$FeCl_2$(**L41**)] and [$FeCl_2$(**L40**)] precatalysts are 0.44, 0.33 and 0.34, respectively, being much lower than those obtained with the analogous alkyl-substituted complexes [15, 52, 103]. Although the *o*-trifluoromethyl monosubstitution leads to ethylene polymerization iron precatalysts [74, 91], the *m*- and *p*-trifluoromethyl-substituted iron complexes are active exclusively toward ethylene oligomerization, showing comparable catalytic activities and *K* values [55].

Ionkin et al. [76] introduced the nitro group into the BIP ligand **L15**, which bears a *o*-methyl substituent on the aryl ring, and found that all the corresponding nitro-substituted iron complexes, [$FeCl_2$(**L79**)], [$FeCl_2$(**L80**)] and [$FeCl_2$(**L162**)], show lower activities when compared to that of the parent iron complex [$FeCl_2$(**L15**)], but give higher *K* values. Conversely, the introduction of the diethylamino group ([$FeCl_2$(**L68**)]) results in the decrease in both activity and *K* values [76]. Likewise, the same group [75] synthesized a nitrile *o*-aryl monosubstituted derivative of the BIP ligand, the resulting complex [$FeCl_2$(**L76**)] showing to be inactive toward ethylene oligomerization. Nevertheless, the nitrile *p*-substituted aryl ring, either in symmetrical or unsymmetrical BIP ligands, results in decreased activities and increased *K* values produced by their iron complexes [$FeCl_2$(**L77**)], [$FeCl_2$(**L154**)], [$FeCl_2$(**L161**)], [$FeCl_2$(**L181**)] [75]. Among them, the symmetrical iron complex [$FeCl_2$(**L77**)] gives the highest *K* value of 0.69, and the remaining unsymmetrical iron complexes lead to lower *K* values of ca. 0.62.

3.3 Mono(arylimino)pyridine-Based Ligands and Their Corresponding Iron and Cobalt Complexes

3.3.1 Syntheses of Ligands and Complexes

The 2-mono(arylimino)pyridine ligands mentioned in this section are bi- or tridentate chelated species, differing from those presented in the previous Sect. 3.2 in that they contain a single imino arm attached to the pyridine fragment.

Table 3.3 lists the 2-mono(arylimino)pyridine-based ligands reported in the literature. Like the 2,6-bis(arylimino)pyridine, these *N,N,E* ligands (E = N, O, S) are prepared by the condensation reaction of either a 6-substituted 2-pyridinecarboxaldehyde or a 6-substituted 2-acetylpyridine with one equivalent of the appropriate aniline, using an alcohol (e.g., methanol [105–107] and ethanol [78, 108]) as solvent, in the presence of a protic acid (e.g., formic [105–107] or acetic [78, 108] acids) as catalyst (Scheme 3.14). In order to improve the yield, the water formed in the reaction is usually concomitantly removed by the addition of water absorbents (e.g., molecular sieves [55, 78, 109] and sodium sulfate [110]). In some cases, the alcohol is replaced by a nonpolar or weakly polar solvent (e.g., benzene

Table 3.3 2-Mono(arylimino)pyridine-based ligands, 2-(ArN=C(R))-6-R′-C_5H_3N

	Ar	R	R′	Refs.[a]
L184	Phenyl	Me	H	[291]
L185	Phenyl	Me	Br	[302]
L186	2-methylphenyl	Me	Phenyl	[303]
L187	2,6-dimethylphenyl	Me	Phenyl	[303]
L188	2,6-dimethylphenyl	Me	Br	[302]
L189	2,6-dimethylphenyl	Me	2-quinolinyl	[111]
L190	2,6-dimethylphenyl	Me	Acetyl	[35, 42, 118]
L191	2,6-dimethylphenyl	Me	Ethoxycarbonyl	[109, 112]
L192	2,4,6-trimethylphenyl	Me	Acetyl	[34, 42]
L193	2,4,6-trimethylphenyl	Me	Ethoxycarbonyl	[112]
L194	2,4,6-trimethylphenyl	Me	2-quinolinyl	[111]
L195	2,4,6-trimethylphenyl	Me	2-hydroxy-2-propyl	[114]
L196	2,6-diethylphenyl	Me	Acetyl	[42]
L197	2,6-diethylphenyl	Me	Ethoxycarbonyl	[109, 112]
L198	2,6-diethylphenyl	Me	2-quinolinyl	[111]
L199	2-*iso*-propylphenyl	Me	Acetyl	[78]
L200	2,6-di-*iso*-propylphenyl	H	2-benzo[*b*]thienyl	[105]
L201	2,6-di-*iso*-propylphenyl	H	6-(2,2′-bipyridyl)	[108]
L202	2,6-di-*iso*-propylphenyl	H	Hydroxymethyl	[304]
L203	2,6-di-*iso*-propylphenyl	H	Acryloyloxymethyl	[304]
L204	2,6-di-*iso*-propylphenyl	Me	Phenyl	[107]
L205	2,6-di-*iso*-propylphenyl	Me	2-naphthyl	[105]
L206	2,6-di-*iso*-propylphenyl	Me	Br	[106]
L207	2,6-di-*iso*-propylphenyl	Me	2-quinolinyl	[111]
L208	2,6-di-*iso*-propylphenyl	Me	2-pyridyl	[106]
L209	2,6-di-*iso*-propylphenyl	Me	6-methyl-2-pyridyl	[106]
L210	2,6-di-*iso*-propylphenyl	Me	6-(2,2′-bipyridyl)	[108]
L211	2,6-di-*iso*-propylphenyl	Me	2-furanyl	[107]
L212	2,6-di-*iso*-propylphenyl	Me	2-thienyl	[107]
L213	2,6-di-*iso*-propylphenyl	Me	3-thienyl	[106]
L214	2,6-di-*iso*-propylphenyl	Me	5-ethyl-2-thienyl	[107]
L215	2,6-di-*iso*-propylphenyl	Me	2-benzo[*b*]thienyl	[105]
L216	2,6-di-*iso*-propylphenyl	Me	3-benzo[*b*]thienyl	[106]
L217	2,6-di-*iso*-propylphenyl	Me	2-hydroxy-2-propyl	[114]
L218	2,6-di-*iso*-propylphenyl	Me	Acetyl	[110, 118]
L219	2,6-di-*iso*-propylphenyl	Me	Ethoxycarbonyl	[109]
L220	2,6-di-*iso*-propylphenyl	Et	2-benzo[*b*]thienyl	[105]
L221	2-*tert*-butylphenyl	Me	Acetyl	[35]
L222	2-cyclopentylphenyl	Me	Br	[302]
L223	2-cyclopentyl-6-methylphenyl	Me	Br	[302]
L224	2-cyclopentyl-4,6-dimethylphenyl	Me	Br	[302]
L225	2,6-dicyclopentylphenyl	Me	Br	[302]

(continued)

Table 3.3 (continued)

	Ar	R	R′	Refs.[a]
L226	2-cyclohexylphenyl	Me	Br	[302]
L227	2-cyclohexyl-6-methylphenyl	Me	Br	[302]
L228	2-cyclohexyl-4,6-dimethylphenyl	Me	Br	[302]
L229	2,6-dicyclohexylphenyl	Me	Br	[302]
L230	2,6-difluorophenyl	Me	2-quinolinyl	[111]
L231	2,6-difluorophenyl	Me	Ethoxycarbonyl	[109]
L232	3-trifluoromethylphenyl	Me	Acetyl	[55]
L233	2,6-dichlorophenyl	Me	2-quinolinyl	[111]
L234	2,6-dichlorophenyl	Me	Ethoxycarbonyl	[109]
L235	2,6-dibromophenyl	Me	2-quinolinyl	[111]
L236	2,6-dibromophenyl	Me	Ethoxycarbonyl	[109]
L237	4-methoxyphenyl	Me	H	[291]
L238	8-quinolinyl	H	Methyl	[115]
L239	2-methyl-8-quinolinyl	H	H	[115]
L240	2-methyl-8-quinolinyl	H	Methyl	[115]
L241	2-*iso*-propyl-8-quinolinyl	H	H	[115]
L242	2-*iso*-propyl-8-quinolinyl	H	Methyl	[115]
L243	2-cyclohexyl-8-quinolinyl	H	H	[115]
L244	2-cyclohexyl-8-quinolinyl	H	Methyl	[115]

[a] Only references reporting compounds for the first time are given; references are also given when different reaction conditions were used

Scheme 3.14 Synthesis of 2-mono(arylimino)pyridine ligands

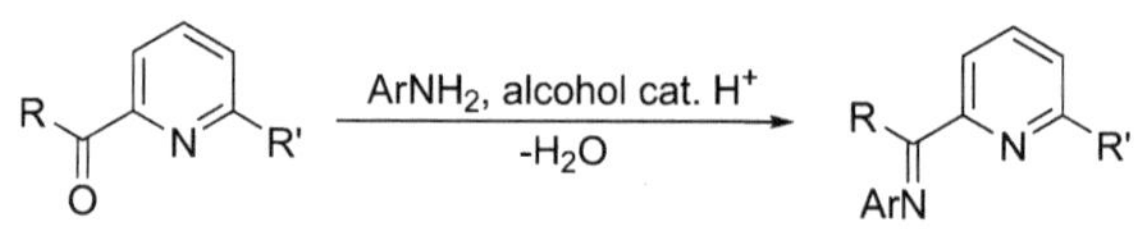

[55] and toluene [34, 109, 111]), combined with the addition of *p*-toluenesulfonic acid as catalyst and removal of water using the Dean–Stark trap technique.

The nature of the substituent R′ and the aryl group (Ar) (Scheme 3.14) has a marked influence on the condensation reaction. When the substituent R′ is ethoxycarbonyl, Su et al. were unsuccessful in the preparation of ligands **L191**, **L193** and **L197** by condensation of ethyl 6-acetylpicolinate with the corresponding anilines, in refluxing ethanol, and in the presence of acetic acid as catalyst [112, 113]. The use of toluene as solvent, *p*-toluenesulfonic acid as catalyst and a Dean–Stark trap to remove the water formed gave very low yields (about 10–15%). Nevertheless, the same authors found that microwave-assisted condensation reactions afford the desired ligands in satisfactory yields. Analogous to the 2,6-bis(arylimino)pyridine ligands, the use of strong electron-withdrawing substituents on the aryl ring (e.g., CF_3 [55]) commonly results in a relatively low yield. Due to the reduced nucleophilicity of the corresponding anilines, forcing conditions are required to improve the reaction yield. For instance, in the case of ligand **L232**, bearing a *m*-trifluoromethyl substituent on the aryl ring [55], a satisfactory result (70%) was obtained only when benzene was employed as solvent, molecular sieves as water absorbent and prolonged reaction times were used (up to 80 days).

Bianchini et al. [105–107] described a new protocol to introduce aryl or heterocyclic ring substituents, e.g., 2-furanyl, 2-thienyl and 2-pyridyl groups, into the 6-position of the 2-mono(arylimino)pyridine ligand, leading to a new family of 6-substituted 2-(arylimino)pyridine neutral ligands (e.g., **L204**, **L211**, **L212** and **L216**) (Scheme 3.15). This method consisted in the condensation of 6-bromo-2-pyridinecarboxaldehyde or 2-acetyl-6-bromopyridine with the appropriate anilines to form the 6-bromo-2-(arylimino)pyridines, followed by the palladium catalyzed Stille coupling of the latter compounds with the appropriate stannanes. Alternatively, introduction of the aryl substituents, e.g., phenyl and 2-naphthyl, into the 6-position of the 2-mono(arylimino)pyridine ligand can be achieved by the palladium catalyzed Suzuki coupling of 6-bromo-2-pyridinecarboxaldehyde or 2-acetyl-6-bromopyridine with the appropriate substituted boronic acid, followed by the condensation of the resulting 6-substituted 2-pyridinecarboxaldehyde or 2-acetylpyridine with the desired anilines (Scheme 3.15).

Treatment of 6-acetyl-2-(arylimino)pyridine ligands with two equivalents of trimethylaluminum affords the air-sensitive bimetallic 2-arylacetiminopyridinyl-6-(propan-2-olate) aluminum complexes [2-(ArN=CMe)-6-{$C(CH_3)_2O(AlMe_3)$}-C_5H_3N]$AlMe_2$ (Ar = 2,6-$^i Pr_2C_6H_3$ or 2,4,6-$Me_3C_6H_2$), which, followed by hydrolysis, generate a new class of tridentate ligands such as **L195** and **L217** (Scheme 3.16) [114].

The iron and cobalt complexes based on the 2-mono(arylimino)pyridine ligands are readily prepared by the reactions of the ligands with hydrated or dehydrated iron and cobalt dichloride, in ethanol or tetrahydrofuran [42, 78, 105, 106, 109, 112]. However, the molecular structures of the corresponding iron and cobalt complexes

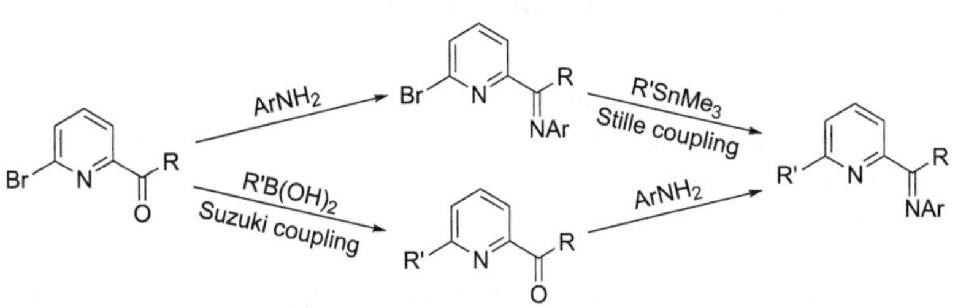

Scheme 3.15 Syntheses of 6-substituted 2-mono(arylimino)pyridine ligands reported by Bianchini et al. [105–107]

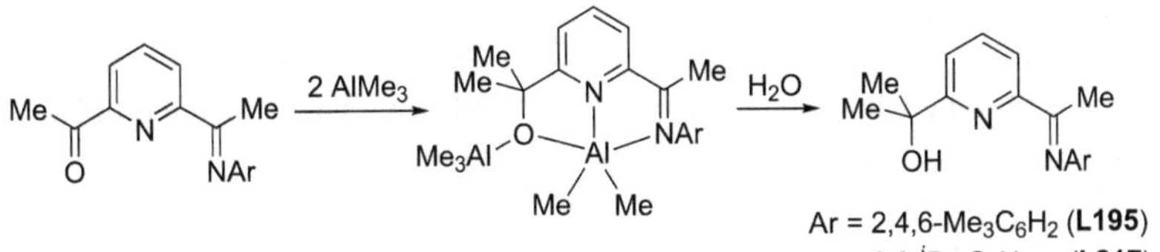

Scheme 3.16 Syntheses of ligands **L195** and **L217**

are largely dependent upon the nature of the substituents (R′) at the 6-positions of the pyridyl ring. The 2-acetyl R′ substituent exclusively gives rise to the N^N^O tridentate chelated iron and cobalt complexes [42, 78, 110], the molecular structures of which adopt square pyramidal geometries around the metal centers. Two nitrogen atoms, one oxygen atom and one of the chlorine atoms form the basal plane, the remaining chlorine occupying the apical position, e.g., [$FeCl_2$(**L218**)] [78]. The 2-ethoxycarbonyl substitution at the 6-position of the pyridyl ring gives rise to either N^N^O(carbonyl) tridentate chelates of iron and cobalt, e.g., [$FeCl_2$(**L197**)] [109] and [$CoCl_2$(**L197**)] [112], or to N^N bidentate chelated complexes, e.g., [$FeCl_2$(**L219**)], [$FeCl_2$(**L191**)] and [$CoCl_2$(**L236**)] [109]. The tridentate chelated complexes show distorted trigonal bipyramidal geometries around the metal centers, with the pyridyl nitrogen atom and the two chlorine atoms forming the equatorial plane, while the bidentate chelated complexes display distorted tetrahedral geometries around the metal centers. The aryl- or heterocycle-substitution at the pyridyl ring 6-position also leads to either N^N bidentate chelated complexes with tetrahedral geometries around the metal centers, e.g., [$CoCl_2$(**L204**)] and [$CoCl_2$(**L214**)] [107], or to N^N^N tridentate chelated complexes in the case of the 2-pyridyl [106] and 2-quinolinyl [111] derivatives (e.g., [$FeCl_2$(**L189**)], [$CoCl_2$(**L208**)] and [$CoCl_2$(**L207**)]). In addition, the nature of the heterocyclic rings also largely influences the molecular structures of the complexes. Replacement of the phenyl iminic ring by a 8-quinolinyl ring exclusively affords the N^N^N tridentate chelated complexes, e.g., [$FeCl_2$(**L238**)], [$FeCl_2$(**L240**)] and [$CoCl_2$(**L243**)] [115].

3.3.2 Oligo- and Polymerization of Ethylene

Treatment of the 2-mono(arylimino)pyridine-based iron and cobalt complexes with MAO (or MMAO) generates ethylene oligo- and polymerization catalysts. The substituents (R′) at the 6-position of the pyridyl ring and the aryl iminic substituents have marked influences on the catalytic performances of the complexes.

The 6-acetyl substitution on the 2-(arylimino)pyridine ligand bearing sterically congested *ortho* aryl substituents (e.g., 2,6-dimethyl (**L190**), 2,6-diethyl (**L196**) and 2,6-di-*iso*-propyl (**L217**)) exclusively leads to highly active ethylene polymerization iron and cobalt catalysts [42, 110]. The highest activity can reach 1.32×10^7 g/(mol h bar) ([$FeCl_2$(**L217**)] [110]), which is comparable with those of 2,6-bis(arylimino)pyridine iron catalysts [14, 53]. Notably different from the 2,6-bis(arylimino)pyridine iron catalysts, branched polyethylenes, instead of linear ones, were obtained by this kind of iron complexes. The cobalt complexes are substantially less active than their iron analogues [42]. The iron complex [$FeCl_2$(**L232**)], containing a relatively less bulky *ortho* aryl substituent exclusively affords oligomers, upon activation with MAO [55].

The 6-ethoxycarbonyl-2-(arylimino)pyridine-based iron and cobalt complexes show moderate activities, in the order of 10^4 g/(mol h bar), in the oligo- and polymerization of ethylene, upon activation with MAO, affording a mixture of oligomers and polymers [109, 112]. The oligomers predominantly consist of butenes and hexenes and small amounts of higher oligomers, following the Schulz–Flory distribution, with *K* values varying in the range of 0.73–0.96, while the polymers comprise relatively low molecular weight polyethylenes (M_n = 500–4,700). The aryl substituents have marked influences on the catalytic activities. The *o*-alkyl substitution gives an identical influence on the activities of iron and cobalt complexes, i.e., the catalytic activity decreases with the size of the *o*-substituents (Me < Et < ^{i}Pr), e.g., [$FeCl_2$(**L191**)] > [$FeCl_2$(**L197**)] > [$FeCl_2$(**L219**)], which is consistent with the trends observed for the 2,6-bis(arylimino)pyridine-based iron and cobalt complexes [14, 53]. In the case of the *o*-halo substitution, the activity varies in the order of chloro- ([$FeCl_2$(**L234**)]) > fluoro- ([$FeCl_2$(**L231**)]) > bromo- ([$FeCl_2$(**L236**)]) for iron complexes, and bromo- ([$CoCl_2$(**L236**)]) > chloro- ([$CoCl_2$(**L234**)]) > fluoro- ([$CoCl_2$(**L231**)]) for cobalt complexes. It is worth to note that the catalytic activities of cobalt complexes are, in the latter cases, not lower, but remarkably higher than those of their iron analogues.

The 2-quinolinyl substitution at the 6-position leads exclusively to iron and cobalt catalysts that exhibit high activities in the oligomerization of ethylene (ca. 10^5 g/(mol h bar)), under a low pressure of ethylene (1 bar), yielding a mixture of ethylene oligomers and polymer, under higher pressures of ethylene (e.g., 10 bar). The oligomers formed mainly consist of butenes and hexenes [111]. The cobalt complexes are slightly less active than the corresponding iron analogues. The substituents on the aryl groups have no relevant effects on the catalytic behavior of the system.

The heterocycle substitution (e.g., 2-thienyl (**L212**), 2-furanyl (**L211**) and 2-benzo[*b*]thienyl (**L215**)) at the 6-position of the 2-(arylimino)pyridine ligands results in ethylene oligomerization cobalt catalysts [105–107]. The substitution of 3-thienyl (**L213**), 3-benzo[*b*]thienyl (**L216**), 2-furanyl (**L211**), phenyl (**L204**) and 2-naphthyl (**L205**) give rise to comparable catalytic activities in the order of 10^6 g/(mol h bar), which are 5–10 times lower than those obtained with the 2-thienyl (**L212**), 5-ethyl-2-thienyl (**L214**) and 2-benzo[*b*]thienyl (**L215**) substituted derivatives. It is worth noting that the 3-thienyl- and 3-benzo[*b*]thienyl-substituted cobalt complexes [$CoCl_2$(**L213**)] and [$CoCl_2$(**L216**)] produce exclusively 1-butene, while the 2-thienyl-, 5-ethyl-2-thienyl- and 2-benzo[*b*]thienyl-substituted ones (respectively, [$CoCl_2$(**L212**)], [$CoCl_2$(**L214**)] and [$CoCl_2$(**L215**)]) afford C_4–C_{14} α-olefins with very low *K* values in the range of 0.06–0.21.

The 2-hydroxy-2-propyl-substituted iron complexes [$FeCl_2$(**L195**)] and [$FeCl_2$(**L217**)] display moderate activities upon treatment with MAO, affording highly linear polymers along with some oligomeric products [114].

Replacement of the iminic phenyl by the 8-quinolinyl ring leads to highly active ethylene oligomerization iron and cobalt catalysts [115]. The butenes are the major products with good selectivity for 1-butene. Both the substituent (R′) at the 6-position of the pyridyl ring and the substitution with the 8-quinolinyl ring have pronounced effects on the catalytic activities. Generally, the bulkier the substituents, the higher the activities. The cobalt complexes show comparable activities with their iron analogues, and all cobalt complexes exhibit good selectivity for 1-butene, regardless of the substituent bulkiness, a behavior that is quite different from the corresponding iron complexes.

3.4 Mono- or Bis(alkylimino)pyridine-Based Ligands and Their Corresponding Iron and Cobalt Complexes

3.4.1 Syntheses of Ligands and Complexes

To date, just a few mono- or bis(alkylimino)pyridine ligands and their iron and cobalt derivatives have been reported (Table 3.4). Compared with the 2,6-bis(arylimino)pyridine, the 2,6-bis(alkylimino)pyridine ligands are more readily

Table 3.4 Mono- or bis(alkylimino)pyridine-based ligands, (R′N=C(R))-6-(R″N=C(R))C_5H_3N

	R′	R″	R	Refs.[a]
Symmetrical ligands				
L245	*iso*-propyl	*iso*-propyl	Me	[61]
L246	*n*-hexyl	*n*-hexyl	Me	[204]
L247	Cyclohexyl	Cyclohexyl	Me	[34, 36, 61]
L248	2,6-dimethylcyclohexyl	2,6-dimethylcyclohexyl	Me	[120]
L249	*n*-octadecyl	*n*-octadecyl	Me	[204]
L250	Benzyl	Benzyl	H	[116]
L251	Benzyl	Benzyl	Me	[116]
L252	Benzhydryl	Benzhydryl	Me	[100]
L253	Cyclohexylmethyl	Cyclohexylmethyl	Me	[116]
L254	*cis*-myrtanyl	*cis*-myrtanyl	H	[116]
L255	*cis*-myrtanyl	*cis*-myrtanyl	Me	[58]
L256	(*R*)-bornyl	(*R*)-bornyl	H	[116]
L257	9*H*-fluoren-9-yl	9*H*-fluoren-9-yl	Me	[100]
Unsymmetrical ligands				
L258	Cyclohexyl	2,6-dimethylphenyl	Me	[118]
L259	*n*-butyl	2,6-di-*iso*-propylphenyl	Me	[119]
L260	Cyclohexyl	2,6-di-*iso*-propylphenyl	Me	[118]
L261	Benzyl	2,6-di-*iso*-propylphenyl	Me	[101]
L262	(*R*)-1-phenylethyl	2,6-di-*iso*-propylphenyl	Me	[118]
L263	(*S*)-1-naphthalen-2-yl-ethyl	2,6-di-*iso*-propylphenyl	Me	[118]
L264	2-propenyl	2,6-di-*iso*-propylphenyl	Me	[119]
L265	3-butenyl	2,6-di-*iso*-propylphenyl	Me	[119]
L266	4-pentenyl	2,6-di-*iso*-propylphenyl	Me	[119]
L267	5-hexenyl	2,6-di-*iso*-propylphenyl	Me	[119]

[a] Only references reporting compounds for the first time are given; references are also given when different reaction conditions were used

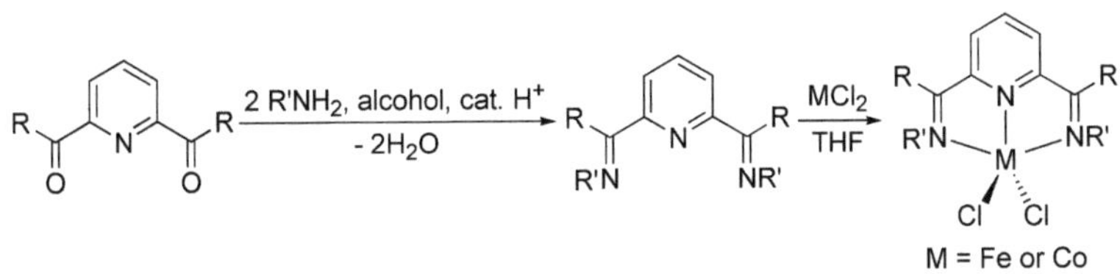

Scheme 3.17 Syntheses of symmetrical 2,6-bis(alkylimino)pyridine ligands and their iron and cobalt complexes

accessible by the typical condensation reaction of pyridine-2,6-dicarboxaldehyde or 2,6-diacetylpyridine with the appropriate aliphatic amines. This is possibly due to the electron-donating nature of the alkyl groups, which lead to enhanced nucleophilic reactivities of the corresponding aliphatic amines. Generally, the use of an alcohol as solvent (e.g., methanol [116] and ethanol [34]), and a protic acid as catalyst (e.g., formic [116] or acetic [34] acids), in the absence of water absorbents, is sufficient to afford a satisfactory yield (Scheme 3.17). Treatment of the 2,6-bis(alkylimino)pyridine ligands with equimolar amounts of iron or cobalt dichloride, in tetrahydrofuran, gives the corresponding iron and cobalt complexes in good yields (Scheme 3.17). Addition of an excess of ligand leads to bis-chelating ligand derivatives. For instance, reaction of hydrated cobalt dichloride with an equivalent of ligand **L247**, bearing cyclohexyl groups, in methanol, generates the monoligand cobalt complex [$CoCl_2$(**L247**)], whereas addition of the ligand in a 2:1 or 5:1 ligand-to-metal ratio suppresses the formation of [$CoCl_2$(**L247**)], and the cation $[Co(\mathbf{L247})_2]^{2+}$ is observed using UV/vis spectroscopy [117]. In contrast, treatment of the ligand **L247** with hydrated cobalt tetrafluoroborate exclusively afforded the bis-chelating ligand complex $[Co(\mathbf{L247})_2][BF_4]_2$, regardless of the ligand-to-metal ratio employed [117].

Bianchini and Erker groups [101, 118, 119] reported a new family of 2-(alkylimino)-6-(arylimino)pyridine ligands **L258–L267**, which were easily prepared by treatment of 2,6-diacetylpyridine with an equimolar quantity of an aromatic amine, followed by reaction of the resulting 6-acetyl-2-(arylimino)pyridine with an equimolar amount of an aliphatic amine (Scheme 3.18). Reaction of these ligands with anhydrous iron or cobalt dichloride, in boiling *n*-butanol, gave rise to the corresponding iron and cobalt complexes (Scheme 3.18).

Similar to the 2,6-bis(arylimino)pyridine-based iron and cobalt complexes, bearing less bulky *o*-substituents on the aryl rings (e.g., [$FeCl_2$(**L17**)] [52] and [$FeCl_2$(**L41**)] [63]), the molecular structures of the 2,6-bis(alkylimino)pyridine-based iron and cobalt complexes also feature a distorted trigonal bipyramidal geometry around the metal center, with the pyridyl nitrogen atom and two chlorine atoms positioned in the equatorial plane and the two imino nitrogen atoms located at the axial positions, e.g., [$CoCl_2$(**L247**)] [117] and [$CoCl_2$(**L256**)] [116], due to the presence of the less sterically congested alkyl substituents, except [$FeCl_2$(**L248**)] containing the bulkier 2,6-dimethylcyclohexyl, which adopted a distorted square pyramidal geometry around the metal center. On the contrary, iron

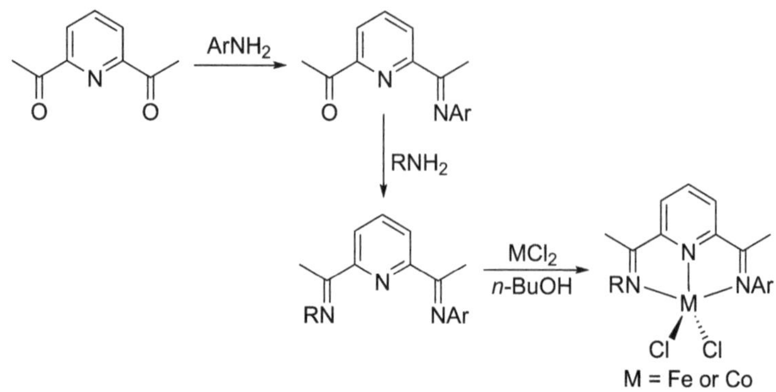

Scheme 3.18 Syntheses of the 2-(alkylimino)-6-(arylimino)pyridine unsymmetrical ligands and their iron and cobalt derivatives

and cobalt complexes based on the unsymmetrical 2-(alkylimino)-6-(arylimino) pyridine ligands present structures with geometries in between square pyramidal and trigonal bipyramidal, by virtue of the co-contributions of the less sterically congested alkyl and the more sterically congested aryl substituents [118, 119]. A pronounced feature in the unsymmetrical ligand-based complexes is that the bond distance of $Co–N_{alkylimino}$ is dramatically shorter than that of $Co–N_{arylimino}$, 2.185(6) Å vs. 2.348(6) Å, respectively, in [$CoCl_2$(**L258**)] [118], likely due to the relatively higher nucleophilicity and smaller size of the alkyl substituents.

3.4.2 Oligo- and Polymerization of Ethylene

The majority of iron and cobalt complexes based on the symmetrical bis(alkylimino)pyridine ligands are inactive toward the oligo- or polymerization of ethylene, upon treatment with MAO, possibly due to the lack of steric protection of the active species' metal center, leading to the deactivation of the catalyst. Just a few cases exhibit low ethylene oligo- or polymerization activities, e.g., [$FeCl_2$(**L255**)] with the *cis*-myrtanyl group [58], [$CoCl_2$(**L248**)] with the 2,6-dimethylcyclohexyl group [120] and [$FeCl_2$(**L257**)] with the 9*H*-fluoren-9-yl group [100]. When activated with MAO, the first two catalysts give very low ethylene polymerization activities, while the latter shows a moderate activity toward ethylene oligomerization (ca. 10^5 g/(mol h bar)).

The catalytic performance of iron and cobalt complexes containing unsymmetrical 2-(alkylimino)-6-(arylimino)pyridine backbones is determined by the combined effects of the alkyl and aryl groups. Bianchini et al. [118] studied the catalytic behavior of a series of such complexes, in which the alkyl groups involve cyclohexyl, (*R*)-1-phenylethyl and (*S*)-1-naphthalen-2-yl-ethyl, the aryl groups being either 2,6-dimethylphenyl or 2,6-di-*iso*-propylphenyl, on the oligo- or polymerization of ethylene, and found that all the iron and cobalt complexes are active in the oligomerization of ethylene, upon activation with MAO, affording α-olefins with Schulz–Flory distributions ($K = 0.61$–0.91). The catalytic activity increases with the bulkiness of the alkyl substituents, e.g., [$FeCl_2$(**L260**)] (cyclohexyl) < [$FeCl_2$(**L262**)] ((*R*)-1-phenylethyl) < [$FeCl_2$(**L263**)] ((*S*)-1-naphthalen-2-yl-ethyl), and decreases with the size of the aryl substituents, e.g., [$FeCl_2$(**L258**)] (2,6-dimethylphenyl) > [$FeCl_2$(**L260**)] (2,6-di-*iso*-propylphenyl). The cobalt complexes are approximately one order of magnitude less active than their iron analogues.

Wallenhorst et al. [119] revealed that the presence of a pendant alkenylimino substituent (instead of a normal alkylimino group) in the 2-(alkylimino)-6-(arylimino)pyridine ligands (e.g., **L264–L267**) leads to a unique catalytic behavior of the corresponding iron and cobalt complexes in the oligo- or polymerization of ethylene. Iron complexes [$FeCl_2$(**L264**)]–[$FeCl_2$(**L267**)] afford a mixture of linear polyethylenes and low molecular weight oligomers with low polydispersities ($M_w/M_n = 1.15$–1.31), whereas the corresponding cobalt complexes exclusively yield linear polyethylenes. All the cobalt complexes give relatively lower activities than their iron analogues. The same author ascribed this unusual catalytic behavior to the reversible involvement of the alkenyl pendant group in the polymerization process, as an internal comonomer.

3.5 Derivatization of 2,6-Bis(imino)pyridine Ligands and Their Corresponding Iron and Cobalt Complexes

Besides the aforementioned modified BIP ligands, a wealth of other new BIP derivatives have to date been achieved by further in-depth modification of the parent BIP ligand, and their corresponding iron and cobalt complexes have been tested for the oligo- or polymerization of ethylene. These modifications can be, in principle, divided into three aspects: (1) modifications on the pyridine moiety; (2) modifications on the imino arms; and (3) modifications both on the pyridine moiety and on the imino arms.

3.5.1 Modifications on the Pyridine Moiety

The simplest modification on the pyridine moiety was achieved by alkylation of the pyridyl ring 4-position of the BIP ligand [56, 121, 122]. The reaction of 2,6-bis(arylimino)pyridine with dialkylmanganese gives a thermally unstable dialkylmanganese species that undergoes a spontaneous rearrangement involving the migration of one of the alkyl groups from the metal center to the 4-position of the heterocyclic ring, which looses a hydrogen to the solution, affording the 4-alkylated BIP manganese(I) alkyl complex (Scheme 3.19). Treatment of these

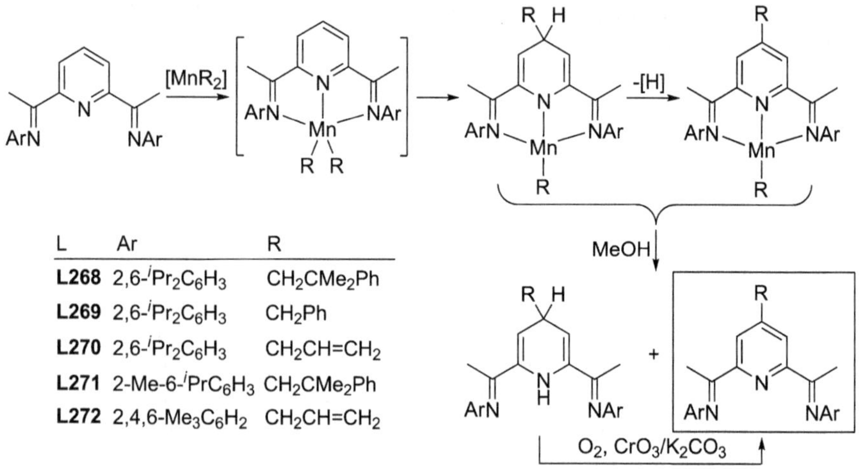

L	Ar	R
L268	2,6-$^{i}Pr_2C_6H_3$	CH_2CMe_2Ph
L269	2,6-$^{i}Pr_2C_6H_3$	CH_2Ph
L270	2,6-$^{i}Pr_2C_6H_3$	$CH_2CH{=}CH_2$
L271	2-Me-6-$^{i}PrC_6H_3$	CH_2CMe_2Ph
L272	2,4,6-$Me_3C_6H_2$	$CH_2CH{=}CH_2$

Scheme 3.19 Syntheses of 4-alkyl-2,6-bis(imino)pyridine ligands

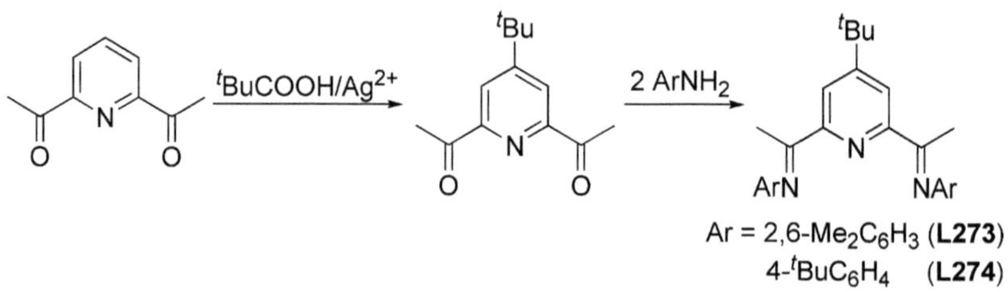

Scheme 3.20 Substitution of *tert*-butyl group at the 4-position of the pyridyl ring of the BIP ligand

manganese species with methanol leads to a mixture of the free 4-alkylated BIP ligand and 4-alkyl-2,6-bis(imino)-1,4-dihydropyridine. The latter is slowly transformed into the former on exposure to air, a reaction that is promoted by the addition of potassium carbonate and a catalytic amount of chromium trioxide, leading to the desired 4-alkylated BIP ligand (e.g., **L268–L272** in Scheme 3.19) as the main product. Alternatively, the introduction of a *tert*-butyl group into the 4-position of the pyridyl ring can be achieved by radical nucleophilic substitution on 2,6-diacetylpyridine with ^{t}Bu· radicals, using Minisci conditions, followed by the typical condensation reaction with the appropriate anilines, e.g., **L273** and **L274**, (Scheme 3.20) [123]. In addition, the 4-chloro-substituted BIP ligands are readily prepared by the condensation reaction of 2,6-diacetyl-4-chloropyridine with the appropriate anilines, e.g., **L275** (4-Cl-2,6-(ArN=C(Me))$_2C_5H_2$N, Ar = 2,6-$^{i}Pr_2C_6H_3$) [120]. Iron and cobalt complexes based on the 4-substituted BIP ligands are readily accessible by treatment of the corresponding ligands with iron or cobalt dichlorides. These iron and cobalt complexes are structurally comparable with the corresponding parent BIP ligands, e.g., [$FeCl_2$(**L268**)] vs. [$FeCl_2$(**L87**)] (both show distorted square pyramidal geometries around the metal centers).

The 4-substituted BIP ligand-based iron and cobalt complexes are active toward the polymerization of ethylene, upon activation with MAO, being the iron complexes generally more active than the cobalt analogues. Compared to iron and cobalt complexes based on the corresponding parent BIP ligands, 4-alkyl substitution has an insignificant effect on the catalytic activity, whereas the 4-chloro substitution results in a pronounced decrease in the activity [120, 122].

Ligands related to BIP but bearing alternative heterocyclic backbones have also been synthesized. This is the case of **L276** and **L277**, which are easily obtained by the condensation of 2,3,7,8-tetrahydro-1*H*,6*H*-acridine-4,5-dione with the proper anilines. Coordination of these ligands to iron dichloride affords the corresponding iron derivatives (Scheme 3.21) [31, 124]. On activation with MMAO, precatalysts [$FeCl_2$(**L276**)] and [$FeCl_2$(**L277**)] are active toward ethylene oligomerization, with activities in the order of 10^4 g/(mol h bar).

Replacement of the pyridine ring by further heteroaryl rings, such as pyrazine [125, 126], pyrimidine [127], furan [127] and thiazole [128], affords a large family of novel BIP derivatives.

The condensation of 2,6-diacetylpyrazine with the proper anilines yields the 2,6-bis(imino)pyrazine ligands, which, upon coordination to iron dichloride, in tetrahydrofuran, give rise to the corresponding iron complexes, e.g., [$FeCl_2$(**L278**)] and [$FeCl_2$(**L280**)] (Scheme 3.22) [125, 126]. Treatment of these complexes with MMAO leads to active ethylene polymerization catalysts showing activities in the order of 10^4 g/(mol h bar). When compared with the corresponding BIP

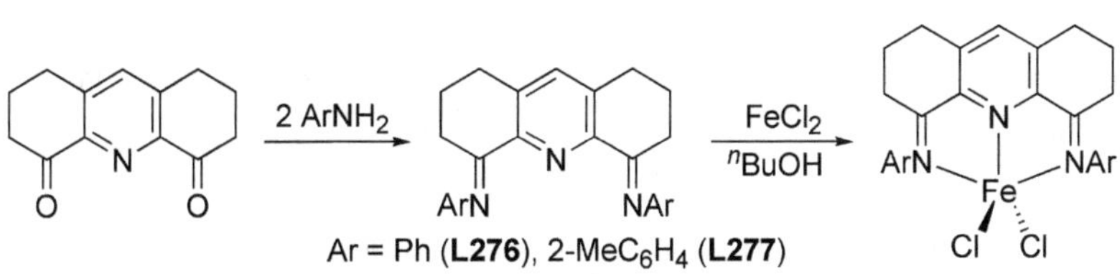

Scheme 3.21 Syntheses of BIP ligands with a cyclic backbone and their iron complexes

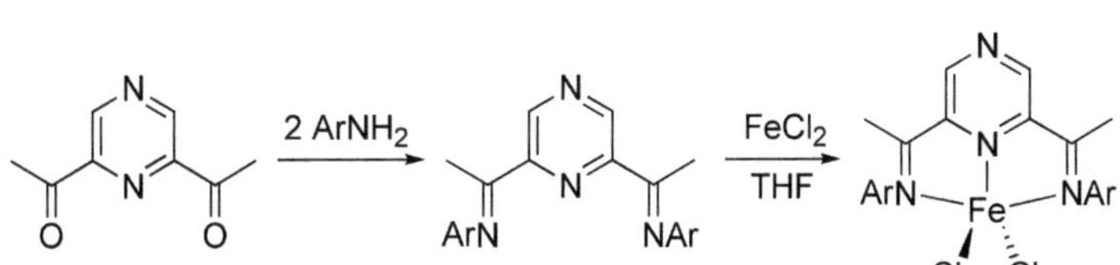

Scheme 3.22 Syntheses of 2,6-bis(imino)pyrazine ligands and their iron complexes

precatalysts, the replacement of pyridine by the pyrazine ring causes remarkable decreases both in the activity and in the molecular weight of the resulting polyethylenes.

Using an analogous reaction protocol, the 4-methyl-2,6-bis(imino)pyrimidine ligands and their corresponding iron and cobalt complexes were synthesized (Scheme 3.23) [127]. It is worth to mention that attempts to prepare the cobalt complexes [$CoCl_2$(**L285**)] were unsuccessful, possibly as a consequence of the increased steric bulk of the *iso*-propyl groups in combination with an intrinsically weaker metal–ligand interaction compared to iron. Further introduction of a nitrogen atom into the pyrimidine ring leads to two ligand derivatives **L287** and **L288** containing a 1,3,5-triazine core (Scheme 3.24) [127]. Attempts to coordinate these two molecules to iron or cobalt dichloride were unsuccessful. The high isomerization barriers to the conformer suited to coordination, as well as the reduced basicities of these triazine ligands, likely prevent the formation of metal complexes. Upon activation with MAO, the complexes [MCl_2(**L283**)], [MCl_2(**L284**)] and [MCl_2(**L285**)] (M = Fe or Co), bearing the bulky 2,6-dimethyl, 2,4,6-trimethyl and 2,6-di-*iso*-propyl groups on the aryl rings, respectively, convert ethylene to polymers, in which the former two show comparable catalytic activities and yield similar polymer molecular weights, and the latter displays a reduced activity and affords higher polymer molecular weights. The complexes [$FeCl_2$(**L286**)] and [$CoCl_2$(**L286**)], bearing the less bulky 2-methyl groups on the aryl rings, convert the ethylene to oligomers with Schulz–Flory distribution (K values of 0.69 and 0.83, respectively), when activated by MAO, thereby following the same trend as that observed for the BIP ligand-based iron and cobalt catalysts [127].

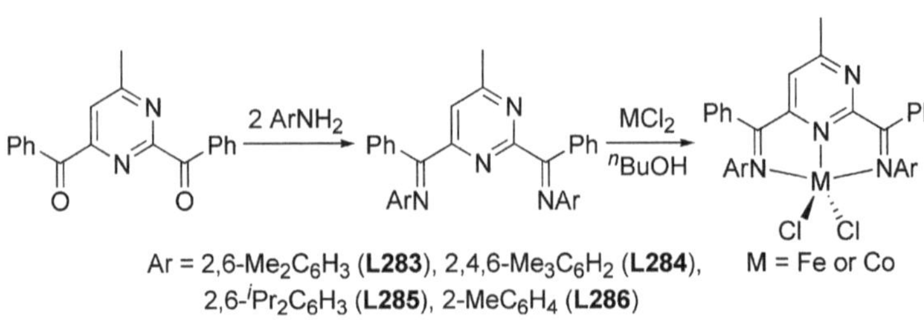

Scheme 3.23 Syntheses of 2,6-bis(imino)pyrimidine ligands and their iron and cobalt complexes

Scheme 3.24 Potential ligands bearing a 1,3,5-triazine core

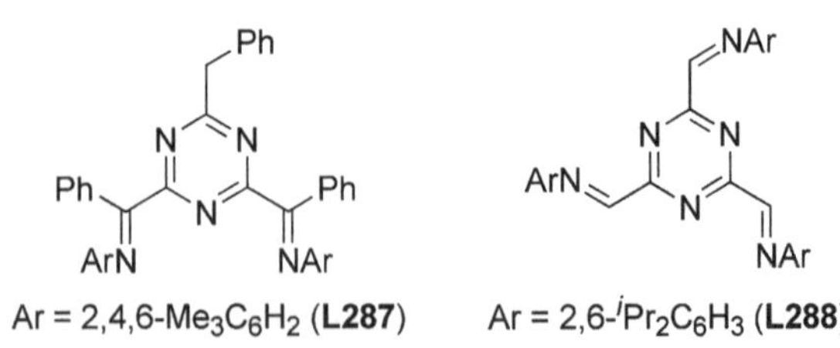

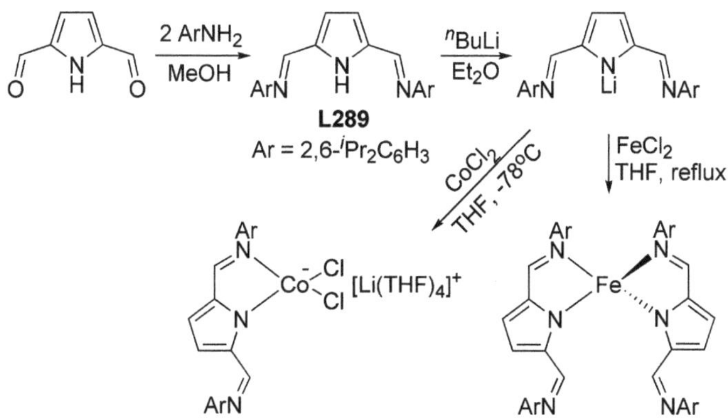

Scheme 3.25 Synthesis of the 2,5-bis(imino)pyrrole ligand precursor **L289** and its iron and cobalt complexes

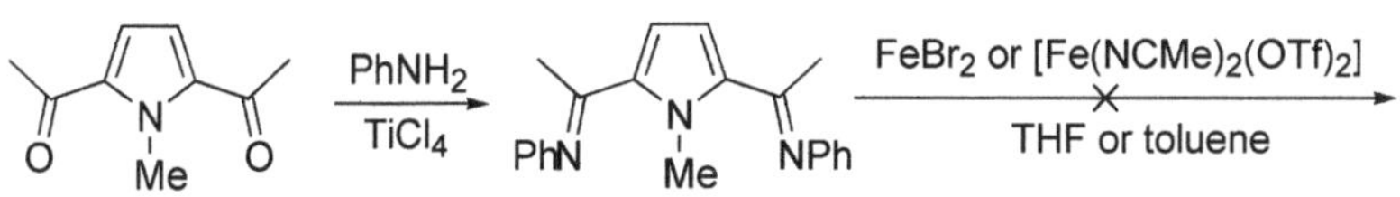

Scheme 3.26 Synthesis of the potential ligand 1-methyl-2,5-bis(imino)pyrrole **L290**

Of a slightly different nature from BIP, the 2,5-bis(imino)pyrrole 2,5-$(ArN{=}CH)_2C_4H_2NH$ (Ar = 2,6-$^iPr_2C_6H_3$) (**L289**), which is a potential precursor to a monoanionic tridentate ligand, can be prepared by condensation of pyrrole-2,5-dicarboxaldehyde with 2,6-di-*iso*-propylaniline, in methanol. Reaction of the lithium salt of **L289** with iron dichloride, in refluxing tetrahydrofuran, gives the neutral bisligand complex [Fe(**L289**)$_2$], showing a distorted tetrahedral geometry around the metal center, whereas the reaction with cobalt dichloride generates the ionic complex $[Li(THF)_4]^+[CoCl_2($**L289**$)]^-$, which is converted into $[Li(THF)_2]^+$-$[CoCl_2($**L289**$)]^-$ on pumping in vacuum (Scheme 3.25) [129]. It is apparent that **L289** acts exclusively as a monoanionic bidentate ligand in both complexes, but not as a tridentate ligand. When activated with MAO, [Fe(**L289**)$_2$] is unreactive toward ethylene, but the cobalt derivative can convert ethylene to oligomers with an activity of ca. 3.1×10^4 g/(mol h bar). The resulting oligomers are a mixture of terminal and internal olefins, and branched products. Using the analogous condensation protocol, 1-methyl-2,5-bis(phenylimino)pyrrole (**L290**) was prepared, but attempts to coordinate it to iron failed (Scheme 3.26) [128].

Replacement of the pyridine ring of the BIP ligand by furan or thiophene heterocycles to achieve potentially new neutral tridentate ligands can be accomplished by condensation of furan-2,5-dicarboxaldehyde or thiophene-2,5-dicarboxaldehyde with the appropriate anilines, e.g., **L291–L293** (Scheme 3.27) [127, 128]. However, attempts to prepare the tridentate iron or cobalt complexes of

Scheme 3.27 Syntheses of the potential ligands 2,5-bis(imino)furan and 2,5-bis(imino)thiophene

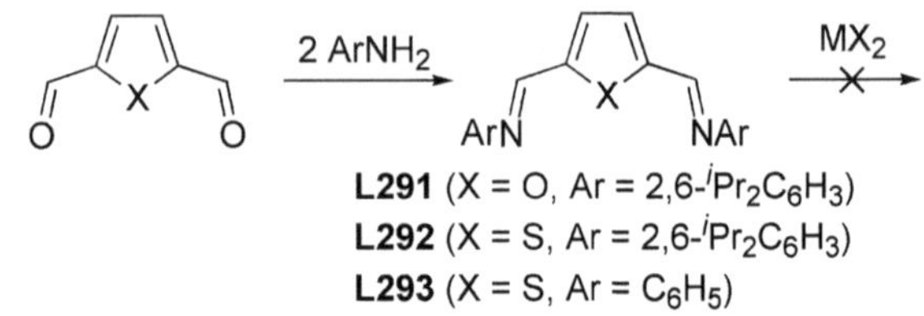

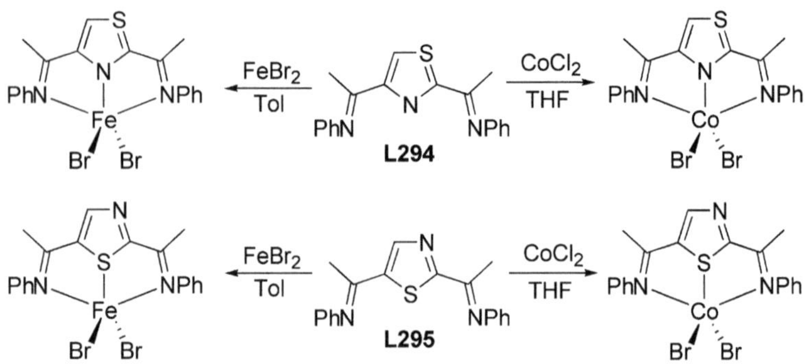

Scheme 3.28 Syntheses of the bis(imino)thiazole iron and cobalt complexes

these ligands failed. Conversely, the bis(imino)thiazole ligands **L294** and **L295**, which are easily accessible by condensation of the corresponding 2,5-diacetylthiazole and 2,4-diacetylthiazole with anilines, can be coordinated to iron and cobalt, leading to the corresponding tridentate derivatives (Scheme 3.28) [128]. Treatment of these iron and cobalt derivatives with MMAO results in ethylene oligomerization catalysts, yielding predominantly butenes and hexenes.

Considering the facts that ligands bearing five-membered heterocyclic cores, such as pyrrole (**L289**), furan (**L291**) and thiophene (**L292** and **L293**) are not able to form the corresponding tridentate chelating iron and cobalt complexes, presumably due to insufficient imino arm length and metal radii, an extra carbon atom linker between the imino moiety and the central heterocyclic core was introduced to overcome this problem, as depicted in Scheme 3.29 [127, 130]. The bis(imino)carbazole monoanionic ligand precursors **L296–L299** were prepared by condensation of 1,8-diformylcarbazole with the appropriate amines. The lithium or sodium salts of these ligands, which can be obtained by deprotonation of these precursors, in tetrahydrofuran, using *n*-butyllithium or sodium hydride, were treated with iron or cobalt dichloride to afford the complexes of the type [MClL] (L = **L296** or **L298**, M = Fe or Co), presenting a severely distorted and flattened tetrahedral geometry around the metal center. Similar reaction was performed with iron trichloride to generate complexes of the type [$FeCl_2L$] (L = **L296–L299**), which have slightly distorted trigonal bipyramidal geometries. All the iron and cobalt complexes obtained were found to be inactive toward the oligo- or polymerization of ethylene in combination with MAO. With the same objective, the dibenzofuran and the diphenyl ether and thioether ligand derivatives were prepared

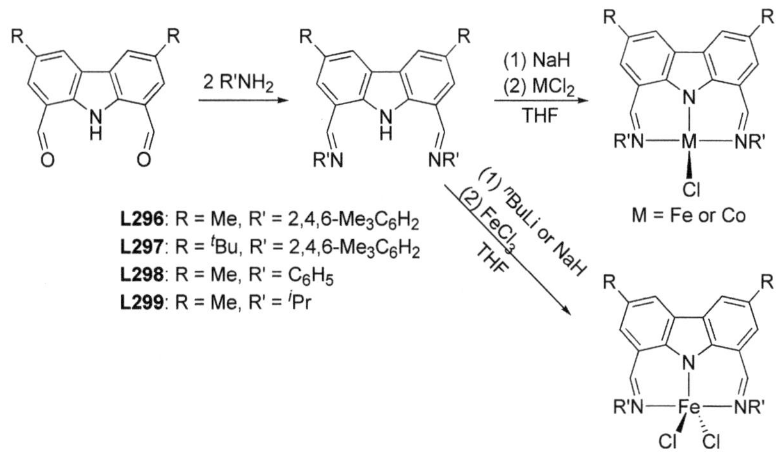

Scheme 3.29 Syntheses of the bis(imino)carbazole ligands and their iron and cobalt complexes

Scheme 3.30 Syntheses of potential ligands derived from dibenzofuran, diphenyl ether and diphenyl thioether

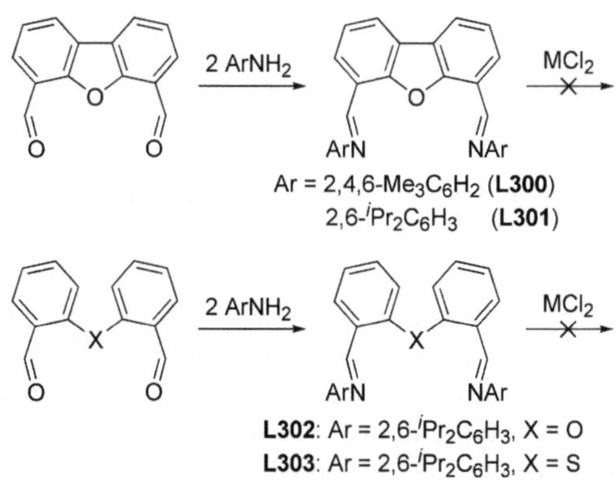

by condensation of the corresponding dicarboxaldehydes with the appropriate anilines (Scheme 3.30) [127]. However, like in the cases of bis(imino)furan and bis(imino)thiophene, these ligands fail to coordinate to iron and cobalt, not forming the corresponding tridentate chelated complexes.

Sun and coworkers [131, 132] prepared a series of 2,6-bis(imino)phenol ligands by condensation of 2-hydroxy-5-methylisophthalaldehyde or 5-*tert*-butyl-2-hydroxyisophthalaldehyde with the appropriate anilines (Scheme 3.31). The reaction of iron trichloride with 4-methyl-2,6-bis(imino)phenol bearing *ortho* aryl substituents (e.g., **L307–L309**), in acetonitrile, resulted in the formation of salts of the type $[2,6\text{-}(Ar(H)N{=}CH)_2\text{-}4\text{-}MeC_6H_2O]^+[FeCl_4]^-$ (where the phenol group is deprotonated, while the C=N groups are protonated), which adopt a columnar helical structure through the formation of hydrogen bonding between the anion and cation counterparts and intermolecular π–π interactions. When the *ortho* aryl groups are hydrogen atoms (e.g., **L310**), the adduct $FeCl_3$·**L310** is formed

Scheme 3.31 Syntheses of 2,6-bis(imino)phenol ligands and their iron and cobalt complexes

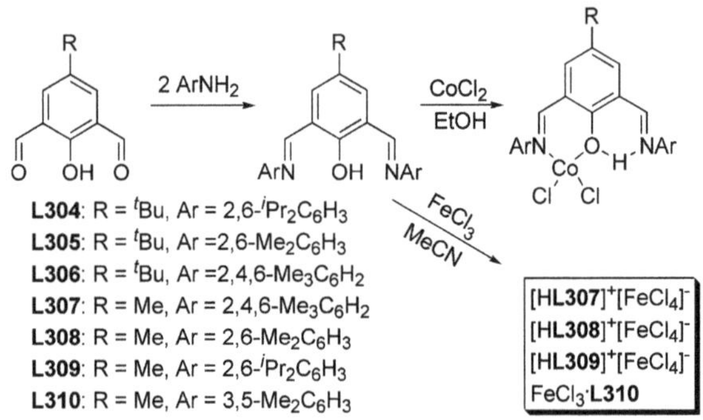

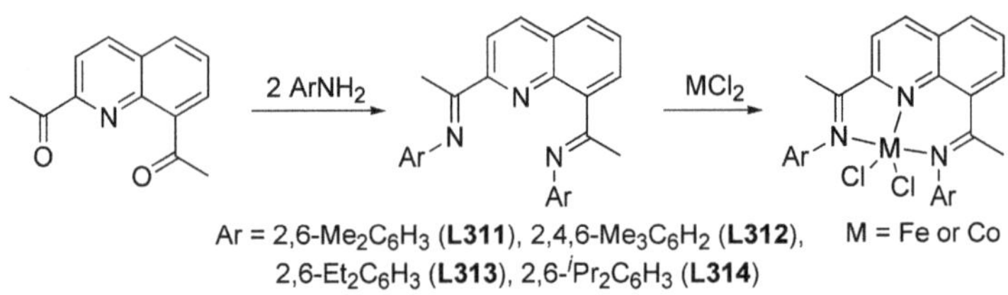

Scheme 3.32 Syntheses of 2,8-bis(imino)quinoline ligands and their iron and cobalt complexes

(Scheme 3.31). Conversely, reaction of cobalt dichloride with 4-*tert*-butyl-2,6-bis(imino)phenols such as **L304–L306** gives the corresponding bidentate chelated cobalt complexes, which adopt distorted tetrahedral geometries around the metal centers. Upon treatment with MAO, cobalt complexes afford exclusively butenes with very low catalytic activities (ca. 10^3 g/(mol h bar)).

The 2,8-bis(imino)quinoline ligands are readily accessible by condensation of 2,8-diacetylquinoline with the corresponding anilines, which, followed by treatment with iron or cobalt dichloride, yield their iron and cobalt derivatives (Scheme 3.32) [133]. It is worth to mention that the coordination reaction to iron and cobalt chlorides is limited to the ligands (**L311** and **L312**) bearing *o*-methyl substituents on the aryl rings, whereas those with the bulkier *o*-ethyl or *o*-*iso*-propyl substituents (**L313** and **L314**) do not react. Treatment of the resulting iron and cobalt complexes with MAO leads to highly active ethylene polymerization catalysts. The activities of cobalt complexes are comparable with those of their iron analogues. Maximum catalytic activities of 7.61×10^6 g/(mol h) and 6.93×10^6 g/(mol h) can be reached for [$FeCl_2$(**L311**)] and [$CoCl_2$(**L311**)], respectively, under 30 bar of ethylene. The most pronounced feature of these iron and cobalt complexes is that their catalytic activities remarkably increase with the polymerization temperature up to 100 °C, along with a decrease in the molecular weight.

The bis(imino)bipyridine ligand **L315** can be easily prepared by condensation of 2,2′-bipyridine-6,6′-dicarboxaldehyde with 2,4,6-trimethylaniline, in ethanol, in the presence of acetic acid as catalyst (Scheme 3.33) [134]. Treatment of **L315**

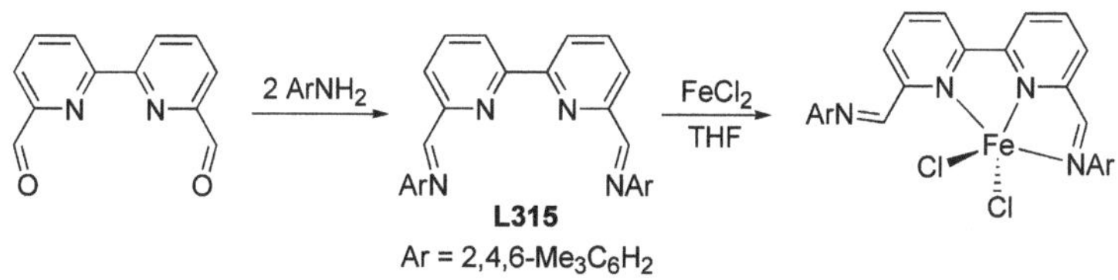

Scheme 3.33 Syntheses of a bis(imino)bipyridine ligand and the corresponding iron complex

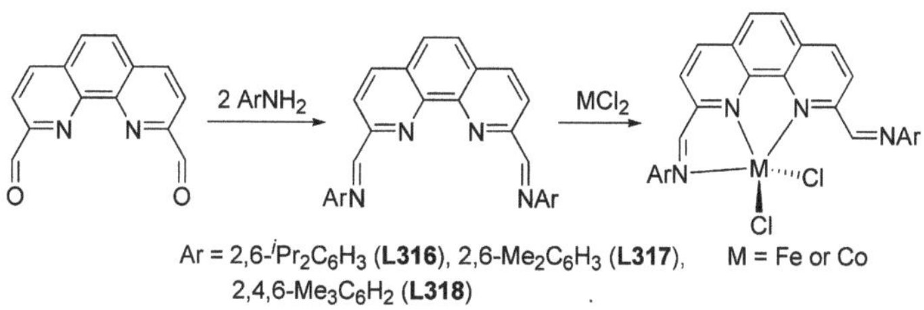

Scheme 3.34 Syntheses of 2,9-bis(imino)-1,10-phenanthroline ligands and their iron and cobalt complexes

with iron dichloride, in tetrahydrofuran, affords a tridentate chelated iron complex [$FeCl_2$(**L315**)], which demonstrated to be inactive toward the oligo- or polymerization of ethylene, in the presence of MAO.

The 2,9-bis(imino)-1,10-phenanthroline ligands (**L316–L318**) are prepared by condensation of 1,10-phenanthroline-2,9-dicarboxaldehyde with the corresponding anilines, which, followed by treatment with iron or cobalt dichloride, afford their iron and cobalt complexes that adopt a distorted trigonal bipyramidal geometry around the metal center (Scheme 3.34) [134, 135]. Upon treatment with MAO, the iron complexes are not active toward the oligo- or polymerization of ethylene, under various pressures of ethylene (1, 10 and 20 bar), while the cobalt complexes are as well inactive under 1 bar of ethylene, but can serve as ethylene oligomerization catalysts, at pressures above 10 bar of ethylene, with activities in the order of 10^5 g/(mol h).

3.5.2 Modifications on the Imino Arms

In order to place a catalytic site at the core of a dendrimer, and thus create a controllable microenvironment around it, Moss and coworkers [136] prepared a series of the BIP ligands containing dendritic wedges, as shown in Scheme 3.35. These dendritically functionalized BIP ligands are prepared by reaction of BIP derivatives containing 4-hydroxy groups on the aryl rings, which are easily synthesized by the condensation of 2,6-diacetylpyridine with the corresponding 4-anilines, with the appropriate bromoalkyl wedge, using the Williamson ether

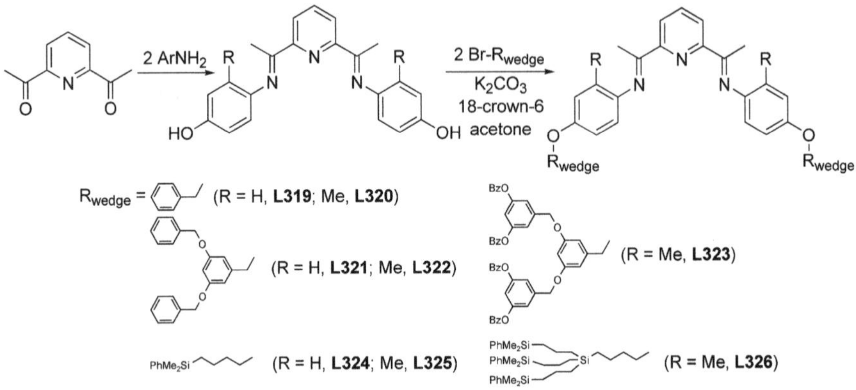

Scheme 3.35 Syntheses of BIP ligands bearing dendritic wedges

synthesis protocol. Treatment of these ligands with iron dichloride, in tetrahydrofuran, leads to the corresponding iron derivatives [$FeCl_2$(**L319**)]–[$FeCl_2$(**L326**)]. Upon treatment with MAO, all the iron complexes, except [$FeCl_2$(**L321**)] and [$FeCl_2$(**L324**)], can serve as ethylene oligomerization catalysts with activities in the order of 10^5 g/(mol h bar). In comparison with the corresponding unsubstituted BIP iron complex [$FeCl_2$(**L15**)], bearing 2-methyl groups on the aryl rings, the dendritically functionalized iron complexes show higher activities and *K* values. In addition, the dendritically substituted complexes [$FeCl_2$(**L322**)], [$FeCl_2$(**L323**)] and [$FeCl_2$(**L326**)] are evidently more active than their nondendritic analogues [$FeCl_2$(**L320**)] and [$FeCl_2$(**L325**)], respectively, suggesting that the dendritic functionalization at the BIP ligand can efficiently improve the catalytic activities of the corresponding iron complexes toward ethylene oligomerization.

Replacement of the aryl rings of the 2,6-bis(arylimino)pyridine ligands by azolyl rings, such as pyrrolyl, indolyl, carbazolyl and triazolyl rings, leads to a new family of BIP ligands with *N*–heterocyclic groups **L327–L338**, which are readily accessible by condensation of 2,6-diacetylpyridine with the corresponding *N*-amino azoles (Scheme 3.36) [137]. The unsymmetrical ligand **L339** is prepared by stepwise condensation of pyridine-2,6-carboxaldehyde with *N*-aminocarbazole and *N*-aminopyrrole, respectively. Treatment of these ligands with the appropriate metal halides ($FeCl_2$, $CoCl_2$ and $FeCl_3$) gives the corresponding iron and cobalt derivatives in good yields, except for **L330** and **L333**, bearing exclusively phenyl substituents on the *N*-pyrrolyl rings, possibly due to a high steric hindrance and/or electron-withdrawing effect imposed by two pairs of *ortho* phenyl groups. Upon treatment with MAO, all these iron and cobalt complexes, except [$FeCl_2$(**L336**)] containing triazolyl groups, show remarkable catalytic activities in the oligo- or polymerization of ethylene, comparable with that of the 2,6-bis(arylimino)pyridine iron complex [$FeCl_2$(**L87**)]. The highest activity is achieved by [$FeCl_2$(**L328**)], bearing 2,5-dimethylpyrrolyl groups, with almost twice of the productivity as that of [$FeCl_2$(**L87**)]. Iron complexes are generally more active by a factor of 10–100

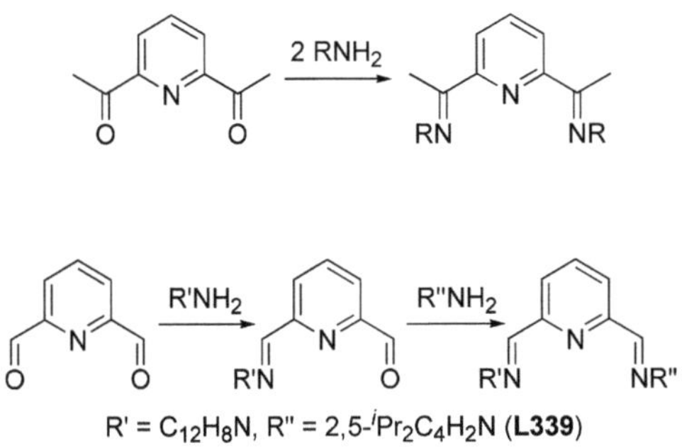

L327 R = C_4H_4N
L328 R = 2,5-$Me_2C_4H_2N$
L329 R = 2,5-$^iPr_2C_4H_2N$
L330 R = 2,5-$Ph_2C_4H_2N$
L331 R = 2-Me-5-$^iPrC_4H_2N$
L332 R = 2-Me-5-PhC_4H_2N
L333 R = 2,3,5-Ph_3C_4HN
L334 R = 2-Me-5-(2-MePh)C_4HN
L335 R = $C_{12}H_8N$
L336 R = 3,5-$Me_2C_2N_3$
L337 R = C_8H_6N
L338 R = 2-Me-C_8H_5N

Scheme 3.36 Syntheses of *N*-azolyl BIP ligands

than their corresponding cobalt analogues. Similar to the trend observed for the 2,6-bis(arylimino)pyridine iron and cobalt complexes, increasing the steric hindrance of the *N*-azolyl groups leads to lower catalytic activities and higher polymer molecular weights, while the opposite effect is observed as the steric hindrance decreases. The molecular weights of the resulting polyethylenes are remarkably lower than those obtained with the normal 2,6-bis(arylimino)pyridine-based iron and cobalt complexes. It is worth to note that the oligomers produced with the *N*-pyrrolyl complexes possess a significant degrees of branching, being substantially different from the strictly linear oligomers obtained with the *N*-aryl systems. For instance, an NMR analysis shows the polyethylene formed by [$FeCl_2$(**L335**)] contains 1.1% of vinyl, 0.2% of vinylidene, 4.3% of *cis*/*trans* double bonds, with a branching of 19 methyl, 71 ethyl, 11 butyl and 80 alkyl $\geq C_6$ per 1000 carbon atoms.

Gibson and coworkers prepared a series of new aminoiminopyridine and bis(amino)pyridine ligands [138, 139]. The aminoiminopyridine ligands (**L339**–**L341**) are prepared by reaction of 2,6-bis(arylimino)pyridine with trimethylaluminum, in which one of the imino arms is transformed into an amino group via nucleophilic attack of this alkylating agent on only one of the imino carbons, followed by hydrolysis (Scheme 3.37). Treatment of these ligands with iron or cobalt dichlorides, in *n*-butanol, at elevated temperature, leads to their corresponding metal complexes. The bis(amino)pyridine ligand (**L342**) is prepared by reaction of 2,6-bis(bromomethyl)pyridine with lithium amide LiNHAr (Ar = 2,6-di-*iso*-propylphenyl) (Scheme 3.37) [140]. However, its iron derivative could not be prepared by reaction with anhydrous iron dichloride, in *n*-butanol or tetrahydrofuran. Alternatively, the reaction of **L342** with the iron dichloride tetrahydrofuran adduct [$FeCl_2(THF)_{1.5}$], in refluxing toluene, gives the desired complex in high yield [138]. When activated with MAO, these iron and cobalt complexes based on the aminoiminopyridine and bis(amino)pyridine ligands are active in the polymerization of ethylene, with the activities in the order of 10^4 g/(mol h bar), which are dramatically lower than those obtained with the corresponding bis(arylimino)pyridine-based metal complexes. The catalytic activity of [$FeCl_2$(**L342**)]

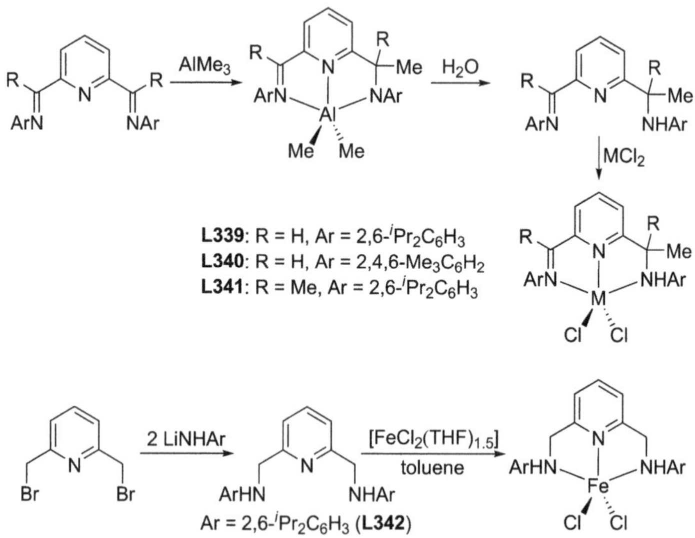

Scheme 3.37 Syntheses of aminoiminopyridine and bis(amino)pyridine ligands and their iron and cobalt complexes

is remarkably lower than those of [FeCl$_2$(**L339**)]–[FeCl$_2$(**L341**)], suggesting that the substitution of both imino groups by two amino functionalities is not favorable. Using a similar reaction protocol, Britovsek et al. [141] prepared a couple of non-protic bis(amino)pyridine ligands 2,6-(R(Me)NCH$_2$)$_2$C$_5$H$_3$N (R = Me (**L343**) or Ph (**L344**)). The corresponding iron complexes [Fe(OTf)$_2$(**L343**)] and [FeCl$_2$-(**L344**)], which are easily prepared by reaction of the ligands with the iron(II) triflate acetonitrile adduct ([Fe(OTf)$_2$(NCMe)$_2$]) and iron dichloride tetrahydrofuran adduct ([FeCl$_2$(THF)$_{1.5}$]), respectively, were found to be active for the oxidation of cyclohexane with hydrogen peroxide.

The biphenyl-bridged bis(iminopyridine) ligands **L345–L348** are prepared by the reaction of biphenyl-2,2′-diamines with the appropriate 6-substituted 2-acetylpyridine or pyridine-2-carboxaldehyde (Scheme 3.38) [142]. Treatment of the ligands bearing no alkyl substituents at the pyridyl rings 6-positions (R″ = H) with metal halides (FeCl$_2$, FeBr$_2$, and CoCl$_2$) affords the tetradentate chelated iron and cobalt complexes, which adopt distorted octahedral geometries around the metal centers, whereas the reaction of **L347**, bearing a 6-methyl substituent, with iron dichloride gives the cationic complex [FeCl(**L347**)]$^+$Cl$^-$, which shows a square pyramidal geometry around the metal center. These iron or cobalt complexes show extremely low or no activities toward ethylene polymerization, upon activation with MAO.

The 2-benzoxazolyl-6-(arylimino)pyridine ligands **L349–L355** are accessible by condensation of 2-(2-benzoxazolyl)-6-acetylpyridine with the appropriate anilines, which, followed by treatment with iron or cobalt dichloride, give rise to the corresponding iron and cobalt derivatives (Scheme 3.39) [143–145].

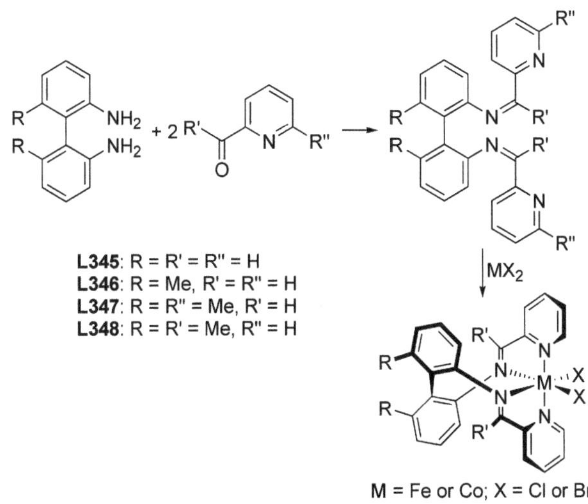

Scheme 3.38 Syntheses of biphenyl-bridged bis(iminopyridine) ligands and their iron and cobalt complexes

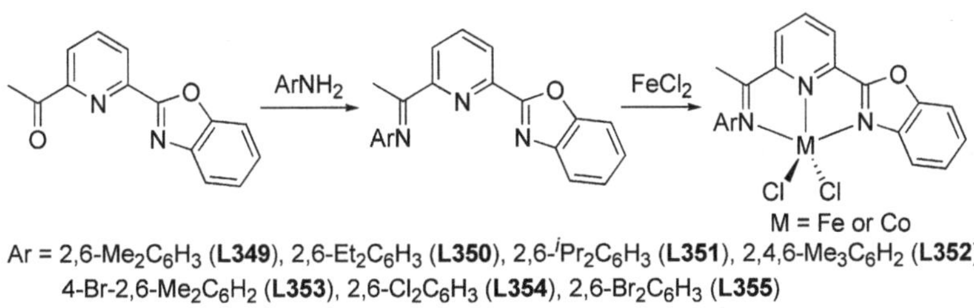

Scheme 3.39 Syntheses of 2-benzoxazolyl-6-(arylimino)pyridine ligands and their iron and cobalt complexes

The complexes [$FeCl_2$(**L349**)], [$FeCl_2$(**L351**)], [$CoCl_2$(**L349**)] and [$CoCl_2$(**L350**)], bearing *o*-methyl, *o*-ethyl or *o*-*iso*-propyl groups on the aryl rings, adopt distorted trigonal bipyramidal geometries around the metal centers, whereas [$FeCl_2$(**L355**)] and [$CoCl_2$(**L355**)], bearing *o*-bromo groups, show distorted square pyramidal geometries [143, 144]. Upon exposure to ethylene, the iron complexes [$FeCl_2$(**L349**)]–[$FeCl_2$(**L355**)], in combination with MMAO, afford oligomers with Schulz–Flory distributions, characterized by *K* values of 0.55–0.78, along with waxy oligomers, consisting mainly of linear α-olefins, whereas the cobalt complexes [$CoCl_2$(**L349**)]–[$CoCl_2$(**L355**)] exclusively generate butenes and traces of hexenes. The oligomerization activities of the iron complexes (in the order of 10^5 g/(mol h bar)) are one order of magnitude higher than those of their cobalt analogues. The substituents on the aryl rings substantially, but not identically, influence the catalytic performance of the iron and cobalt complexes. For example, in the case of [$FeCl_2$(**L350**)], containing *o*-ethyl groups, the highest oligomerization activity of 1.02×10^6 g/(mol h bar) is obtained. The presence of halogen substituents on the *ortho* positions of the aryl rings ([$FeCl_2$(**L354**)] and [$FeCl_2$(**L355**)]) leads to activities similar to those of the *o*-alkyl-substituted complexes ([$FeCl_2$(**L349**)]–[$FeCl_2$(**L351**)]). Compared with the parent iron complex [$FeCl_2$(**L349**)], which is 4-methyl substituted, precatalyst [$FeCl_2$(**L352**)] gives rise to an increase in the activity, whereas the 4-bromo substituted [$FeCl_2$(**L353**)] leads to a marked decrease in the activity [144]. On the contrary, for cobalt complexes, the precatalyst [$CoCl_2$(**L355**)], containing *o*-bromo atoms, exhibits the highest oligomerization activity of 7.52×10^4 g/(mol h bar). Compared with the *o*-alkyl-substituted complexes [$CoCl_2$(**L349**)]–[$CoCl_2$(**L351**)], the *o*-halo substituted [$CoCl_2$(**L354**)] and [$CoCl_2$(**L355**)] give rise to a significant increase in the activity. Both 4-methyl and 4-bromo substitutions result in remarkable decrease in the activity with respect to the parent complex [$CoCl_2$(**L349**)] [143].

The 2-benzimidazolyl-6-(arylimino)pyridine ligands can be prepared routinely by the condensation of 2-benzimidazolyl-6-acetylpyridine with the appropriate substituted anilines (Scheme 3.40) [146–149]. Treatment of these ligands with iron or cobalt dichloride gives the desired metal complexes of the type [MCl_2L] (M = Fe or Co), except for **L365** that contains the less bulky *o*-fluoro groups on the aryl rings, and leads to the bisligand complexes [Fe(**L365**)$_2$]$^{2+}$[$FeCl_4$]$^{2-}$ [150], analogous to the 2,6-bis(arylimino)pyridine-based iron complex [Fe(**L39**)$_2$]$^{2+}$ [$FeCl_4$]$^{2-}$ (**L39** = 2,6-(ArN=C(Me))$_2C_5H_3N$, Ar = 2-fluorophenyl) [60, 72]. The coordination geometry of these [MCl_2L] complexes is dependent upon the bulkiness of the substituents on both the aryl and the benzimidazolyl rings. The less bulky substituents (H or Me) on the benzimidazolyl ring favor distorted

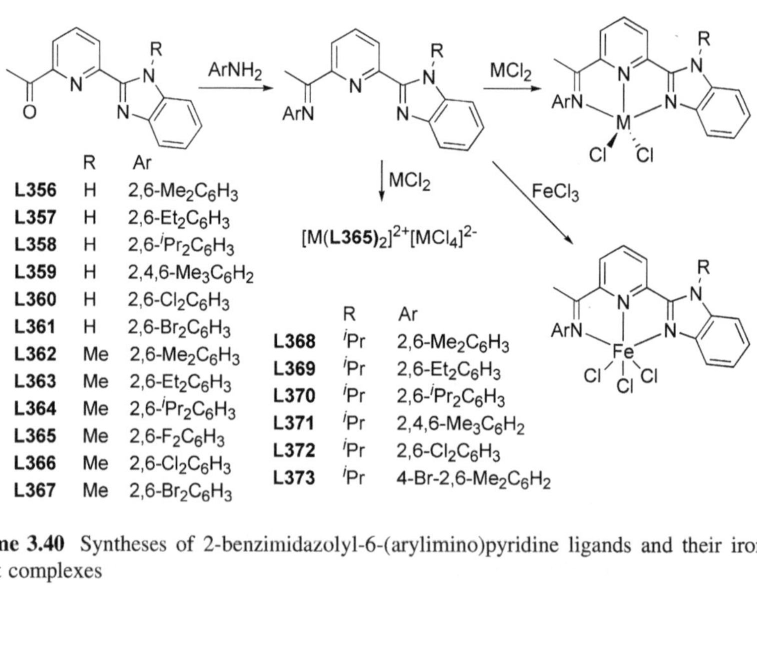

Scheme 3.40 Syntheses of 2-benzimidazolyl-6-(arylimino)pyridine ligands and their iron and cobalt complexes

trigonal bipyramidal geometries, e.g., [$FeCl_2$(**L358**)], [$CoCl_2$(**L357**)], [$FeCl_2$(**L362**)], [$CoCl_2$(**L362**)] and [$FeCl_2$(**L366**)], regardless of the substituents on the aryl ring, whereas when the substituent on the benzimidazolyl ring is *iso*-propyl, less bulky substituents on the aryl ring lead to trigonal bipyramidal geometries, e.g., [$FeCl_2$(**L371**)] and [$CoCl_2$(**L371**)], bearing *o*-methyl groups, while much bulkier substituents on the aryl ring result in the square pyramidal geometry, such as in the case of [$CoCl_2$(**L370**)] that contains *iso*-propyl groups. Treatment of the ligands with iron trichloride gives iron(III) complexes of the type [$FeCl_3$L]. Except cobalt complexes bearing the *iso*-propyl groups on the benzimidazolyl rings ([$CoCl_2$(**L368**)]–[$CoCl_2$(**L373**)]), which show higher catalytic activities in combination with diethylaluminum chloride ($AlEt_2Cl$) than with MAO or MMAO, iron and cobalt complexes of the type [MCl_2L] (M = Fe and Co) display the highest activities for ethylene oligomerization, when activated with MAO or MMAO, forming predominantly oligomers with Schulz–Flory distributions, along with a small amount of waxy polymers. The latter were proved to be linear α-olefins.

Iron complexes are generally much more active than the corresponding cobalt analogues. The substituents on both the aryl and benzimidazolyl rings have remarkable influences on the catalytic performance of the metal complexes. In general, the bulkier the substituent (R) on the benzimidazolyl rings, the lower the catalytic activity. For instance, when the size of the R group increases in the order of H < Me < ^{i}Pr, the activities of [$FeCl_2$(**L356**)] (R = H), [$FeCl_2$(**L362**)] (R = Me) and [$FeCl_2$(**L368**)] (R = ^{i}Pr), all bearing 2,6-dimethylphenyl groups, decrease in the order of [$FeCl_2$(**L356**)] (1.35×10^6 g/(mol h)) > [$FeCl_2$(**L362**)] (9.2×10^5 g/(mol h)) > [$FeCl_2$(**L368**)] (2.34×10^5 (mol h)), at 10 bar of ethylene. For complexes containing a hydrogen atom or a methyl group on the benzimidazolyl rings (R = H or Me), the bulkier *o*-alkyl substituents on the aryl rings lead to higher activities, e.g., [$FeCl_2$(**L358**)] (Ar = 2,6-di-*iso*-propylphenyl) (2.62×10^6 g/(mol h)) > [$FeCl_2$(**L357**)] (Ar = 2,6-diethylphenyl) (1.57×10^6 g/(mol h)) > [$FeCl_2$(**L356**)] (Ar = 2,6-dimethylphenyl) (1.35×10^6 g/(mol h)), at 10 bar of ethylene, whereas for complexes containing an *iso*-propyl group on the benzimidazolyl rings (R = ^{i}Pr), the reverse trend is observed, i.e., [$FeCl_2$(**L370**)] (Ar = 2,6-di-*iso*-propylphenyl) (0.48×10^5 g/(mol h)) <[$FeCl_2$-(**L369**)] (Ar = 2,6-diethylphenyl) (1.62×10^5 g/(mol h)) < [$FeCl_2$(**L368**)] (Ar = 2,6-dimethylphenyl) (2.34×10^5 g/(mol h)), at 10 bar of ethylene. For all the complexes, *o*-halo substituents on the aryl rings cause a dramatic decrease in the activity when compared with those of the *o*-alkyl-substituted derivatives. It is worth to mention that the cocatalyst also has a remarkable effect on the catalytic behavior of these complexes. For instance, in the case of complexes bearing a methyl group on the benzimidazolyl rings (R = Me), in the presence of MAO, the catalytic activity varies in the order of [$FeCl_2$(**L362**)] (with Me) < [$FeCl_2$(**L363**)] (with Et) < [$FeCl_2$(**L364**)] (with ^{i}Pr) and [$FeCl_2$(**L366**)] (with Cl) < [$FeCl_2$(**L367**)] (with Br), whereas in the presence of MMAO, the reverse trend is observed, [$FeCl_2$(**L362**)] (with Me) > [$FeCl_2$(**L363**)] (with Et) > [$FeCl_2$(**L364**)] (with ^{i}Pr) and [$FeCl_2$(**L366**)] (with Cl) > [$FeCl_2$(**L367**)] (with Br). Upon activation with

MAO, iron(III) complexes of the type [$FeCl_3$L] can serve as highly active ethylene oligomerization catalysts, with activities of ca. 10^6 g/(mol h), at 10 bar of ethylene. The catalytic activity varies in the order of [$FeCl_3$(**L362**)] (with Me) > [$FeCl_3$(**L363**)] (with Et) > [$FeCl_3$(**L364**)] (with ^{i}Pr) and [$FeCl_3$(**L365**)] (with F) > [$FeCl_3$(**L366**)] (with Cl) > [$FeCl_3$(**L362**)] (with Br), which is consistent with the trend observed for the corresponding iron(II)/MMAO system, but opposite to that obtained in the corresponding iron(II)/MAO system.

Sun et al. [149] synthesized a series of 2-(benzimidazolyl)pyridine derivatives, starting from *o*-phenylenediamine and 2,6-dimethylpyridine, as shown in Scheme 3.41. Reaction of *o*-phenylenediamine with 2,6-dimethylpyridine in the presence of sulfur, as oxidant, gives 2-(2-benzimidazolyl)-6-methylpyridine (**L374**), which, followed by treatment with selenium dioxide, affords 6-(2-benzimidazolyl)pyridine-2-carboxylic acid. The latter is esterified with ethanol, in the presence of a catalytic amount of sulfuric acid, to form 2-(ethoxycarbonyl)-6-(2-benzimidazolyl)pyridine (**L376**). *N*-methylation of **L374** and **L376** with methyl iodide, in the presence of potassium carbonate, in acetonitrile, gives 2-(1-methyl-2-benzimidazolyl)-6-methylpyridine (**L375**) and 2-(ethoxycarbonyl)-6-(1-methyl-2-benzimidazolyl)pyridine (**L377**), respectively. The latter can be transformed into 6-(2-benzimidazolyl)-2-acetylpyridine (**L378**) through a Claisen reaction with ethyl acetate, in the presence of sodium ethoxide. Treatment of these ligands with iron or cobalt dichloride (MCl_2, M = Fe or Co) gives the desired N^N bidentate chelated complexes [MCl_2(**L374**)]–[MCl_2(**L377**)] and N^N^O tridentate chelated complexes [MCl_2(**L378**)]. Upon activation with MAO, MMAO or $AlEt_2Cl$, these iron or cobalt complexes give moderate ethylene oligomerization activities (ca. 10^4 g/(mol h bar)). For cobalt complexes, the cocatalyst $AlEt_2Cl$ gives higher activities than MAO and MMAO. Compared with the complexes [MCl_2(**L374**)]

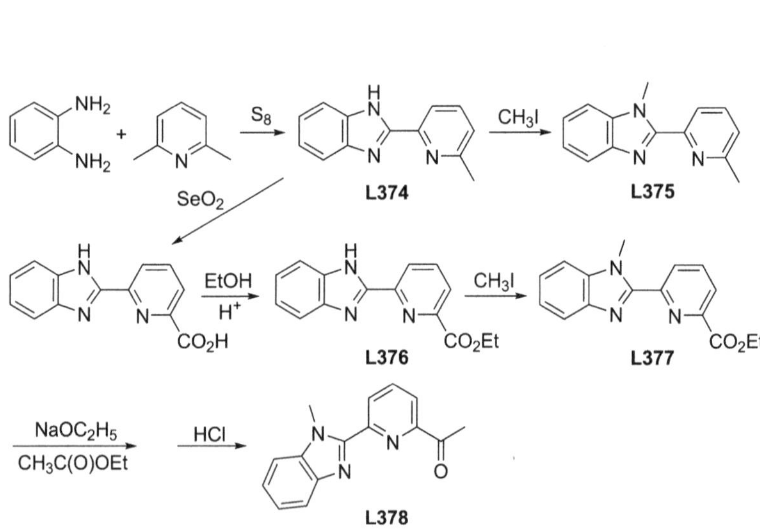

Scheme 3.41 Syntheses of 2-(benzimidazolyl)pyridine derivatives

and [MCl_2(**L376**)], their corresponding *N*-methylated complexes [MCl_2(**L375**)] and [MCl_2(**L377**)] show lower activities. Iron complexes [MCl_2(**L376**)] and [MCl_2(**L377**)], containing an ethoxycarbonyl group at the 6-position of the pyridyl ring, exhibit higher activities than their corresponding iron analogues [MCl_2(**L374**)] and [MCl_2(**L375**)], containing a methyl group at the 6-position of the pyridyl ring.

The 2,6-di(1*H*-pyrazol-1-yl)pyridine ligands (**L379–L382**) are prepared by treatment of the pyrazole sodium salt with 2,6-dibromopyridine, while the 2,6-bis((1*H*-pyrazol-1-yl)methyl)pyridine ligands (**L383** and **L384**) can be synthesized by phase transfer catalyzed reactions of 2,6-bis(chloromethyl)pyridine with the pyrazole salt (Scheme 3.42) [128, 151]. Treatment of these ligands with iron or cobalt dihalide (MX_2, M = Fe or Co, X = Cl or Br), in tetrahydrofuran or *n*-butanol, leads to the corresponding tridentate chelated iron and cobalt derivatives. It is noteworthy that iron complexes [$FeBr_2$(**L381**)] [128] and [$FeCl_2$(**L381**)] [152], bearing the 3,5-dimethyl groups, adopt a distorted trigonal bipyramidal geometry around the metal center, whereas the cobalt complex [$CoCl_2$(**L380**)] [151], containing 3,4,5-trimethyl groups, shows a distorted square pyramidal geometry. Iron and cobalt complexes [$FeCl_2$(**L379**)], [$FeCl_2$(**L380**)], [$CoCl_2$-(**L370**)], [$CoCl_2$(**L380**)], [$FeCl_2$(**L383**)] and [$FeCl_2$(**L384**)], containing unsubstituted or 3,4,5-trimethyl-substituted pyrazole rings, are active toward ethylene polymerization, when activated with MAO, forming linear polyethylenes with high molecular weights, whereas they give very low or no activities in the polymerization of ethylene, when activated with MMAO. Iron complexes are more active than their cobalt analogues, and the introduction of a methylene linker between the pyrazole and pyridine rings causes a decrease in activity, e.g., [$FeCl_2$(**L379**)] (9.7×10^4 g/(mol h bar)) > [$FeCl_2$(**L383**)] (6.6×10^4 g/(mol h - bar)). On the contrary, iron complexes [$FeBr_2$(**L383**)] and [$FeBr_2$(**L382**)], containing 3,5-dimethyl- or 3,5-diphenyl-substituted pyrazole rings, are active in the

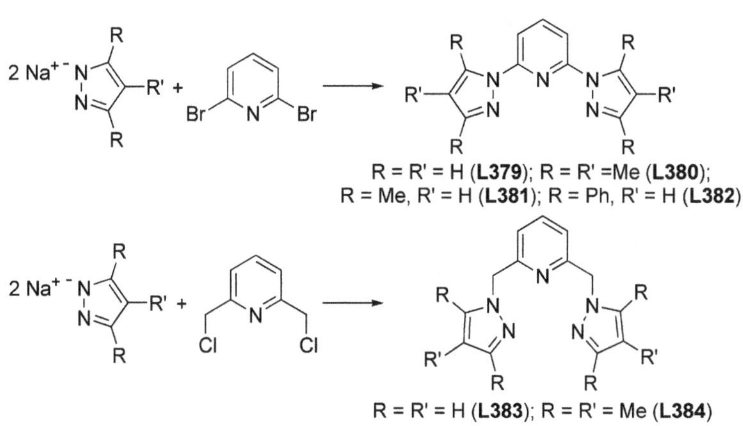

Scheme 3.42 Syntheses of 2,6-bis(pyrazolyl)pyridine ligands

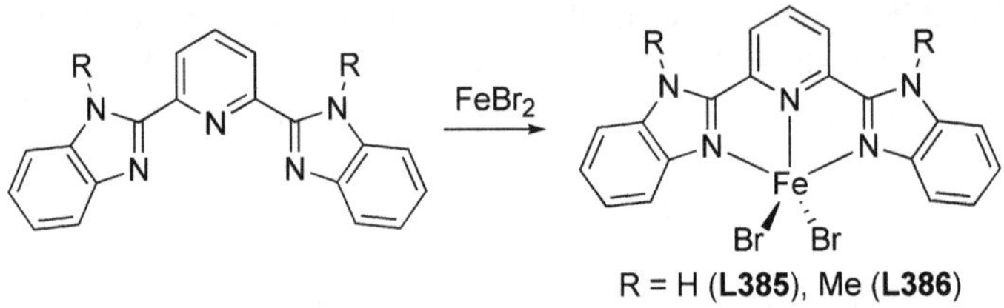

Scheme 3.43 Syntheses of 2,6-bis(2-benzimidazolyl)pyridine-based iron complexes

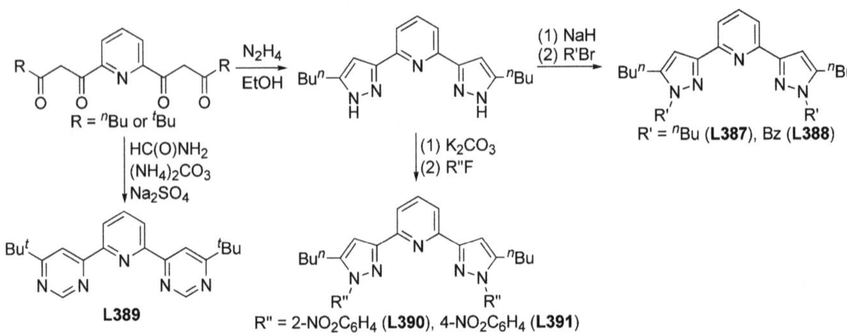

Scheme 3.44 Syntheses of 2,6-bis(pyrazolyl)pyridine and 2,6-bis(pyrimidinyl)pyridine ligands

oligomerization of ethylene, when activated with MMAO, generating predominantly butenes and a small amount of hexenes.

The 2,6-bis(2-benzimidazolyl)pyridine-based iron complexes [$FeBr_2$(**L385**)] and [$FeBr_2$(**L386**)] are prepared by reaction of iron dibromide with the commercially available 2,6-bis(2-benzimidazolyl)pyridine and 2,6-bis(1-methyl-2-benzimidazolyl)pyridine, respectively, in tetrahydrofuran (Scheme 3.43) [128]. The latter can be easily synthesized by *N*-methylation of 2,6-bis(2-benzimidazolyl)pyridine with methyl iodide. Upon activation with MMAO, these two iron complexes are active toward ethylene oligomerization, also giving rise predominantly to butenes and a small amount of hexenes.

Starting from the tetraketones, 2,6-bis(5-*n*-butylpyrazol-3-yl)pyridine is prepared by a ring closure reaction with hydrazine, while 2,6-bis(6-*tert*-butylpyrimidin-4-yl)pyridine (**L389**) is obtained by treatment of the tetraketones with formamide, in the presence of ammonium carbonate and sodium sulfate (Scheme 3.44) [153]. Deprotonation of the former compound with sodium hydride, in tetrahydrofuran, generates the corresponding pyrazolides, which react with butyl or benzyl bromide to give the alkylated derivatives **L387** and **L388**. Similarly, treatment of the same compound with *ortho*- or *para*-nitro-substituted fluorobenzene results in the formation of the arylated ligands **L390** and **L391**. Treatment of ligands **L387–L391** with iron or cobalt dichloride gives the corresponding iron and cobalt complexes. Upon exposure to ethylene, all the iron and cobalt complexes, except [$FeCl_2$(**L391**)] and [$CoCl_2$(**L391**)], are inactive in the

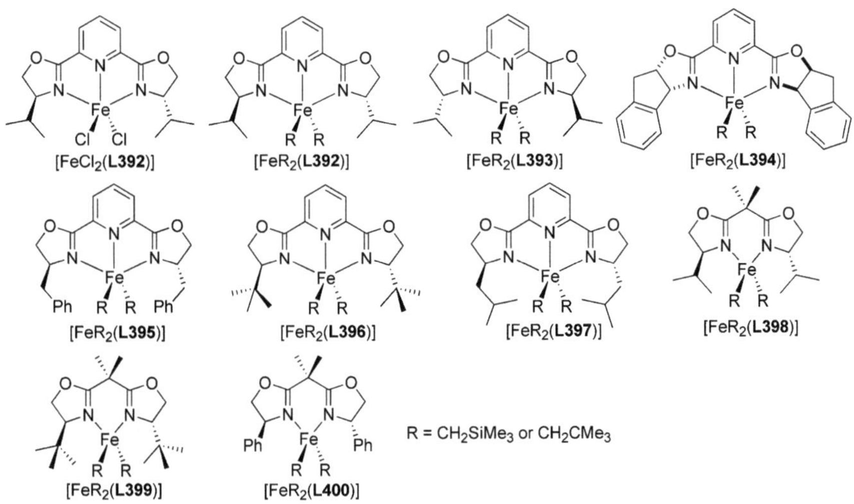

Scheme 3.45 Iron complexes based on 2,6-bis(oxazolinyl)pyridine- or 2,2-bis(oxazolinyl)propane ligands

presence of a large excess of MAO. At 10 bar of ethylene, [$FeCl_2$(**L391**)] and [$CoCl_2$(**L391**)] show moderate activities in the order of 10^5 g/(mol h), forming polyethylenes with high molecular weights.

Imanishi and Nomura [154] synthesized the iron complex [$FeCl_2$(**L392**)] by treatment of the commercially available 2,6-bis[(4*S*)-(–)-*iso*-propyl-2-oxazolin-2-yl]pyridine with iron dichloride, in dichloromethane (Scheme 3.45). This complex exhibits very low activities (ca. 10^3 g/(mol h), at 8 bar of ethylene) toward ethylene polymerization, in the presence of MAO, giving polyethylenes with high molecular weights (ca. 10^5–10^6 g/mol). Similarly, Chirik and coworkers [155] prepared a series of 2,6-bis(oxazolinyl)pyridine or 2,2-bis(oxazolinyl)propane iron dialkyl complexes [FeR_2(**L392**)]–[FeR_2(**L400**)] (R = CH_2SiMe_3 or CH_2CMe_3), either by displacement of pyridine in [$FeR_2(py)_2$] (py = pyridine) with commercially available 2,6-bis(oxazolinyl)pyridine or 2,2-bis(oxazolinyl)propane ligands or by direct dialkylation of the corresponding iron dichloride complexes. These iron dialkyl complexes were found to be active in the catalytic hydrosilylation of various ketones.

The reaction of iron dichloride with 2,2′:6′,2″-terpyridine (**L401**) gives the bisligand iron complex $[Fe(\mathbf{L401})_2]^{2+}[FeCl_4]_2^{2-}$, whereas the reactions with **L402** and **L403**, bearing bulky substituents at the *o*-positions, afford the monoligand iron derivatives [$FeCl_2$(**L402**)] and [$FeCl_2$(**L403**)], respectively (Scheme 3.46) [156]. Treatment of **L401** and **L404** with iron trichloride leads to the iron(III) complexes [$FeCl_3$(**L401**)] and [$FeCl_3$(**L404**)], respectively. It was found that, except in the cases of [$FeCl_2$(**L402**)] and [$FeCl_2$(**L403**)], which show extremely low activities for ethylene polymerization in the presence of MMAO (respectively, 90 and 70 g/(mol h bar)), all the other complexes are totally inactive toward ethylene polymerization.

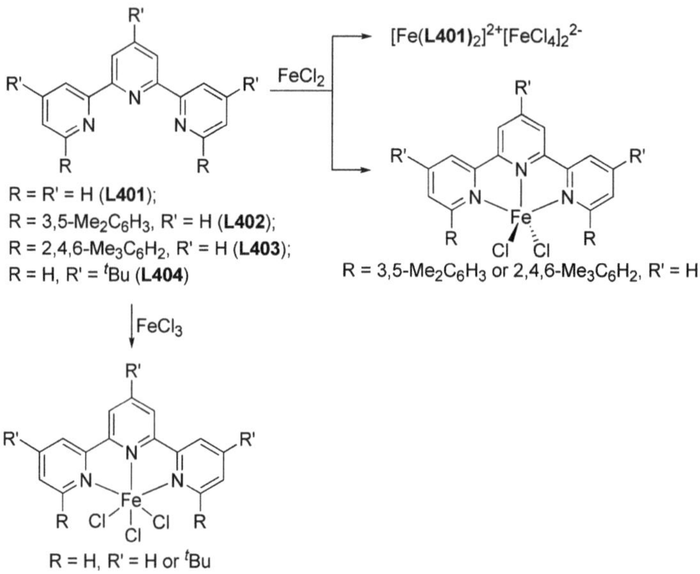

Scheme 3.46 Syntheses of terpyridine-based iron complexes

Treatment of iron dibromide with the 2,6-bis(*N*-heterocyclic carbene)pyridine **L405**, a tridentate chelating ligand containing the *iso*-propyl substituent, in tetrahydrofuran, gives the bisligand iron complex $[Fe(\mathbf{L405})_2]^{2+}[FeBr_4]^{2-}$, whereas the reaction with **L406**, bearing the bulkier 2,6-di-*iso*-propylphenyl substituent, generates the monoligand iron complex $[FeBr_2(\mathbf{L406})]$, which can be alternatively synthesized by the direct metalation reaction from 2,6-bis(arylimidazolium)pyridine dibromide (aryl = 2,6-di-*iso*-propylphenyl) and bis[bis(trimethylsilyl)amido]iron(II) ($[Fe\{N(SiMe_3)_2\}_2]$), in tetrahydrofuran (Scheme 3.47) [157, 158]. Likewise, the reaction of 2,6-bis(arylimidazolium)pyridine dibromide (aryl = 2,6-di-*iso*-propylphenyl) with bis[bis(trimethylsilyl)amido]cobalt(II) ($[Co\{N(SiMe_3)_2\}_2]$), in tetrahydrofuran, gives quantitative yields of $[CoBr_2(\mathbf{L406})]$ [159]. These 2,6-bis(carbene)pyridine-based iron and cobalt complexes were found to be totally inactive toward the oligo- or polymerization of ethylene, in the presence of MAO.

Treatment of the 2,6-bis(iminophosphoranyl)pyridine **L407–L410** and bis(iminophosphoranyl)ethane ligands **L416–L419** with iron or cobalt dihalide (MX_2, M = Fe or Co, X = Cl or Br) give, respectively, the corresponding tridentate or bidentate iron and cobalt complexes (Scheme 3.48) [160]. Deprotonation of bis(iminophosphoranyl)methane ligand precursors with *n*-butyllithium, in tetrahydrofuran, followed by reaction with cobalt dichloride, affords the cobalt complexes [CoCl(L)] (**L411–L415**). All the iron and cobalt complexes were tested toward ethylene polymerization, in the presence of MAO, the cobalt complexes based on the 2,6-bis(iminophosphoranyl)pyridine and bis(iminophosphoranyl)alkane ligands showing moderate catalytic activities, with typical values in

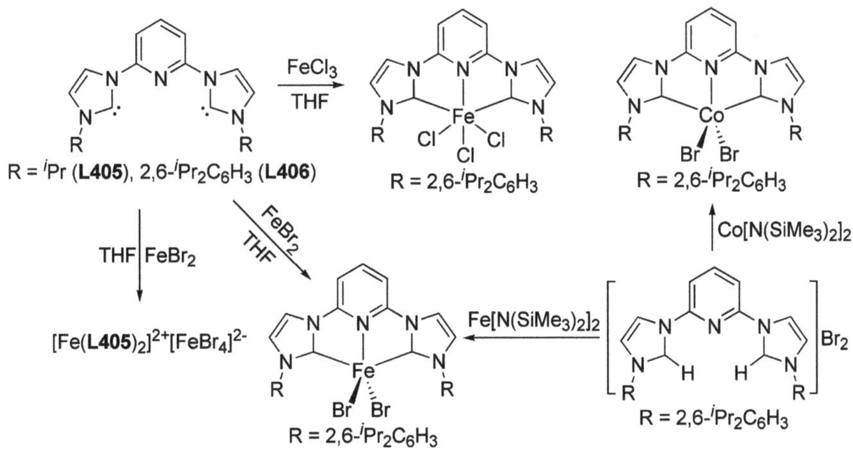

Scheme 3.47 Syntheses of 2,6-bis(*N*-heterocyclic carbene)pyridine ligands and their iron and cobalt complexes

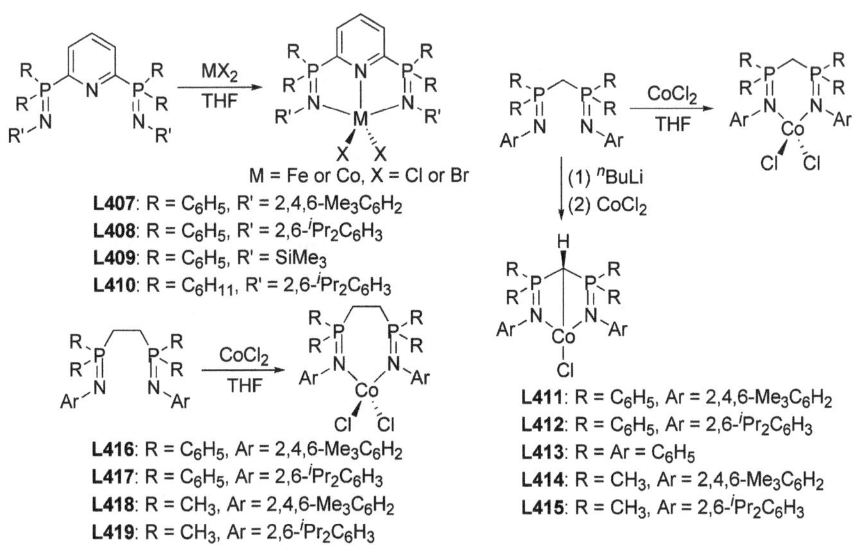

Scheme 3.48 Syntheses of 2,6-bis(iminophosphoranyl)pyridine- and bis(iminophosphoranyl) alkane-based iron and cobalt complexes

the range of 5–30 × 10^3 g/(mol h bar), and giving rise to polyethylenes with high molecular weights (M_w = 1.5–4 × 10^5 g/mol), while the 2,6-bis(iminophosphoranyl)pyridine-based iron complexes exhibited very low or no activities.

The 2,6-bis((diarylphosphino)methyl)pyridine ligands **L420–L422** are prepared by treatment of diarylphosphines with 2,6-bis(chloromethyl)pyridine in the presence of potassium *tert*-butoxide. These tridentate ligands react with iron or cobalt

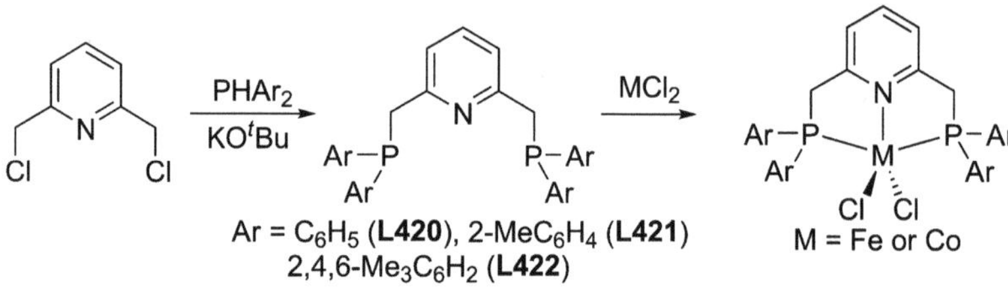

Scheme 3.49 Syntheses of 2,6-bis((diarylphosphino)methyl)pyridine ligands and their iron and cobalt complexes

dichloride, in ethanol, to afford the corresponding metal complexes (Scheme 3.49) [161]. Upon activation with MAO, iron complexes [$FeCl_2$(**L420**)]–[$FeCl_2$(**L422**)] are not active toward the polymerization of ethylene, with a fast decomposition of the system being observed, while cobalt complexes [$CoCl_2$(**L420**)]–[$CoCl_2$(**L422**)] exhibit very low activities of 74–225 g/(mol h), at 60 bar of ethylene, forming polyethylenes with high molecular weights (ca. 3×10^5 g/mol).

3.5.3 Modifications on Both Pyridine and Imino Moieties

Sun and Solan groups [162–167] recently reported a large family of 2-imino-1,10-phenanthrolinyl ligands, which were prepared by the condensation of the 1,10-phenanthrolinyl ketones or aldehydes with substituted anilines (Scheme 3.50). Depending on the difference in reactivities between aldehydes or ketones and alkyl- or halo-substituted anilines, the reaction conditions were tuned, including solvent (e.g., ethanol, toluene), dehydrating agent (e.g., tetraethyl orthosilicate), catalyst (e.g., acetic acid and *p*-toluenesulfonic acid), and reaction temperature and time, in order to improve the product yields.

The 2-imino-1,10-phenanthrolinyl ligands can be classified according to the nature of the R^1 substituent at the imino carbon as: methyl-ketimine (R^1 = Me, **L423**–**L447** and **L472**–**L474**), aldimine (R^1 = H, **L448**–**L459**), ethyl-ketimine (R^1 = Et, **L460**–**L466**) and phenyl-ketimine (R^1 = Ph, **L467**–**L471** and **L475**–**L477**). Reaction of the ligands with anhydrous or hydrated iron or cobalt dichloride, in tetrahydrofuran or ethanol, gave the corresponding metal complexes. Analogous to the 2,6-bis(arylimino)pyridine metal complexes, the coordination geometries of the 2-imino-1,10-phenanthrolinyl iron and cobalt complexes are flexible, varying from trigonal bipyramidal to square pyramidal, and are greatly dependent on the bulkiness of the aryl rings *o*-substituents. Generally, the complexes containing the less bulky *o*-substituents on the aryl rings adopt trigonal bipyramidal geometries, in which the nitrogen atom of the phenanthrolinyl group (next to the imino carbon) and two chlorides form the equatorial plane, and the other nitrogen atom of the phenanthrolinyl group and the imino nitrogen atom occupy the axial positions, e.g., [$FeCl_2$(**L424**)] (with Et), [$FeCl_2$(**L427**)]

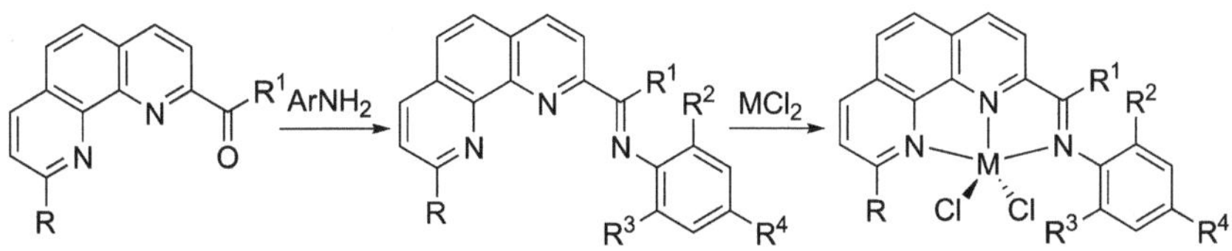

	R	R¹	R²	R³	R⁴
L423	H	Me	Me	Me	H
L424	H	Me	Et	Et	H
L425	H	Me	^{i}Pr	^{i}Pr	H
L426	H	Me	^{i}Pr	^{i}Pr	^{i}Pr
L427	H	Me	Me	Me	Me
L428	H	Me	Me	Me	Br
L429	H	Me	F	F	H
L430	H	Me	Cl	Cl	H
L431	H	Me	Br	Br	H
L432	H	Me	Br	Br	Me
L433	H	Me	Br	Br	Cl
L434	H	Me	Br	Br	Br
L435	H	Me	Me	H	H
L436	H	Me	Et	H	H
L437	H	Me	^{i}Pr	H	H
L438	H	Me	Me	Et	H
L439	H	Me	Br	H	H
L440	H	Me	Br	H	Me
L441	H	Me	Br	H	F

	R	R¹	R²	R³	R⁴
L442	H	Me	Br	H	Br
L443	H	Me	F	F	F
L444	H	Me	Cl	Cl	Cl
L445	H	Me	Br	F	F
L446	H	Me	Br	Cl	F
L447	H	Me	Br	F	Cl
L448	H	H	Me	Me	H
L449	H	H	Et	Et	H
L450	H	H	^{i}Pr	^{i}Pr	H
L451	H	H	^{i}Pr	^{i}Pr	Br
L452	H	H	^{i}Pr	^{i}Pr	CN
L453	H	H	^{i}Pr	^{i}Pr	^{i}Pr
L454	H	H	F	F	H
L455	H	H	Cl	Cl	H
L456	H	H	Br	Br	H
L457	H	H	Me	H	H
L458	H	H	Me	Me	Me
L459	H	H	Me	Me	Br

	R	R¹	R²	R³	R⁴
L460	H	Et	Me	Me	H
L461	H	Et	Et	Et	H
L462	H	Et	^{i}Pr	^{i}Pr	H
L463	H	Et	Me	Me	Me
L464	H	Et	Et	Et	Me
L465	H	Et	F	F	H
L466	H	Et	Cl	Cl	H
L467	H	Ph	Me	Me	H
L468	H	Ph	Et	Et	H
L469	H	Ph	^{i}Pr	^{i}Pr	H
L470	H	Ph	Me	Me	Me
L471	H	Ph	Me	Me	Br
L472	Ph	Me	Me	Me	H
L473	Ph	Me	Et	Et	H
L474	Ph	Me	^{i}Pr	^{i}Pr	H
L475	Ph	Ph	Me	Me	H
L476	Ph	Ph	Et	Et	H
L477	Ph	Ph	^{i}Pr	^{i}Pr	H

Scheme 3.50 Syntheses of 2-imino-1,10-phenanthrolinyl ligands and their iron and cobalt complexes

(with Me), [$FeCl_2$(**L430**)] (with Cl), [$FeCl_2$(**L431**)] (with Br), [$CoCl_2$(**L449**)] (with Et), [$CoCl_2$(**L468**)] (with Et), [$FeCl_2$(**L471**)] (with Me) and [$FeCl_2$(**L473**)] (with Et), while the complexes bearing the much bulkier *o*-substituents on the aryl rings exhibit square pyramidal geometries with the three nitrogen atoms and one of the chlorine atoms forming the base, and the remaining chlorine atom occupying the apical position, e.g., [$CoCl_2$(**L452**)] (with ^{i}Pr) and [$CoCl_2$(**L455**)] (with ^{i}Pr).

All the iron and cobalt complexes, when activated with MAO or MMAO, display high catalytic activities toward ethylene oligomerization, with high selectivity for α-olefins, and the distribution of oligomers obtained in all cases follows the Schulz–Flory rules. The catalytic activities of iron complexes (ca. 10^6 g/(mol h bar)) are generally one order of magnitude higher than those obtained by their cobalt analogues. The various substituents (R, R^1, R^2, R^3 and R^4, Scheme 3.50) of the ligand backbone are found to have remarkable influences on the catalytic performances of the metal complexes. For the variation of R, the phenyl substitution (R = Ph) gives rise to a substantial decrease in the activity, in the case of iron complexes, and a slight increase for the cobalt ones. For instance, the phenyl-substituted iron complexes [$FeCl_2$(**L472**)]–[$FeCl_2$(**L474**)] show activities of 40.8, 29.1 and 0.62 × 10^4 g/(mol h), respectively, at 10 bar of ethylene, being strikingly lower than those of 38.9, 49.1 and 9.42 × 10^6 g/(mol h), observed for the corresponding unsubstituted iron analogues [$FeCl_2$(**L423**)]–[$FeCl_2$(**L425**)] (R = H), while the cobalt complexes [$CoCl_2$(**L472**)]–[$CoCl_2$(**L474**)] give slightly higher activities of 6.85, 6.71 and 5.23 × 10^5 g/(mol h) (at 1 bar of ethylene) than those of 1.93, 2.34 and 2.57 × 10^5 g/(mol h) for [$CoCl_2$(**L423**)]–[$CoCl_2$(**L425**)]. Additionally, it is noteworthy that the phenyl-substituted iron complexes [$FeCl_2$(**L472**)]–[$FeCl_2$(**L477**)] exclusively form 1-butene, whereas the corresponding unsubstituted iron analogues ($FeCl_2$(**L423**)]–[$FeCl_2$(**L425**)] and $FeCl_2$(**L467**)]–[$FeCl_2$(**L469**)]) afford oligomers with *K* values in the range of 0.50–0.67, along with a small amount of waxy polyethylenes. On the contrary, the oligomers obtained with the phenyl-substituted and unsubstituted cobalt complexes display comparable molecular weight distributions, the latter possessing a higher content of butenes.

Concerning the influence of the imino carbon substituents (R^1 = H, Me, Et, and Ph, Scheme 3.50), the catalytic activity varies in the order: methyl-ketimine > phenyl-ketimine > ethyl-ketimine ≈ aldimine, for the iron complexes having no alkyl or aryl substituents on the 9-position of the 1,10-phenanthrolinyl ring (R = H), whereas among the iron complexes with phenyl substituents on the 9-position (R = Ph), the phenyl-ketimine is slightly more active than the methyl-ketimine one. For all the cobalt complexes, the aldimine and methyl-ketimine show comparable activities, which are slightly lower than those of the phenyl-ketimine.

The influence of the aryl substituents (R^2, R^3 and R^4, Scheme 3.50) on the catalytic performances of metal complexes is of significance and varies according to the nature of the complexes (e.g., aldimine, methyl-ketimine or phenyl-ketimine). For the 2,6-dialkyl-substituted methyl-ketimine iron complexes ([$FeCl_2$(**L423**)]–[$FeCl_2$(**L425**)] and [$FeCl_2$(**L472**)]–[$FeCl_2$(**L474**)]), the bulkier precatalysts lead to reduced catalytic activities, e.g., [$FeCl_2$(**L424**)] (Et, 4.91 × 10^7 g/(mol h)) > [$FeCl_2$(**L423**)] (Me, 3.89 × 10^7 g/(mol h)) > [$FeCl_2$(**L425**)] (^{i}Pr, 9.42 × 10^6 g/(mol h)) [162] and [$FeCl_2$(**L472**)] (Me, 4.08 × 10^5 g/(mol h)) > [$FeCl_2$(**L473**)] (Et, 2.91 × 10^5 g/(mol h)) > [$FeCl_2$(**L474**)] (^{i}Pr, 0.62 × 10^4 g/(mol h)), at 10 bar of ethylene [166]. Furthermore, the bulkier the substituents, the smaller the *K* values, and the smaller the amount of low-molecular-weight waxy polyethylene produced. The same trend is observed for the dialkyl-substituted aldimine, ethyl-ketimine and phenyl-ketimine iron complexes ([$FeCl_2$(**L448**)]–[$FeCl_2$(**L450**)], [$FeCl_2$(**L460**)]–[$FeCl_2$(**L462**)], [$FeCl_2$(**L467**)]–[$FeCl_2$(**L469**)] and [$FeCl_2$(**L475**)]–[$FeCl_2$(**L477**)]) and the monoalkyl-substituted methyl-ketimine iron complexes ([$FeCl_2$(**L435**)]–[$FeCl_2$(**L437**)]). For the 2,6-dihalo-substituted methyl-ketimine iron complexes ([$FeCl_2$(**L429**)]–[$FeCl_2$(**L431**)]), a reverse tendency is observed since bulkier substituents result in higher catalytic activities and *K* values (bromo > chloro > fluoro). However, in the case of the 2,6-dihalo-substituted aldimine iron complexes ([$FeCl_2$(**L454**)]–[$FeCl_2$(**L456**)]), the precatalyst [$FeCl_2$(**L456**)], with *o*-bromo substituents on the aryl ring, gives the lowest activity (1.08 × 10^6 g/(mol h)), at 10 bar of ethylene, while the 2,6-dichloro-substituted [$FeCl_2$(**L455**)] displays the highest one (7.30 × 10^6 g/(mol h)), with some waxy polymer formed. The 2,6-difluoro-substituted [$FeCl_2$(**L454**)] shows a catalytic activity similar to the latter (6.99×10^6 g/(mol h)), with a lower K value [162]. For all the dialkyl- or dihalo-substituted cobalt complexes, regardless of their aldimine ([$CoCl_2$(**L448**)]–[$CoCl_2$(**L450**)] and [$CoCl_2$(**L454**)]–[$CoCl_2$(**L456**)]), methyl-ketimine ([$CoCl_2$-(**L423**)]–[$CoCl_2$(**L425**)] and [$CoCl_2$(**L429**)]–[$CoCl_2$(**L431**)]) or phenyl-ketimine ([$CoCl_2$(**L467**)]–[$CoCl_2$(**L469**)]) nature, the catalytic activity increases with the size of the aryl ring *o*-substituents (Me < Et < ^{i}Pr and F < Cl < Br).

The 2-(benzimidazol-2-yl)-1,10-phenanthrolyl ligands **L478–L487** can be prepared by treatment of *o*-phenylenediamine either with 2,9-dimethyl-1,10-phenanthroline, using sulfur as oxidant, or with 1,10-phenanthroline-2-carboxylic acid, in the presence of polyphosphoric acid (ppa), under microwave radiation, followed by *N*-alkylation with an alkyl halide, in the presence of potassium carbonate (Scheme 3.51) [168]. Reaction of the ligands with iron or cobalt dichloride, in ethanol, gives the corresponding metal complexes. These complexes, when activated with MMAO, are active in the oligomerization of ethylene, affording oligomers with predominance of butenes. The highest activity can reach 3.51×10^6 g/(mol h) in the case of [$FeCl_2$(**L483**)], and 2.32×10^6 g/(mol h) for [$CoCl_2$(**L483**)], at 30 bar of ethylene. In general, the iron complexes are more active than their cobalt analogues. The backbone substituents (R and R′, Scheme 3.51) show marked influences on the catalytic performance of the pre-catalysts. The introduction of a methyl group on the 9-position of the phenanthrolinyl ring (R = Me) results in a decrease in the activity and a slight increase in the α-C_4 selectivity. Replacement of the active proton attached to the benzimidazolyl nitrogen by an alkyl group leads to a decrease both in the activity and in the α-C_4 selectivity. The steric effect of these alkyl groups on the catalytic activity does not follow regular patterns.

Zhang et al. [169] prepared several 2-oxazolinyl/benzoxazolyl-1,10-phenanthrolinyl ligands (**L488-L491** and **L492-L495**, respectively), using a multistep

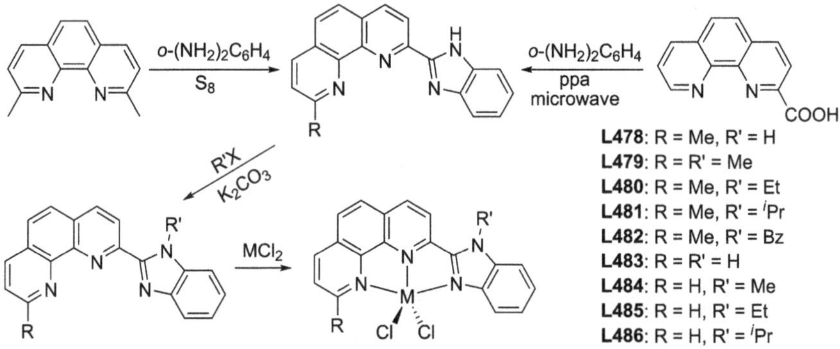

Scheme 3.51 Syntheses of 2-(benzimidazol-2-yl)-1,10-phenanthrolyl ligands and their iron and cobalt complexes

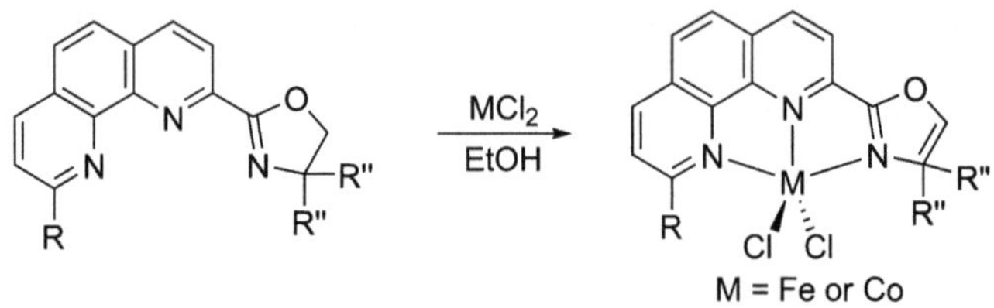

L488: R = R" = H; **L489**: R = H, R" = Me; **L490**:R = Ph, R" = H; **L491**: R = Ph, R" = Me

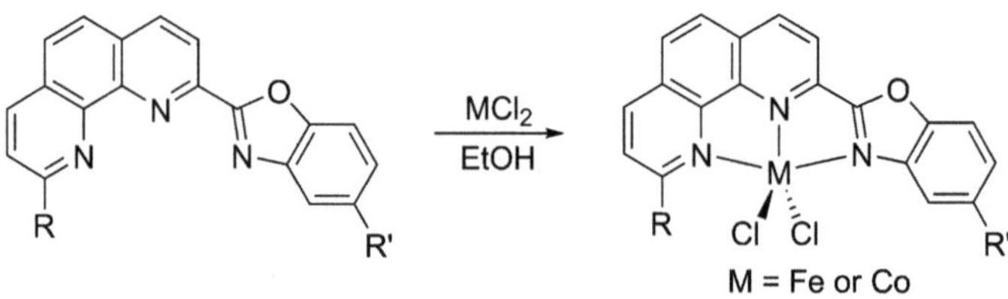

L492: R = R' = H; **L493**: R = H, R' = Me; **L494**: R = H, R' = ^{t}Bu; **L495**: R= Ph, R' = H

Scheme 3.52 Syntheses of 2-oxazolinyl/benzoxazolyl-1,10-phenanthrolinyl iron and cobalt complexes

reaction protocol starting from 2-methyl-1,10-phenanthroline. Reaction of the resulting ligands with iron or cobalt dichloride, in ethanol, gives the corresponding iron and cobalt complexes (Scheme 3.52). The substituent (R′) on the benzoxazolyl ring greatly influences the coordination geometry of complexes. For instance, [$CoCl_2$(**L492**)], bearing a hydrogen atom on the benzoxazolyl ring (R′ = H) displays a distorted trigonal bipyramidal geometry around the metal center, whereas [$CoCl_2$(**L494**)], with a *tert*-butyl group (R′ = ^{t}Bu) exhibits a square pyramidal geometry. Treatment of these complexes with MMAO gives rise to highly active ethylene oligomerization catalysts, with activities of ca. 10^5 g/(mol h), at 10 bar of ethylene. The maximum activity reaches 1.51×10^6 g/(mol h) for cobalt complexes ([$CoCl_2$(**L492**)]), and 1.89×10^6 g/(mol h) for iron complexes ([$FeCl_2$(**L492**)]). The resulting oligomers are mainly butenes and hexenes. The substitution of the phenyl group on the 9-position of the phenanthrolinyl ring (R = Ph) and the alkyl-substitution on the 2-oxazolinyl/benzoxazolyl rings (R′ = Me or ^{t}Bu, R″ = Me) give rise to a decrease in the activity and an increase in the selectivity for α-olefins. Complexes bearing the benzoxazolyl ring show higher activity and selectivity than their oxazolinyl-containing analogues. All the iron complexes show slightly higher activities than their cobalt analogues.

The stoichiometric reaction of *N*-(2-pyridylmethyl)-2-hydroxy-3,5-di-*tert*-butylbenzaldimine **L496** with iron dichloride, in tetrahydrofuran, gives the bidentate complex [$FeCl_2$(**L496**)], in high yield (90%), in which **L496** is a neutral ligand, while the reaction of the lithium salt of **L496** with iron dichloride, in tetrahydrofuran, affords a complex of the type [FeClL], in moderate yield (46%), in which L is the monoanionic tridentate ligand generated by the deprotonation of **L496**

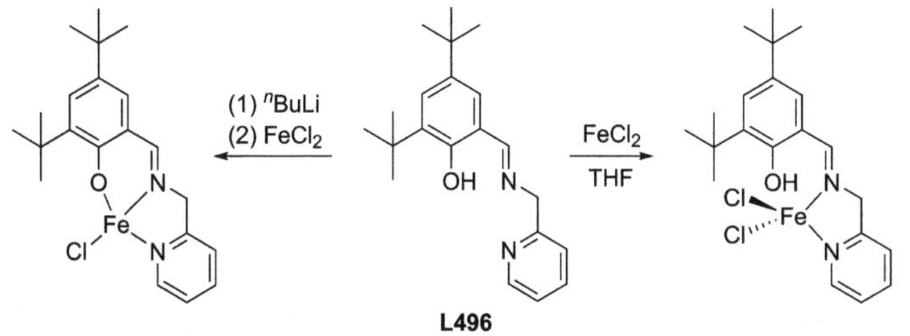

Scheme 3.53 Syntheses of *N*-(2-pyridylmethyl)-2-hydroxy-3,5-di-*tert*-butylbenzaldimine-based iron complexes

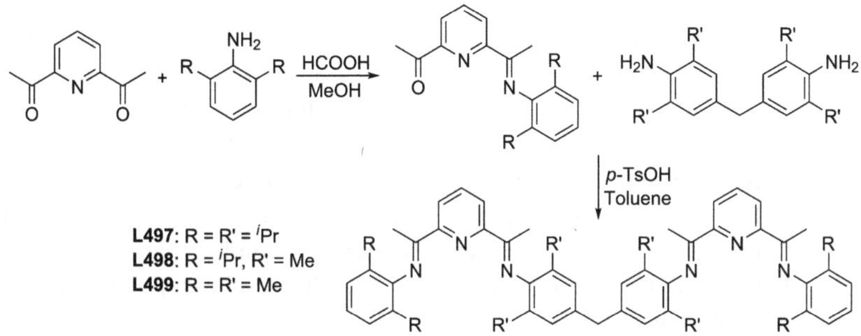

Scheme 3.54 Syntheses of methyl-bridged binuclear BIP ligands

(Scheme 3.53) [156]. None of these complexes is active toward ethylene polymerization, in the presence of MMAO.

3.5.4 Polynuclear Metal Complexes

Incorporating the BIP ligand, as a unit, into a macrocycle or a polymeric chain may lead to polydentate ligands, and their corresponding homo- or heteropolynuclear metal complexes are expected to show some particular features as precatalysts in the oligo- or polymerization of ethylene. Indeed, such homo- or heteropolynuclear metal complexes may produce unconventional mixtures of polyethylenes and/or α-olefins, improve the catalytic activities compared with the parent mononuclear BIP complex, and tailor the molecular weight and microstructure of the polymers.

The simplest methyl-bridged BIP ligands **L497–L499** are prepared by condensation of the bridged diamine with 2 equivalents of 6-acetyl-2-(arylimino)pyridine, which is formed by reaction of 2,6-diacetylpyridine with an equimolar amount of substituted aniline, either in methanol, catalyzed by formic acid, or in toluene, in the presence of *p*-toluenesulfonic acid (Scheme 3.54) [170, 171].

Treatment of these ligands with iron dichloride, in tetrahydrofuran, gives the corresponding binuclear iron complexes [$(FeCl_2)_2$(**L497**)]–[$(FeCl_2)_2$(**L499**)]. Upon treatment with tri-*iso*-butylaluminum (Al^iBu_3), [$(FeCl_2)_2$(**L498**)] and [$(FeCl_2)_2$(**L499**)] are highly active toward ethylene polymerization, forming high molecular weight polyethylenes with broad/bimodal molecular weight distributions, while [$(FeCl_2)_2$(**L497**)] is almost inactive, but it can be activated by triethylaluminum ($AlEt_3$) to generate a highly active ethylene polymerization catalyst. Compared with the parent mononuclear BIP iron complexes ([$FeCl_2$(**L84**)], [$FeCl_2$(**L87**)] and [$FeCl_2$(**L171**)]), the corresponding binuclear modification leads to a remarkable increase both in activity and polymer molecular weight, e.g., [$(FeCl_2)_2$(**L498**)] (activity = 5.36×10^6 g/(mol h), $M_w = 4.60 \times 10^5$ g/mol) vs. [$FeCl_2$(**L171**)] (activity = 0.69×10^6 g/(mol h), $M_w = 3.21 \times 10^4$ g/mol), at 1 bar of ethylene.

The macrocyclic BIP ligand **L500** can be synthesized by condensation of 2,6-diacetylpyridine with 4,4′-methylenebis(2,6-di-*iso*-propylaniline), using *p*-toluenesulfonic acid as catalyst, under pseudo-high-dilution conditions. Reaction of this macrocycle with iron dichloride affords the corresponding trinuclear iron complex [$(FeCl_2)_3$(**L500**)] (Scheme 3.55) [172]. Treatment of [$(FeCl_2)_3$(**L500**)] with MMAO or Al^iBu_3 gives origin to a highly active ethylene polymerization catalyst. Compared with its mononuclear parent complex [$FeCl_2$(**L87**)], this complex exhibits higher catalytic activities and the resulting polymers possess higher molecular weights.

The oligomeric BIP ligand **L501** and **L502** are prepared by condensation of 2,6-diacetylpyridine with equimolar quantities of 4,4′-methylenebis(2-cyclopentyl-6-methylaniline) or 4,4′-methylenebis(2-cyclohexyl-6-methylaniline), in methanol. Further treatment with iron dichloride, in tetrahydrofuran, leads to their iron complexes [$(FeCl_2)_3$(**L501**)] and [$(FeCl_2)_2$(**L502**)] (Scheme 3.56), which are active toward ethylene polymerization, when activated with MAO [173]. Compared with the mononuclear analogues ([$FeCl_2$(**L93**)] and [$FeCl_2$(**L97**)]), the

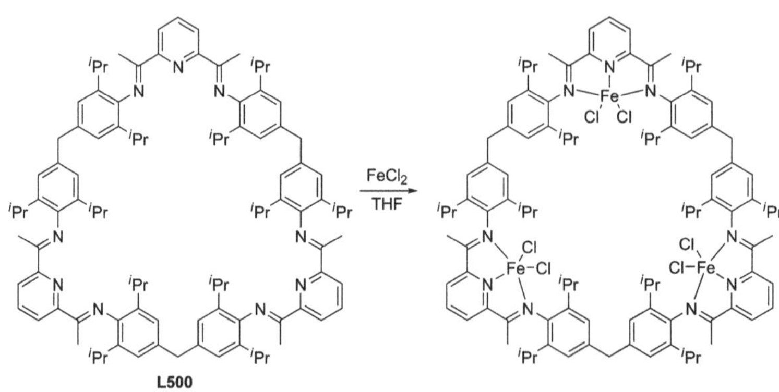

Scheme 3.55 Synthesis of a macrocyclic trinuclear BIP iron complex

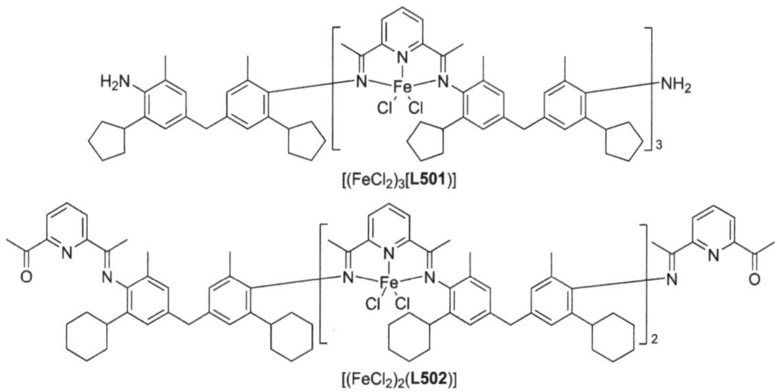

Scheme 3.56 Oligomeric BIP iron complexes

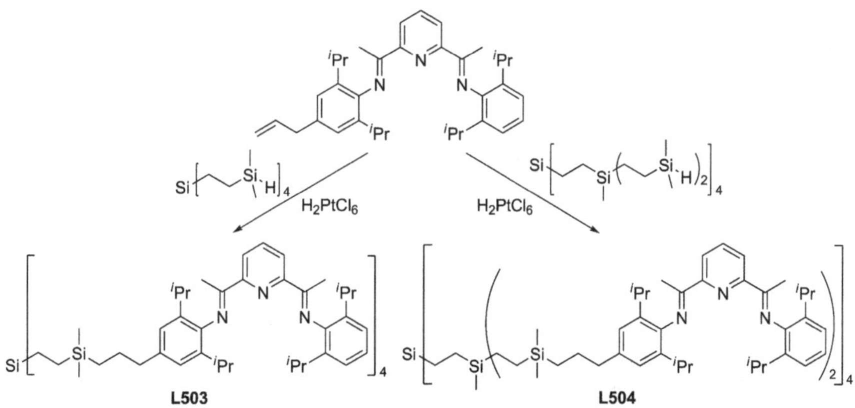

Scheme 3.57 Syntheses of dendritic BIP ligands

polynuclear precatalysts [$(FeCl_2)_3$(**L501**)] and [$(FeCl_2)_2$(**L502**)] show higher activity and stability at elevated temperature.

The dendritic BIP ligands **L503** and **L504** are prepared by platinum-catalyzed hydrosilylation reactions of the allyl-containing BIP with tetra(Si–H)silane or octa(Si–H)silane. The treatment of the resulting ligands with iron dichloride affords the corresponding polynuclear iron complexes [$(FeCl_2)_4$(**L503**)] and [$(FeCl_2)_8$(**L504**)], which are active toward ethylene polymerization upon activation with MMAO (Scheme 3.57) [174]. Compared with the mononuclear analogue ([$FeCl_2$(**L87**)], these polynuclear precatalysts display higher catalytic activities and produce higher molecular weight polyethylenes.

The bis(imino)quaterpyridine ligand **L505** can be easily synthesized by condensation of 2,4,6-tri-*iso*-propylaniline with 6,6‴-diacetylquaterpyridine. Further treatment with metal dihalide ($FeCl_2$, $CoCl_2$ or $CoBr_2$), in *n*-butanol, gives the corresponding metal complexes [M_2X_3(**L505**)]$_2$[MX_4] (M = Fe or Co, X = Cl or Br)

(Scheme 3.58) [175]. Their molecular structures are sensitive to the solvent. Crystallization of $[Co_2Cl_3($**L505**$)]_2[CoCl_4]$ in acetonitrile gives a dinuclear monocation balanced by a half occupancy tetrahedral tetrachlorocobaltate(II) dianion, in which the presence of a molecule of acetonitrile coordinated at only one of the metal centers results in two different coordination geometries, namely octahedral and distorted square pyramidal, while crystallization of $[Co_2Br_3($**L505**$)]_2[CoBr_4]$ in acetonitrile generates two distinct monocations balanced by a discrete tetrabromocobaltate dianion, in which one of the monocations are isostructural with the $CoCl_2$ analogues and the other consists of a $[CoBr_2(N$^N^N$)]$ unit with a square pyramidal geometry and a $[CoBr(N$^N^N$)(NCMe)(OH_2)]$ unit with a distorted octahedral geometry. Upon activation with MAO, the cobalt complexes $[Co_2Cl_3($**L505**$)]_2[CoCl_4]$ and $[Co_2Br_3($**L505**$)]_2[CoBr_4]$ show moderate activities in the oligomerization of ethylene, affording α-olefins with Schulz–Flory distributions (*K* values of 0.76 and 0.74, respectively), whereas the iron complex $[Fe_2Cl_3($**L505**$)]_2[FeCl_4]$ is inactive.

The tetrakis-imino-bis-pyridine ligand **L506** is prepared by condensation of 1,1′,1″,1‴-(4,4’-bipyridine-2,2′,6,6′-tetrayl)tetraethanone with 2,6-di-*iso*-propylaniline. Reaction of ligand **L506** with 2 equivalents of iron or cobalt dichloride affords the corresponding bimetallic complexes $[Fe_2Cl_4($**L506**$)]$ and $[Co_2Cl_4($**L506**$)]$ (Scheme 3.59) [176]. Upon activation with MAO, $[Fe_2Cl_4($**L506**$)]$ shows a catalytic activity comparable with that of its monometallic analogue $[FeCl_2($**L87**$)]$, whereas $[Co_2Cl_4($**L506**$)]$ is strikingly more active than its monometallic analogue $[CoCl_2($**L87**$)]$.

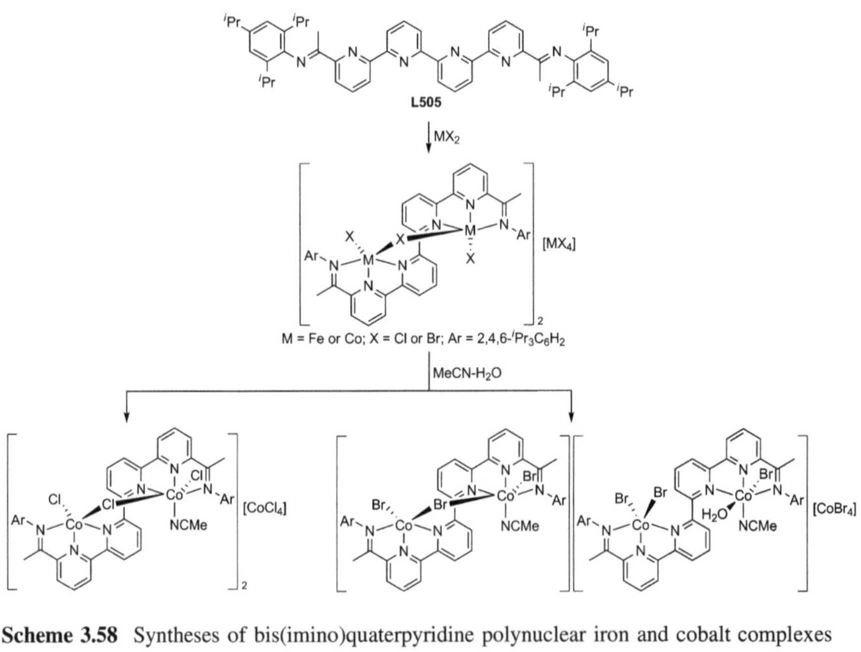

Scheme 3.58 Syntheses of bis(imino)quaterpyridine polynuclear iron and cobalt complexes

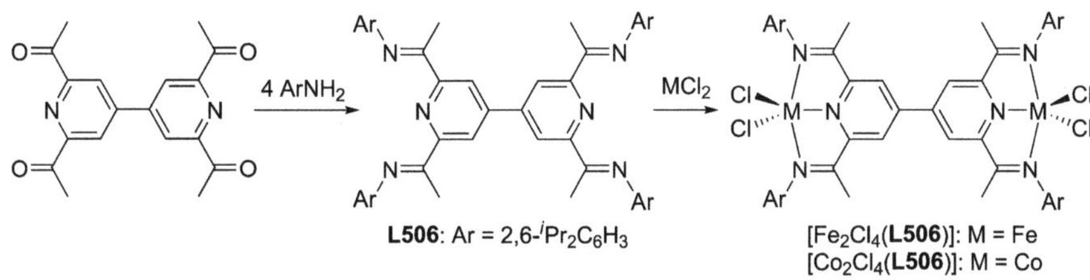

Scheme 3.59 Syntheses of a tetrakis-imino-bis-pyridine ligand and its bimetallic iron and cobalt complexes

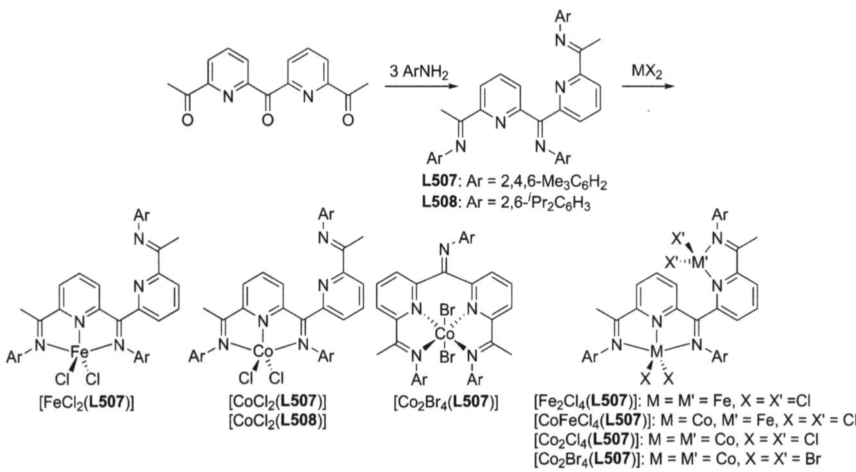

Scheme 3.60 Syntheses of tri-imino-bis-pyridine ligands and their mono- or bimetallic iron and cobalt complexes

The tri-imino-bis-pyridine ligands **L507** and **L508** are prepared by the formic acid-catalyzed condensation of 1,1′-(6,6′-carbonylbis(pyridine-6,2-diyl))diethanone with the appropriate anilines in methanol (Scheme 3.60) [176]. Reaction of these ligands with an equimolar amount of metal dihalide (MX_2, M = Fe or Co, X = Cl or Br) leads to the monometallic complexes [$FeCl_2$(**L507**)], [$CoCl_2$(**L507**)], [$CoCl_2$(**L508**)] and [$CoBr_2$(**L507**)], while reaction with 2 equivalents of metal dihalide, in tetrahydrofuran, generates the homo-bimetallic complexes ([Fe_2Cl_4(**L507**)], [Co_2Cl_4(**L507**)] and [Co_2Br_4(**L507**)]). The hetero-bimetallic complex [$CoFeCl_4$(**L507**)] is prepared by mixing equimolar quantities of iron dichloride with [$CoCl_2$(**L507**)], in dichloromethane. It is noteworthy that the coordination geometries of the monometallic complexes [$FeCl_2$(**L507**)] and [$CoCl_2$(**L508**)] in the solid state and in solution are invariable, being described as a distorted trigonal bipyramid and a square pyramid, respectively, consistent with those of their corresponding BIP complexes [$FeCl_2$(**L85**)] and [$CoCl_2$(**L87**)], whereas [$CoCl_2$(**L507**)] and [$CoBr_2$(**L507**)] are characterized by an equilibrium between the five-coordinate trigonal bipyramidal and the six-coordinate octahedral

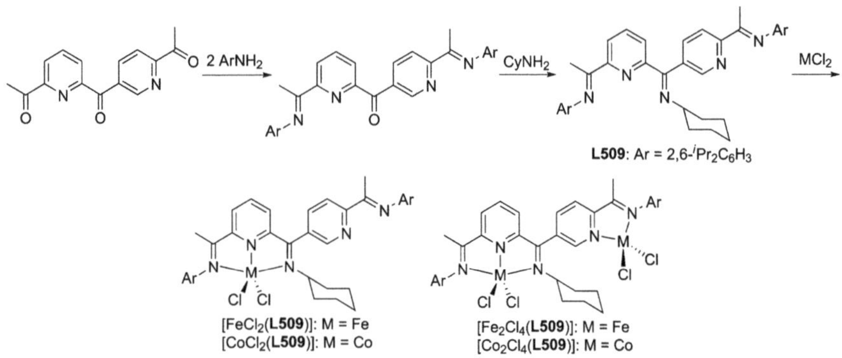

Scheme 3.61 Syntheses of a tetrakis-imino-bis-pyridine ligand and its mono- and bimetallic iron and cobalt complexes

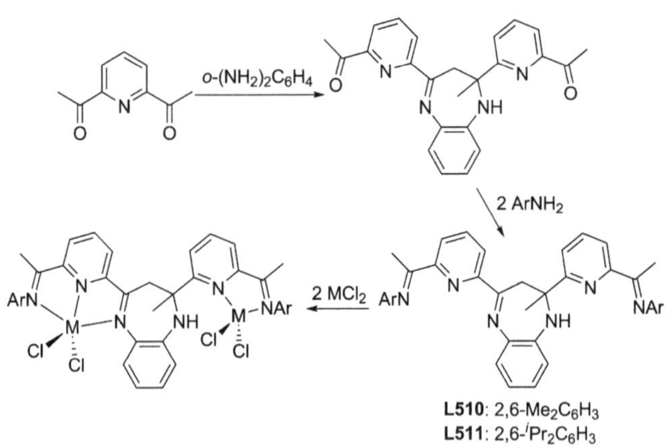

Scheme 3.62 Syntheses of bis-imino-bis-pyridine ligands and their bimetallic iron and cobalt complexes

geometries in solid state and solution. All the mono- and bimetallic iron and cobalt complexes can serve as catalysts for the polymerization of ethylene, when activated with MAO, and, irrespective of the metal, the catalytic activities of the monometallic complexes are significantly higher than those of the corresponding bimetallic derivatives, but lower than those obtained with the typical BIP complexes (e.g., [$FeCl_2$(**L85**)], [$CoCl_2$(**L85**)] and [$CoCl_2$(**L87**)]).

The tetrakis-imino-bis-pyridine ligand **L509** can be accessible by the stepwise condensation of 1-(6-(6-acetylnicotinoyl)pyridin-2-yl)ethanone with 2,6-di-*iso*-propylaniline and cyclohexylamine. Further treatment with 1 or 2 equivalents of iron or cobalt dichloride, in tetrahydrofuran, affords, respectively, the mono- and bimetallic complexes [MCl_2(**L509**)] and [M_2Cl_4(**L509**)] (M = Fe or Co) (Scheme 3.61) [177]. Upon activation with MAO, these complexes are active in

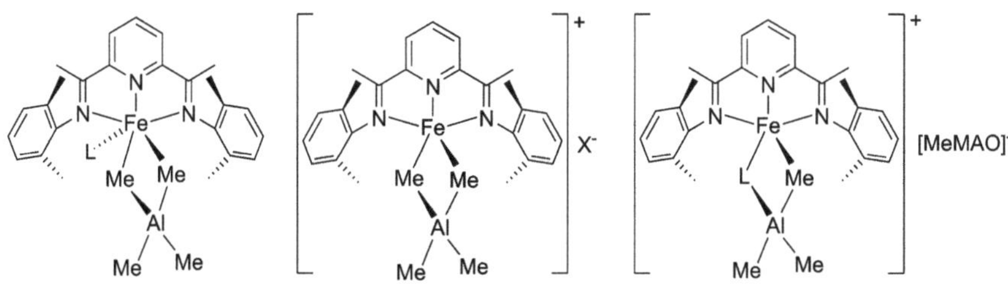

Scheme 3.63 The proposed active species present in the [FeCl$_2$(**L84**)]/various activator systems

the oligomerization of ethylene, forming α-olefins with a Schulz–Flory distribution.

The bis-imino-bis-pyridine ligands **L510** and **511** are prepared by condensation of 3,3-dihydro-2-methyl-2,4-bis(6-acetylpyridin-2-yl)-1*H*-1,5-benzodiazepine, which is synthesized by reaction of *o*-phenylenediamine with 2,6-diacetylpyridine, with the appropriate *o*-substituted anilines (Scheme 3.62) [178]. Further treatment with 2 equivalents of iron or cobalt dichloride, in ethanol, results in the corresponding bimetallic complexes [M$_2$Cl$_4$(**L510**)] and [M$_2$Cl$_4$(**L511**)] (M = Fe or Co), which show high catalytic activities in the oligomerization of ethylene (ca. of 10^5 g/(mol h bar)), when activated with MMAO, affording oligomers with a very high selectivity for α-olefins, along with a small amount of waxy polyethylene.

3.6 Oligo- or Polymerization Mechanisms

3.6.1 Activation and Initiation

3.6.1.1 Iron Complexes

It has been well recognized that the polymerization of olefins initiated by homogeneous single-site catalysts, e.g., metallocenes/MAO systems, starts from a highly reactive mono-methylated metal cation of the type [LM–Me]$^+$, commonly referred to as the active species, which is balanced by a weakly coordinating counteranion such as [X–MAO]$^-$ (X = halide or Me). By analogy, one would tend to consider that the MAO-activated BIP iron and cobalt catalyst systems initiate olefin polymerization starting also from this kind of active species. However, there is a number of controversial experimental and theoretical findings [79–81, 83, 179–181] that render the nature of the active species in the BIP iron and cobalt catalyst systems still a matter of debate.

The initial experimental [86] and theoretical [182–184] results support the assumption of formation of a cationic monoalkyl iron(II) or cobalt(II) species when the BIP iron or cobalt complexes are treated with MAO. Gould and coworkers [183] presented the first theoretical studies on these type of catalyst systems. They performed a full ab initio calculation on the iron complex [$FeCl_2$(**L87**)], and identified the key structures operating for the first monomer insertion within a cationic alkyl mechanism. Based on the same iron complex, combined density functional theory (DFT) and molecular mechanics (MM) calculations were carried out by Ziegler and coworkers [182]. It was found that the rate-determining step, for both propagation and chain transfer, is the capture of ethylene by the iron alkyl cation and that the steric bulk exerted by the aryl substituents suppresses the ethylene capture for the chain transfer step and increases the rate of insertion.

Given the very low activities that characterize the polymerization of propylene with BIP iron complexes/MAO catalyst systems, Babik and Fink [86] activated the iron complex [$FeCl_2$(**L89**)] with triphenylcarbenium tetrakis(pentafluorophenyl)borate ($[Ph_3C]^+[BC_6F_5]_4^-$) and subsequently reacted it with tri-*iso*-butylaluminum or triethylaluminum to generate a very active catalytic system toward propylene polymerization. This system is over 200 times more active than that activated with MAO, and these authors found that the characteristic aliphatic polymer end groups, derived from the different iron alkyl cations, could be identified. Recently, electrospray ionization mass spectrometry (ESI–MS) complemented by UV–visible spectroscopy was used to investigate the activation process of [$FeCl_2$(**L87**)] with MAO [180]. The four-coordinate cationic methyl iron(II) complex [FeMe(**L87**)]$^+$ was observed. In addition, the cationic iron(II) monochloride [FeCl(**L87**)]$^+$, the product resulting from the α-hydrogen transfer from [FeMe(**L87**)]$^+$ to trimethylaluminum, [Fe(CH_2AlMe_2)(**L87**)]$^+$, and the cationic iron hydride complex [FeH(**L87**)]$^+$ were identified.

Talsi et al. [179] employed ^{1}H and ^{2}H NMR spectroscopy to study intermediates formed via the activation of [$FeCl_2$(**L84**)] with not only MAO but also $AlMe_3$, $AlMe_3$/tris(pentafluorophenyl)borane ($B(C_6F_5)_3$) and $AlMe_3/[Ph_3C]$-$[B(C_6F_5)_4]$. They suggested two neutral heterodinuclear species [(**L84**)Fe^{II}(Cl)(μ-Me)$_2$$AlMe_2$] (**A**) and [(**L84**)$Fe^{II}$(Me)(μ-Me)$_2$$AlMe_2$] (**B**) (Scheme 3.63). The former species (**A**) is formed by activation of the precatalyst with MAO at relatively low Al/Fe ratios (less than 50), while the latter (**B**) is generated by activation with MAO at high Al/Fe ratios (more than 500) or by pure $AlMe_3$. Treatment of the precatalyst [$FeCl_2$(**L84**)] with $AlMe_3/B(C_6F_5)_3$ or $AlMe_3/[Ph_3C][B(C_6F_5)_4]$ generates the ion pairs [(**L84**)Fe^{II}(μ-Me)$_2$$AlMe_2$]$^+$$[MeB(C_6F_5)_3]^-$ (**C**) and [(**L84**)Fe^{II}(μ-Me)$_2$$AlMe_2$]$^+$$[B(C_6F_5)_4]^-$ (**D**), respectively. Since comparable catalytic activities were observed in these catalyst systems, Talsi et al. concluded that neutral but not cationic species are active intermediates. Later on, the same authors [185] found that the active species formed by MAO was erroneously assigned on the basis of the accidental similarity of the observed ^{1}H NMR spectra of catalyst systems [$FeCl_2$(**L84**)]/$AlMe_3$, in dichloromethane, and [$FeCl_2$(**L84**)]/MAO, in toluene. When these two systems are studied in the same solvent (toluene), pronounced differences in the ^{1}H NMR spectra are observed, and the new cationic intermediates [(**L84**)Fe^{II}(μ-Cl)(μ-Me)$AlMe_2$]$^+$[Me-MAO]$^-$ (**E**) and [(**L84**)Fe^{II}(μ-Me)$_2$$AlMe_2$]$^+$[Me-MAO]$^-$ (**F**) were suggested for the [$FeCl_2$(**L84**)]/MAO catalyst system. More recently, based on the iron complex [$FeCl_2$(**L87**)], Bryliakov et al. [181] observed similar active species when [$FeCl_2$(**L87**)] is activated with $AlMe_3$/$B(C_6F_5)_3$ or polymeric MAO (PMAO), and, at the same time, they found that these active species decay with time with the concomitant appearance of some new EPR-active iron species ($g = 2.08$, $\Delta\nu_{1/2} = 330$ G). These EPR signals were similar to those detected by Gibson and coworkers [80], which were assigned by the latter authors to iron(III), based on the observed Mössbauer data. On the contrary, Bryliakov et al. considered that these species should be assigned to a low-spin ($S = 1/2$) iron(I) complex, since the transformation of iron(II) precursors to iron(III) is unfeasible under the reductive conditions used (MAO containing $AlMe_3$). In order to explore the latter hypothesis, reactions of [$FeCl_2$(**L87**)] and [FeCl(**L87**)] with $AlMe_3$ were studied, being detected two new species **G** and **H**, as shown in Scheme 3.64. Both reactions yielded the same species, showing ^{1}H NMR resonances in the range of δ +430 to –300 (ppm), which is a much broader range of chemical shifts than that observed for [$FeCl_2$(**L87**)], but close to that of [FeCl(**L87**)]. It is worth to note that, in the latter complex, iron retains the +2 oxidation state, one electron being distributed over the tridentate ligand, to afford a complex of the type [$Fe^{(+)}$Cl(**L87**$^{(-)}$)], which has a $\mu_{\text{eff}} = 3.5$ BM, indicative of a high-spin Fe(II) ion ($S_{\text{Fe}} = 2$) coupled antiferromagnetically to the ligand radical ($S_{\text{L}} = 1/2$). The unpaired electron spin density on the pyridine part of the ligand leads to the paramagnetic contact shifts observed in the ^{1}H NMR spectra of such species. Hence, the species **G** and **H** were assigned to a heterobinuclear complex of the type [(**L87**$^{(-)}$)$Fe^{(+)}$(μ-X))(μ-Me)$AlMe_2$], where X = Cl (**G**) or Me (**H**). The presence of the bridging $AlMe_3$ within the species can be demonstrated by the reaction between [FeMe(**L87**)] and $AlMe_3$. Therefore, the intermediates

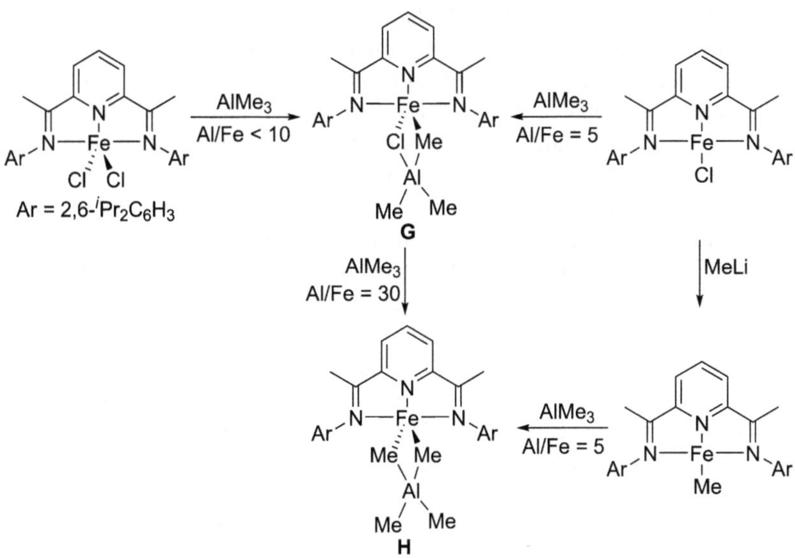

Scheme 3.64 Reactions of [$FeCl_2$(**L87**)] and [FeCl(**L87**)] with $AlMe_3$

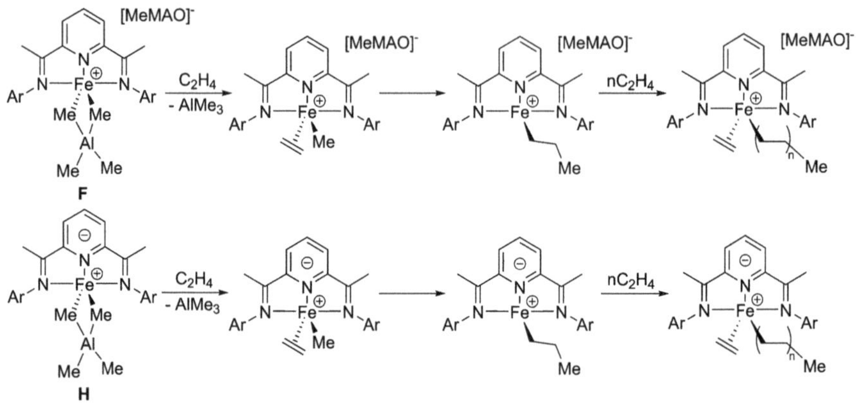

Scheme 3.65 A multicenter model for the initiation and propagation of the polymerization of ethylene promoted by iron BIP complex/MAO or $AlMe_3$ catalyst systems

previously assigned to [(**L84**)Fe^{II}(R)(μ-Me)$_2$AlMe$_2$] (R = Cl (**A**) or Me (**B**)) are actually alkyl-bridged heterobinuclear species [(**L84**$^{(-)}$)$Fe^{(+)}$(μ-Me)$_2$AlMe$_2$]. Based on these experimental results, Bryliakov et al. proposed a multiple active species model for the initiation and propagation steps of the polymerization of ethylene promoted by the BIP iron complex/MAO system, as schematically represented in Scheme 3.65.

The actual oxidation state of the iron center, in the catalytically active species, has been a subject of debate within the area of BIP iron polymerization catalysts, and so far, it remains a "black box". Based on Mössbauer and EPR spectroscopic studies, Gibson and coworkers [80] demonstrated that the divalent iron of the precatalyst is fully oxidized to trivalent iron, upon treatment with MAO. By using global and local descriptors of chemical reactivity and selectivity, Martínez et al. [81] theoretically compared the reactivities of iron(II) and iron(III) toward olefin polymerization, and revealed that the bare iron(III) is more active than iron(II), whereas, upon activation with MAO, iron(III) displays an activity comparable to that of iron(II). The latter presents a stronger delocalization of the charge, making the whole catalytic species of iron(II) much less efficient than that of iron(III). DFT calculations were performed by Raucoules et al. [83] to study the coordination and insertion of the first ethylene molecule, taking into consideration the 10 most reasonable active species which can be formed from the reaction of [$FeCl_2$(**L84**)] with MAO. All the possible coordination and insertion reactions for the two most reactive active species, i.e., the monomethylated iron(III) and iron(II), were evaluated using the B3LYP/LACVP** potential energy surface (PES), taking into account all possible iron spin states and several coordination modes of the ethylene molecule. Consequently, it was found that these reactions occur at the quintet PES of iron(II) and quartet PES of iron(III) species, respectively. Concerning the insertion reaction, the iron(III) species has more favorable reaction and activation enthalpies: iron(II), ΔH(298 K) = −14.1 kcal/mol and

$\Delta H^{\ddagger}$(298 K) = + 21.6 kcal/mol; iron(III), ΔH(298 K) = –22.8 kcal/mol and $\Delta H^{\ddagger}$(298 K) = + 10.0 kcal/mol. Assuming similar insertion barriers for the second insertion reaction (chain propagation), the β-hydrogen transfer reaction to the monomer (chain transfer) ($\Delta H^{\ddagger}$(298 K) = + 11.9 kcal/mol) is favored over further chain growth for the iron(II), whereas for the iron(III) chain growth and chain transfer reactions are apparently in competition ($\Delta H^{\ddagger}$(298 K) = + 10.0 and +12.1 kcal/mol, respectively). Founded on these results, Raucoules et al. concluded that the most active species in the catalytic system is that based on iron(III), which is likely expected to produce low molecular weight polyethylene, consistent with the low molecular weight fraction of the characteristic bimodal molecular weight distribution of the polyethylenes experimentally produced by the BIP iron complex/MAO catalyst system. A similar DFT theoretical study was carried out by Cruz et al. [82], further demonstrating the multicenter nature of the BIP iron catalyst system that exhibits different iron oxidation states. On the contrary, Bryliakov et al. [79] studied the reactions of [$FeCl_2$(**L85**)] and [$FeCl_3$(**L85**)] with MAO, using ^{1}H NMR and EPR spectroscopies, and demonstrated that the iron(III) species in the [$FeCl_3$(**L85**)]/MAO system is completely transformed into iron(II) species, which, according to these authors, are the actual active species in the olefin polymerization process. Their latest studies [181] revealed that iron(II) species can be further reduced to the iron(I) by the $AlMe_3$ present in MAO, and also can serve as active species for olefin polymerization (see above).

Undoubtedly, these ambiguities could be resolved by the preparation of well-defined, single-component BIP iron catalysts, e.g., BIP iron alkyl complexes that may allow the study of the fundamental transformations related to the chain initiation, growth and transfer. Several BIP iron monoalkyl complexes [FeMe(**L85**)]–[FeMe(**L87**)], [Fe(CH_2CMe_3)(**L86**)], [Fe(CH_2CMe_3)(**L87**)] and [FeMe(**L87**)]$^-$[Li(THF)$_4$]$^+$, and dialkyl complexes [Fe(CH_2SiMe_3)$_2$(**L86**)] and [Fe(CH_2SiMe_3)$_2$(**L87**)] have been prepared by Chirik [186, 187] and Gambarotta [188] groups, as shown in Scheme 3.66. Alternatively, Cámpora et al. [189] prepared the BIP iron dialkyl complexes [Fe(CH_2SiMe_3)$_2$(**L84**)], [Fe(CH_2SiMe_3)$_2$(**L85**)] and [Fe(CH_2SiMe_3)$_2$(**L158**)] by reaction of [Fe(CH_2SiMe_3)$_2$(py)$_2$] (py = pyridine) with the appropriate BIP ligands. Scott et al. [190] described a more detailed investigation in the preparation of these complexes and provided evidence for the chemical participation of the BIP ligand in the formation of byproducts. Treatment of [Fe(CH_2SiMe_3)$_2$(**L87**)] with $B(C_6F_5)_3$ or [$PhMe_2NH$][BPh_4] leads to the cationic compounds [Fe($CH_2SiMe_2CH_2SiMe_3$)(**L87**)]$^+$[$MeB(C_6F_5)_3$]$^-$ and [Fe(L)(CH_2SiMe_3)(**L87**)]$^+$[BPh_4]$^-$ (L = Et_2O or THF) [191]. Upon activation with MMAO, Al^iBu_3, $AlMe_3$, $ZnMe_2$ or $AlMe(OAr)_2$ (Ar = 2,4,6-$^tBu_3C_6H_2O$), the iron dialkyl complex [Fe(CH_2SiMe_3)$_2$(**L85**)] show high catalytic activities, whereas the reaction of this complex with $B(C_6F_5)_3$, under the same conditions, is inactive toward the polymerization of ethylene. The complexes [Fe($CH_2SiMe_2CH_2SiMe_3$)(**L87**)]$^+$[$MeB(C_6F_5)_3$]$^-$ and [Fe(Et_2O)(CH_2SiMe_3)(**L87**)]$^+$[BPh_4]$^-$ are single-component catalysts for the polymerization of ethylene, exhibiting activities comparable to that obtained with the corresponding [$FeCl_2$(**L87**)]/MAO system, originating high molecular weight polyethylenes with

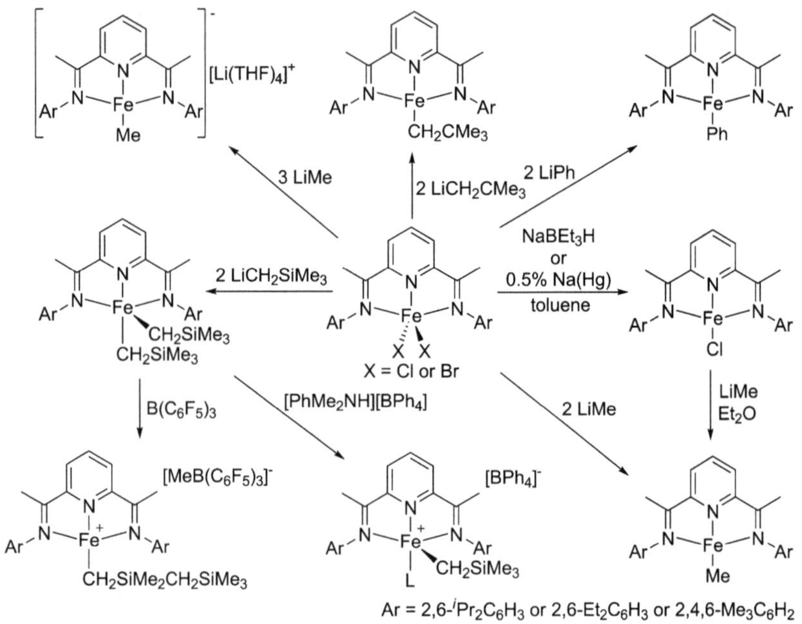

Scheme 3.66 Syntheses of BIP iron alkyl complexes

narrow molecular weight distributions. However, under the same conditions, $[Fe(THF)(CH_2SiMe_3)($**L87**$)]^+[BPh_4]^-$ is inactive, indicating that the dissociation of the donor ligand gives rise to the catalytically active species. The complex $[FeMe($**L87**$)]^-[Li(THF)_4]^+$ alone displays no catalytic activity when exposed to ethylene, but in combination with MAO is a highly active catalyst system. Although the abovementioned iron alkyl complexes are useful to evaluate the stability of the iron–carbon bond and probe the catalytic reactivity toward olefin polymerization, the lack of β-hydrogens in its structure, being used to protect the iron alkyl complex from β-hydrogen elimination, limits its relevance to the propagating species during olefin polymerization. More recently, to expand the number of well-defined single-component iron precatalysts and better mimic the propagating species, several BIP iron alkyl complexes of the type $[Fe(CH_2R)($**L87**$)]$ (R = Me, ^{n}Pr or ^{i}Pr), bearing β-hydrogens, have been synthesized either by salt metathesis reactions or by oxidative addition of alkyl bromide to $[Fe(N_2)_2($**L87**$)]$ (Scheme 3.67) [192]. It was demonstrated that these iron alkyl complexes are unstable and prone to decomposition, by hydrogen transfer from a methyl group belonging to the *iso*-propyl *o*-substituents of the aryl rings. It is worth to mention that a combination of structural, spectroscopic and theoretical studies revealed that all the aforementioned alkyl or chloro iron(I) and neutral iron(0) dinitrogen complexes should be virtually described as high-spin iron(II) species antiferromagnetically coupled to a ligand-centered radical anion or diradical anion, indicating that the corresponding reduction reactions involve a

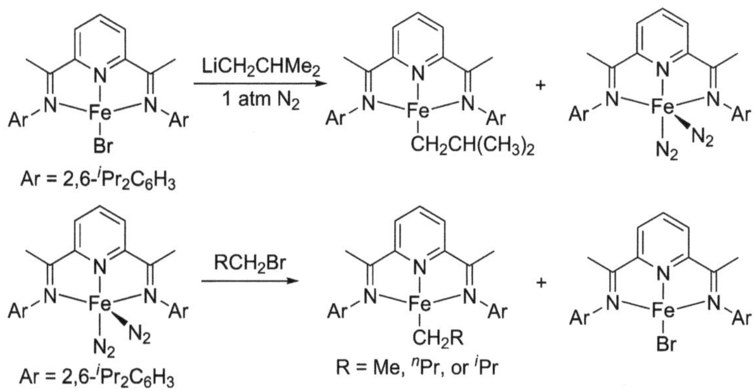

Scheme 3.67 Preparation of BIP iron alkyl complexes containing β-hydrogens

sequential electron transfer to the ligand conjugated π-system rather than to the metal center [43, 193, 194].

3.6.1.2 Cobalt Complexes

Parallel to the investigation on the BIP iron active species in olefin polymerization, the catalytically active species of the BIP cobalt catalysts have also been widely explored [64, 195–202]. By analogy with the theoretical study performed on BIP iron catalysts [182], Ziegler and coworkers [195] disclosed the first theoretical calculations on BIP cobalt catalyst systems. These authors used quantum-mechanics/molecular-mechanics coupling to investigate the mechanisms and energetics of ethylene polymerization promoted by [$CoCl_2$(**L87**)], assuming the polymerization is initiated by a cobalt(II) alkyl cation. Gibson [196], Gal [197] and Erker [64, 198] groups, prepared a series of BIP cobalt monoalkyl complexes [CoMe(**L87**)], [CoBz(**L87**)] (Bz = benzyl) and [Co(CH_2SiMe_3)(**L87**)], which, being diamagnetic, are substantially different from the corresponding iron alkyl complexes, making them suitable to be studied by NMR spectroscopy. They can be prepared either by direct alkylation of [$CoCl_2$(**L87**)] with methylmagnesium bromide (MeMgBr) or by alkylation of [CoCl(**L87**)] with methyllithium (LiMe), dibenzylmagnesium ($MgBz_2$) or ((trimethylsilyl)methyl)lithium ($LiCH_2SiMe_3$) (Scheme 3.68), and their reactivities investigated toward ethylene polymerization. These authors found that all the cobalt monoalkyl complexes are themselves inactive toward this catalytic reaction. On the contrary, upon activation with MAO, all the cobalt alkyl complexes show catalytic activities comparable with that obtained with the [$CoCl_2$(**L87**)]/MAO system. Treatment of [CoMe(**L87**)] and [CoCl(**L87**)] with $B(C_6F_5)_3$ or lithium tetrakis(pentafluorophenyl)borate (Li[$B(C_6F_5)_4$]) results in highly active ethylene polymerization catalysts. It is noteworthy that the reaction of complexes [CoR(**L87**)] (R = Me or Cl) with $B(C_6F_5)_3$, Li[$B(C_6F_5)_4$] or MAO leads to a cobalt(I) cation [Co(**L87**)]$^+$ that does

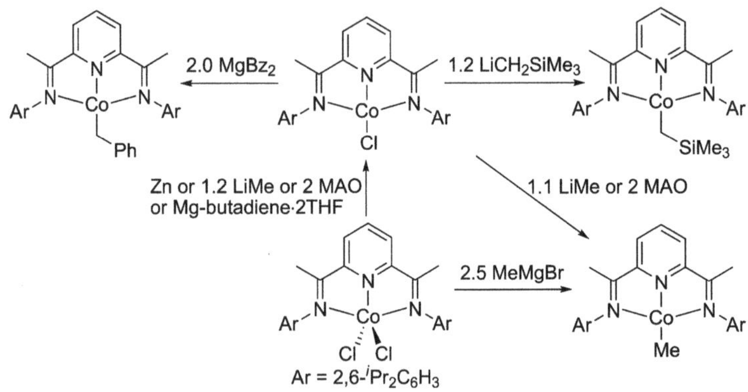

Scheme 3.68 Preparation of BIP cobalt alkyl complexes

not contain a cobalt–C(alkyl) bond. Mechanistic studies revealed that the polymerization initiation from this species involves the incorporation of alkyl groups from the cocatalyst, most likely involving the attack of a methyl anion (from the counteranion) on a cobalt-ethylene species [203].

Although the cobalt alkyl complexes [CoR(**L87**)] (R = ^{n}Pr or ^{n}Bu) are themselves not active toward ethylene polymerization, Gibson et al. [199] revealed they can be reacted with ethylene to give [CoEt(**L87**)] along with the formation of α-olefins, via a stepwise mechanism involving a cobalt-hydride intermediate rather than a concerted process involving direct β-H transfer to the monomer. These reactions gave insight into the chain transfer process commonly encountered in olefin polymerization. The theoretical calculations showed a remarkable agreement with this experimental result [200]. More recently, Talsi and coworkers [201, 202] investigated the activation process of the BIP cobalt complexes by various activators such as MAO, $AlMe_3$ and $AlMe_3$/[Ph_3C][$B(C_6F_5)_4$] using ^{1}H, ^{2}H, ^{19}F NMR and EPR spectroscopies. The results are shown in Scheme 3.69. Treatment of [$CoCl_2$(**L87**)] with MAO or $AlMe_3$/[Ph_3C][$B(C_6F_5)_4$] gives cobalt(II) complex [Co^{II}Me(S)(**L87**)]$^+$X$^-$ (S = solvent or vacancy, X = Me-MAO or $B(C_6F_5)_4$), in which the broad resonance at ca. 150 ppm in the ^{1}H NMR spectrum, is assigned to a Co–CH_3 group. At 20 °C, the complex [Co^{II}Me(S)(**L87**)]$^+$X in toluene gradually transforms into cobalt(I) complexes [Co^I(S)(**L87**)]$^+$X after several days, the ^{1}H NMR spectra of which are consistent with typical diamagnetic cobalt(I) species. On the contrary, the activation of [$CoCl_2$(**L87**)] with $AlMe_3$ rapidly leads to diamagnetic cobalt(I) complexes [(**L87**)Co^I(μ-R)(μ-Me)$AlMe_2$] (R = Me or Cl), which are unstable and rapidly (within several minutes) transform into the paramagnetic aluminum complex [$Al^{(+)}Me_2$(**L87**$^{(\cdot-)}$)] with a radical anion ligand, exhibiting an EPR signal at $g = 2.003$. Interestingly, in the presence of ethylene, reaction of [$CoCl_2$(**L87**)] with $AlMe_3$ yields a diamagnetic ethylene cobalt(I), which is assigned to the ion pair [Co^I(C_2H_4)(**L87**)]$^+$[$AlMe_3Cl$]$^-$. Upon consumption of ethylene, polyethylene is generated and the cobalt complex evolves to [(**L87**)Co^I(μ-Cl)(μ-Me)AlMe(R)] (R = Me or Cl). By analogy with

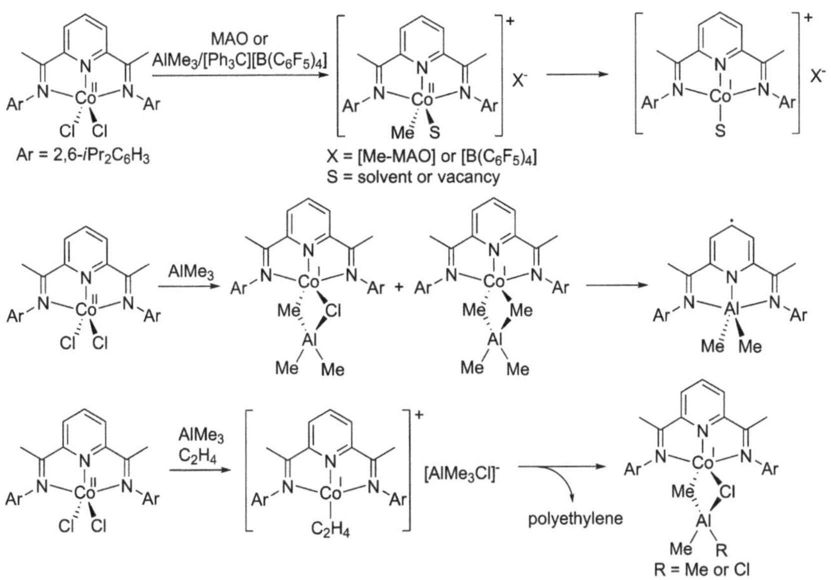

Scheme 3.69 Reactions of the BIP cobalt complex [$CoCl_2$(**L87**)] with various activators

iron(I) or iron(0) complexes [43, 193, 194], cobalt(I) alkyl or chloro complexes (e.g., [CoMe(**L87**)] and [CoCl(**L87**)]) are regarded as a low-spin cobalt(II) species antiferromagnetically coupled to a ligand-centered radical anion [204], whereas neutral cobalt(0) dinitrogen complexes (e.g., [CoN_2(**L84**)] and [CoN_2(**L86**)]), being formed by reduction of the corresponding cobalt(II) dichlorides with sodium amalgam, are virtually low-spin cobalt(I) complexes antiferromagnetically coupled with a ligand-centered radical anion [205].

3.6.2 Propagation and Chain Transfer

Based on a systematic study of the effects of structural modifications of the pre-catalyst and of the reaction conditions on the catalytic activity and polymer properties, an olefin polymerization mechanism for the BIP iron or cobalt catalyst system was put forward by Gibson and coworkers (Scheme 3.70) [53], which actually is not significantly different from those proposed for other classes of olefin polymerization transition-metal catalysts. Assuming that the active species during the olefin polymerization is a cationic metal alkyl complex, a Cossee-type propagation (**a**) is presumed, involving migratory insertion of ethylene into a metal–alkyl (polymer chain) bond, in which the rate of propagation has a first-order dependence on ethylene, in agreement with the observed experimental results. Three chain transfer pathways were proposed. Both β-hydrogen transfer to the monomer (**c**) and β-hydrogen elimination to the metal (**d**) produce the polymer

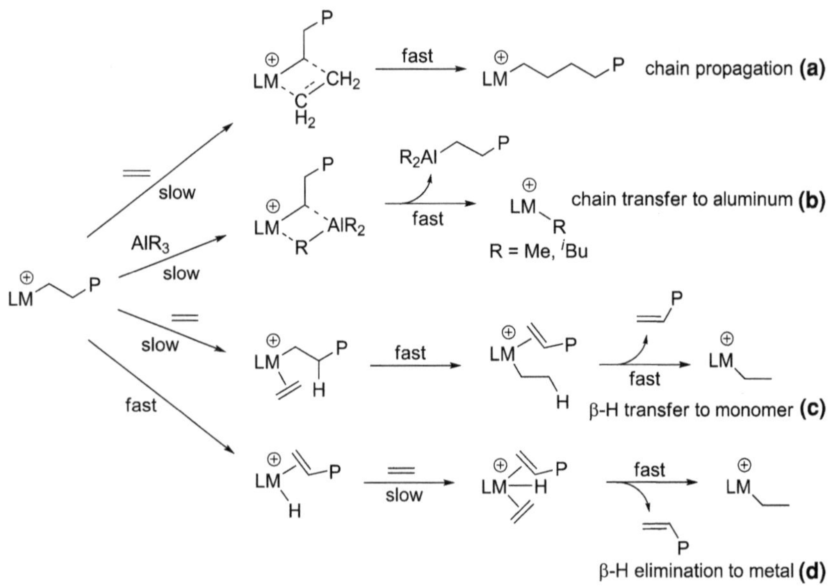

Scheme 3.70 Proposed chain propagation (**a**) and chain transfer (**b–d**) pathways

with unsaturated vinyl end groups and are bimolecular, i.e., first order in monomer and precatalyst, in agreement with the experimental facts. The latter two chain transfer pathways are kinetically indistinguishable.

The chain transfer to aluminum (**b**) involves the formation of an alkyl-bridged Fe–Al species, which induces an exchange of the growing polymer chain for a new alkyl group, generating fully saturated polymer chains. The ^{13}C NMR analysis of the low molecular weight fraction of the typically obtained bimodal polyethylenes, which can be separated from the bulk polymer with toluene Soxhlet extraction, shows it consists of almost fully saturated end groups, indicating that chain transfer to aluminum is responsible for its origin. Combined DFT/MM (QM/MM) theoretical calculations [182] showed that the rate-determining step for both chain transfer and propagation is the uptake of ethylene by the cationic iron(II)–alkyl complex, in which the C_α atom of the alkyl (chain) occupies an axial position above the iron–nitrogen coordination plane and a β-agostic hydrogen binds to the metal center. Chain propagation begins through the backside approach of ethylene, from the same side of the iron–nitrogen coordination plane as C_α, leading to the formation of a π-complex from which the insertion takes place. The dominant chain transfer path is a β-H transfer from the polymer chain to the incoming monomer. Steric bulk helps suppress or reduce the formation of the chain transfer precursor by introducing a coordination barrier, as ethylene would sense the repulsion of the *iso*-propyl groups in its path toward the metal center, from above the iron–nitrogen plane *trans* to C_α (frontside). Gould [183] and Morokuma [184] groups performed two comparable theoretical studies. These results also matched the chain propagation and chain transfer mechanisms proposed in Scheme 3.70.

Recently, Tomov et al. [206] used C_2H_4/C_2D_4 co-oligomerization experiments, which are a powerful tool to differentiate between a metallacyclic and a Cossee-type chain growth mechanism. No H/D scrambling should be observed for a metallacyclic mechanism, whereas for a Cossee-type mechanism, similar rates of chain propagation and chain transfer (β-H elimination) would result in rapid H/D scrambling of the C_2H_4/C_2D_4 feed. These experiments confirmed the Cossee-type mechanism is valid for the BIP iron and cobalt catalysts.

3.6.3 Multiple Active Sites

Compared with the BIP cobalt catalyst system, which commonly yields polyethylenes with narrow molecular weight distributions, one of the most pronounced features of the BIP iron catalyst system is the formation of polymers with bimodal molecular weight distributions, which remarkably varies with the nature of the activators and the molar ratio of activator to iron precatalyst. There are two likely explanations to account for this bimodality. First, there may be multiple active species operating at the same time, and second, the polymerization mechanism involving chain propagation and chain transfer may change during the course of polymerization. A number of experimental or theoretical results [53, 95, 207–212] support these two explanations, thereby the nature of bimodality so far is still unclear. From the kinetic studies, Barabanov et al. [207] observed that two types of active species are present in each of the systems [$FeCl_2$(**L84**)]/MAO and [$FeCl_2$(**L84**)]/Al^iBu_3. The highly active but unstable active species are formed at the early stages of polymerization, yielding low molecular weight polyethylenes, which are transformed into more stable but less active species, giving high molecular weight polyethylenes. Kissin et al. [95] studied the relationship of broad molecular weight distribution produced by the BIP iron catalyst and the active species, by using a Flory–Schulz function fitting, and further supported the abovementioned assumption of active site transformation. Mao and coworkers [208, 209] systematically characterized the compositional heterogeneity in the polyethylenes formed with the BIP iron complex/AlR_3 (R = Et, ^{i}Bu or ^{n}Hex) systems by using gel permeation chromatography/size-exclusion chromatography (GPC/SEC), ^{13}C NMR spectroscopy, solvent gradient elution fractionation (SGEF), crystallization analysis fractionation (CRYSTAF), and differential scanning calorimetry (DSC). The results indicate that at least two different active species are present in the polymerization of ethylene. The aforementioned bridged heterodinuclear Fe–Al active species model, suggested by Talsi and coworkers [179, 181, 185], and the multicenter model with different metal oxidation states, raised by Cruz [82], Martínez [81] and Raucoules [83], are consistent with the assumption of multiple active sites.

In contrast, Gibson et al. [53, 210, 211] considered that the bimodality results from the variation of the polymerization mechanism during the course of the reaction. Studies on the effect of different MAO concentrations on the polymerization behavior of [$FeCl_2$(**L87**)] show that an increase in the MAO concentration leads to an increase in the lower molecular weight fraction of the bimodal polyethylene, which can be separated by toluene Soxhlet extraction. As mentioned in the last subsection, the lower molecular weight fraction consists of fully saturated polyethylene chains, its formation being ascribed to the chain transfer to aluminum, most likely due to the free $AlMe_3$ present in MAO. With a view to a better understanding of this chain transfer process, a number of alkyl aluminum reagents including trialkylaluminum (AlR_3, R = Et, Oct, ^{i}Bu), diethylaluminum ethoxide ($AlEt_2$(OEt)), chlorinated aluminum alkyls (e.g., $AlMe_2Cl$ and $AlEt_2Cl$), and other alkyl metal reagents, including *n*-butyllithium (Li^{n}Bu), di-*n*-butylmagnesium (Mg^{n}Bu$_2$), triethylborane (BEt_3), trialkylgallium (GaR_3, R = Me, Et, ^{n}Bu), tetramethyltin ($SnMe_4$), tetraethyllead ($PbEt_4$) and diethylzinc ($ZnEt_2$), were investigated. The results show $AlEt_3$, $GaMe_3$ and $ZnEt_2$ are the most efficient chain transfer agents, $AlEt_3$ giving lower molecular weight polyethylene with Schulz–Flory distributions, along with a higher molecular weight polyethylene, which indicates a competing chain transfer to aluminum relative to chain propagation. Conversely, $ZnEt_2$ affords low molecular weight product with a Poisson distribution, indicating the rate of chain transfer is far faster than the rate of propagation, and the molecular weight distribution obtained by $GaMe_3$ fit neither a Poisson nor a Schulz–Flory distribution. Therefore, Gibson et al. concluded that alkyl aluminum cocatalysts work as chain transfer agents in the BIP iron catalyst system, leading to a polyethylene with a bimodal molecular weight distribution, due to a competing chain transfer to aluminum relative to chain propagation, while the iron complex/MAO/$ZnEt_2$ system can be regarded as a catalyzed chain growth process, featuring the formation of product with a Poisson distribution. In conjugation with a chain displacement agent ([Ni(acac)$_2$], acac = acetylacetonate), the latter catalyst system can be used to convert ethylene to the technologically desired linear α-olefins with a Poisson distribution, as displayed in Scheme 3.71. Similarly, Wang and coworkers [212–217] studied the effect of oligomeric alkylaluminoxane, formed by partial hydrolysis of trialkylaluminum (e.g., $AlEt_3$, Al^{i}Bu$_3$), on the bimodal molecular weight distribution obtained by [$FeCl_2$(**L84**)]/MAO or [$FeCl_2$(**L87**)]/MAO systems, and found that end group-saturated low molecular weight fraction of the bimodal product is formed at the polymerization early stages and results from chain transfer to aluminum.

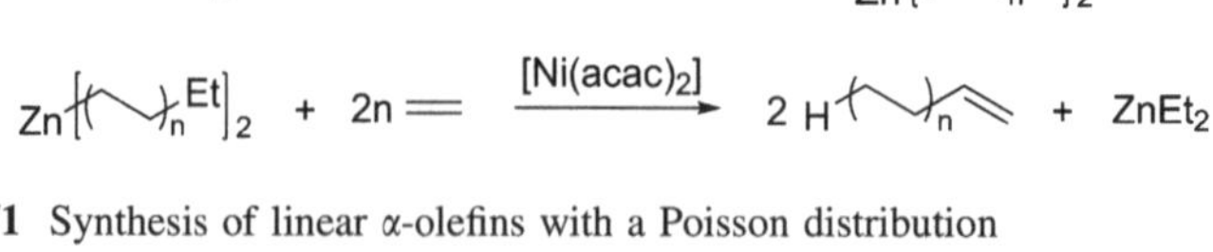

Scheme 3.71 Synthesis of linear α-olefins with a Poisson distribution

3.7 Cocatalysts/Activators

MAO or MMAO have been widely used as cocatalysts to activate the BIP iron and cobalt complexes for the polymerization of olefins, but a number of other alkyl aluminum compounds, including $AlMe_3$ [32, 179, 185, 189, 201, 218, 219], $AlEt_3$ [209, 218, 220], Al^iBu_3 [32, 185, 189, 201, 207, 209, 218–220], Al^nHex_3 [209, 218], Al^nOct_3 [32, 185, 218] and $AlEt_2Cl$ [209], have been screened. By analogy with the roles of MAO or MMAO in catalyst systems, alkyl aluminum compounds can play as alkylating agents, as halide or alkyl abstractors, thus giving rise to cationic species, as reducing agents, or as a combination of the three, in order to activate and stabilize the precatalysts. Additionally, they play key roles in the generation of the active species (e.g., bridged heterodinuclear Fe–Al active species [179, 181, 185]) and in the chain transfer process (e.g., chain transfer to aluminum [53, 210, 211, 216]), which substantially contribute to the formation of broad/bimodal polyethylenes. More recently, some unexpected ligand-participating reactions of alkyl aluminum compounds with the free BIP ligand or BIP iron complexes (such as, ligand alkylation [221], tricyclic reaction [222] and concurrent transmetalation and 4-electron reduction [223]) have attracted interest, and are examples of the complexity of the BIP iron and cobalt catalyst systems. Based on different BIP iron complexes such as [$FeCl_2$(**L87**)], [$FeCl_2$(**L84**)] and [$FeCl_2$(**L85**)], Cramail [218], Bryliakov [185], Semikolenova [32] and Mao [209] groups systematically compared the catalytic behaviors of various alkyl aluminum compounds. In the case of [$FeCl_2$(**L87**)], it was found that the catalytic activity toward ethylene polymerization, at an aluminum-to-iron ratio of 500, increases in the order $AlMe_3 < Al^nOct_3 < Al^iBu_3 < AlEt_3 < Al^nHex_3$, with MAO still displaying the highest activity. The same tendency, i.e., $Al^iBu_3 < AlEt_3 < Al^nHex_3$ was observed for [$FeCl_2$(**L85**)], and $AlEt_2Cl$ revealed to be inactive in this case. For [$FeCl_2$(**L84**)], a reverse tendency, i.e., $Al^nOct_3 > AlMe_3 > Al^iBu_3$, was obtained. Cramail and coworkers [218] also found that in the [$FeCl_2$(**L87**)]/trialkylaluminum catalytic systems, all the primary trialkylaluminum activators, $AlMe_3$, $AlEt_3$, Al^nHex_3 and Al^nOct_3, essentially yield polyethylenes with bimodal molecular weight distributions containing substantial amounts of low molecular weight fractions, except in the case of Al^iBu_3, which leads to polyethylenes with narrow molecular weight distributions ($M_w/M_n = 3.2–4.9$). The occurrence of bimodality was ascribed to extensive primary alkyl-participating chain transfer reactions to aluminum, which was further confirmed by a study on the catalytic behavior of a mixture of $AlMe_3/Al^iBu_3$, in which an enhanced activity was observed relative to each trialkylaluminum compound, and the molecular weight distributions of the resulting polyethylenes narrowed down from bimodal to monomodal with an increase in the amount of Al^iBu_3. In contrast, the [$FeCl_2$(**L84**)]/Al^iBu_3 and [$FeCl_2$(**L84**)]/Al^nOct_3 systems give rise to broad but not bimodal molecular weight distributions of polyethylenes with equimolar saturated and vinyl unsaturated end groups, and Semikolenova et al. [32] considered that a set of different active species are formed in these catalytic systems. Unlike the results observed by Cramail and coworkers [218], Mao and coworkers [209] found that, in the $[FeCl_2($**L85**$)]$/trialkylaluminum catalyst systems, besides the primary trialkylaluminum compounds $AlEt_3$ and Al^nHex_3, also Al^iBu_3 leads to bimodal polyethylenes composed of branched low molecular weight and linear high molecular weight fractions, this bimodality being ascribed to different active species.

Wang and coworkers [212–216, 220, 224, 225] reported a series of tetra-alkylaluminoxanes, $R^1R^2AlOAlR^3R^4$ (R^1, R^2, R^3, R^4 = Et or ^{i}Bu), which are formed by partial hydrolysis of $AlEt_3$ or Al^iBu_3, or a mixture of both, and investigated their catalytic performance in ethylene polymerization, when mixed with precatalysts $[FeCl_2($**L84**$)]$ and $[FeCl_2($**L87**$)]$, as an alternative to the use of MAO. It was shown that tetraalkylaluminoxanes could serve as promising activators to promote ethylene polymerization with high activities comparable with those of MAO, affording polyethylenes with molecular weight distributions varying from monomodal to bimodal. An increase of the amount of Al–Et groups in tetraalkylaluminoxanes results in an increase both in activity and in the low molecular weight fraction of the bimodal distribution, along with a corresponding decrease in the high molecular weight fraction. Additionally, Wang and coworkers [216] found that steric bulk on the aryl substituents of BIP iron complexes also has a substantial effect on the chain transfer reaction to aluminum, a lower steric bulk implying a faster chain transfer to aluminum, e.g., $[FeCl_2($**L84**$)]$, containing methyl *o*-substituents ($k_{trA}/k_p = 2.48$, where k_{trA} and k_p are the rate constants of chain transfer to aluminum and chain propagation, respectively) vs. $[FeCl_2($**L87**$)]$, with *iso*-propyl *o*-substituents ($k_{trA}/k_p = 0.12$). The partial replacement of the aluminoxane aluminum units by boron units, which was carried out by the controlled reaction of trialkylaluminums with boronic acids (e.g., 4-fluorophenylboronic and *n*-butylboronic acids) and water, gives rise to new activators (e.g., BTEAO [214], BEBAO [214] and FB [224]) that showed to dramatically suppress the chain transfer reactions to aluminum and to favor high temperature stability of the catalytic systems, in comparison with MAO.

Semikolenova et al. [201] studied the effect of the free residual $AlMe_3$ of MAO on the catalytic activities of $[FeCl_2($**L84**$)]$ or $[CoCl_2($**L84**$)]$/MAO systems, and found that the highest activity is reached by MAO when completely purified from free $AlMe_3$. The effect of addition of alkyl aluminum compounds to the BIP iron complex/MAO system has also been studied [211]. It was concluded that the addition of $AlMe_3$, $AlEt_3$ and Al^nOct_3 to the $[FeCl_2($**L87**$)]$/MAO catalyst system leads to a slight decrease in activity and a remarkable broadening in molecular weight distribution of the resulting polyethylene, whereas for Al^iBu_3 a reverse trend is observed, i.e., the activity slightly increases and the molecular weight distribution narrows down. Diethyl aluminum ethoxide ($AlEt_2(OEt)$) causes a slight decrease in the activity but has little effect on the molecular weight distribution, whereas chlorinated alkyl aluminum compounds, such as $AlMe_2Cl$ and $AlEt_2Cl$, deactivate the catalyst system.

The catalyst systems $[FeCl_2($**L84**$)]/AlMe_3$ and $[FeCl_2($**L84**$)]/AlMe_3/B(C_6F_5)_3$ showed showed comparable activities toward ethylene polymerization, affording polyethylenes with similar molecular weights [179], whereas the [$FeCl_2$(**L84**)]/Al^iBu_3/[Ph_3C][$B(C_6F_5)_4$] system is one order of magnitude less active for ethylene polymerization than the [$FeCl_2$(**L84**)]/Al^iBu_3 one [32]. Babik and Fink [86] found that activation of [$FeCl_2$(**L89**)] with [Ph_3C][$B(C_6F_5)_4$] and subsequent treatment with Al^iBu_3 or $AlEt_3$ generates a highly active catalyst system toward propylene polymerization, which is two orders of magnitude more active than the [$FeCl_2$(**L89**)]/MAO system.

Deffieux and coworkers [226] reported a series of new activators, which are prepared by the reaction of carboxylic acids, such as acetic acid, benzoic acid and benzoic acid derivatives, with trialkylaluminums (e.g., $AlMe_3$, $AlEt_3$ or Al^iBu_3). In the presence of [$FeCl_2$(**L87**)], these activators show similar or higher catalytic activities toward ethylene polymerization in comparison with trialkylaluminum or MAO, and the highest activity (1.73×10^6 g/(mol h bar) was reached with the cocatalyst obtained by the reaction of benzene-1,3,5-tricarboxylic acid (BTCA) with $AlMe_3$ at a ratio of [$AlMe_3$]/[BTCA] = 6.

3.8 Monomers and Comonomers

With a view to expanding their potential applications, the abilities of the BIP iron and cobalt catalysts to polymerize other alkene monomers, such as propylene, 1-butene, 1-hexene and norbornene, besides ethylene, have been widely investigated.

Brookhart and Small [51] systematically studied the polymerization of propylene promoted by the bulkier 2,6-bis(arylimino)pyridine iron complex/MMAO systems, and the corresponding mechanisms of chain initiation, propagation and transfer were discussed at length. Upon activation with MMAO, the 2,6-bis-(arylimino)pyridine iron complexes (e.g., [$FeCl_2$(**L84**)], [$FeCl_2$(**L87**)], [$FeCl_2$(**L153**)] and [$FeCl_2$(**L171**)]) show moderate catalytic activities toward the polymerization of propylene, forming isotactic polypropylenes with low molecular weights (M_n = 600–7000 g/mol) and typical Schulz–Flory distributions ($M_w/M_n \approx 2.0$). The highest activity (1.64×10^6 g/(mol h bar)) is obtained by [$FeCl_2$(**L171**)], at –20 °C. The aldimine iron complexes are substantially less active than the corresponding ketimine analogues, e.g., [$FeCl_2$(**L10**)] (2.95×10^4 g/(mol h bar), at –20 °C) vs. [$FeCl_2$(**L89**)] (7.28×10^5 g/(mol h bar), at –20 °C), respectively. The precatalyst molecular structure does not show a clear influence on the catalytic activities of propylene polymerization. However, the steric bulk of the *o*-substituents has a similar effect on the molecular weights of the resulting polypropylenes to that shown in the polymerization of ethylene, i.e., an increase in the *o*-substituents bulkiness leads to an increase in molecular weight. For instance, the symmetrical precatalyst [$FeCl_2$(**L87**)], containing 2,6-di-*iso*-propylphenyl groups, affords polypropylene, at –20 °C, with a M_n of 6500 g/mol, while the slightly less bulky unsymmetrical [$FeCl_2$(**L173**)], with 2,6-di-*iso*-propylphenyl and 2-methyl-6-*iso*-propylphenyl groups, gives a polymer with a M_n of 5700 g/mol. A further reduction in the steric bulk yields lower molecular weight polymers, such as in the cases of the symmetrical precatalyst [$FeCl_2$(**L89**)], with two 2-methyl-6-*iso*-propylphenyl groups, and the unsymmetrical [$FeCl_2$(**L171**)], with 2,6-di-*iso*-propylphenyl and 2,6-dimethylphenyl groups, giving rise to polymers with M_n of 4100 and 5600 g/mol, respectively. The results obtained with the symmetrical [$FeCl_2$(**L84**)], with 2,6-dimethylphenyl groups, demonstrates this trend conclusively by yielding polypropylene with a M_n of 1800 g/mol. An end group analysis of these polymers indicates that the polymerization proceeds via a secondary enchainment mechanism followed by β-hydrogen elimination, leading to the exclusive formation of 1-propenyl end groups. In all cases, highly isotactic polypropylenes ([*mmmm*] = 55–67%) are obtained, in which the regioregularity decreases with a decrease in the steric bulk of the aryl rings *o*-substituents. ^{13}C NMR spectroscopy demonstrates that the isotacticity results from a chain-end control mechanism. Similar experimental results were obtained by Pellecchia et al. [88] using the [$FeCl_2$(**L87**)]/MAO catalytic system, at a polymerization temperature as high as 50 °C. Babik and Fink [86] found out that when [$FeCl_2$(**L89**)] is activated with [Ph_3C][$B(C_6F_5)_4$], and subsequently treated with Al^iBu_3 or $AlEt_3$, it shows a remarkably higher catalytic activity toward propylene polymerization than that obtained with MAO. Different aliphatic end groups in the polymer were identified by ^{13}C NMR spectroscopy, which arise from the different alkyl aluminum compounds, as a consequence of a catalytic cycle based on an iron hydride species.

Upon activation with MMAO, iron complexes based on less bulky 2,6-bis-(arylimino)pyridine ligands, e.g., [$FeCl_2$(**L14**)] (phenyl), [$FeCl_2$(**L18**)] (2,4-dimethylphenyl) and [$FeCl_2$(**L21**)] (2-ethylphenyl), convert propylene into oligomers with a predominant composition of dimers and trimers. These oligomers, in particular the dimers and trimers, can be identified using analytical gas chromatography, and structurally characterized by ^1H and ^{13}C NMR spectroscopies after separation by preparative gas chromatography [87]. The structural information about the dimers and trimers provides important evidence for the chain propagation and chain transfer mechanisms. Babik and Fink [87] suggested that the first insertion of propylene into an iron hydride species could occur competitively in a primary (1,2-) or a secondary (2,1-) fashion, the bulkier *o*-substituents on the aryl rings favoring the former, while the second insertion of propylene into the Fe–C bonds is dominated by the secondary enchainment mechanism. Tellmann et al. [74] tested the catalytic activities of the fluorine-containing 2,6-bis(arylimino)pyridine iron and cobalt complexes, e.g., [$FeCl_2$(**L46**)], [$FeCl_2$(**L106**)], [$CoCl_2$(**L46**)] and [$CoCl_2$(**L106**)], in the polymerization of propylene, in the presence of MAO, and found that none of the iron complexes are active, whereas both cobalt complexes are active, being [$CoCl_2$(**L106**)] the most active one (1.14×10^6 g/(mol h bar), at –20 °C). The oligomers formed are mainly the linear dimers 1-hexene and 2-hexene, indicating a 1,2-insertion mechanism starting from a cobalt hydride species, which is followed by 2,1-insertion, and finally a β-hydrogen elimination to regenerate the cobalt hydride active species. Analogous highly linear dimers of propylene (>99%) can also be formed by MMAO-activated cobalt complexes [$CoCl_2$(**L15**)], [$CoCl_2$(**L21**)] and [$CoCl_2$(**L23**)] [90]. When activated with MMAO, [$FeCl_2$(**L23**)], with 2-*iso*-propylphenyl aryl groups, and [$FeCl_2$(**L130**)], with 1-anthracenyl groups, were found to be moderately active toward propylene polymerization with moderate activities (ca. 10^5 g/(mol h bar)), giving atactic and isotactic-enriched oligomers, respectively [78].

Treatment of [$FeCl_2$(**L15**)] with MAO or MMAO leads to an active catalyst for the oligomerization of 1-butene, affording predominantly dimers (> 80%), the majority of which are linear, along with a small amount of methyl-branched products [85]. A DFT theoretical study demonstrated [89] that these linear dimers are formed by a primary insertion of 1-butene into an iron hydride species followed by a secondary insertion. Compared with the iron compounds, the cobalt complexes [$CoCl_2$(**L15**)], [$CoCl_2$(**L21**)] and [$CoCl_2$(**L23**)] show much lower catalytic activities toward 1-butene oligomerization, but a much higher content of linear dimers (>97%) are obtained, upon activation with MMAO. Unlike the iron catalysts, in which only a small percentage of the initial amount of 1-butene is isomerized to 2-butene, the cobalt catalysts show a considerably higher yield in this reaction, as evidenced by the approximately equal levels of dimers and 2-butene achieved when 1-butene is oligomerized [90]. It was found that the isomerization occurs with an initial secondary (2,1-) insertion of olefin into a cobalt hydride species, followed by β-hydrogen elimination occurring from the β-methylene group. For iron catalysts, initial 2,1-insertions tend to produce branched dimers, indicating that the propagation is preferred, whereas cobalt catalysts undergo chain transfer after a 2,1-insertion, resulting in high amounts of both linear dimers and 2-butene.

Souane et al. [84] tested the catalytic abilities of several iron and cobalt complexes based on bulky 2,6-bis(arylimino)pyridine ligands such as **L84**, **L26**, **L109** and **L115** toward the polymerization of 1-hexene, in combination with MAO, revealing that all these catalyst systems are inactive. Further copolymerization tests of ethylene with 1-hexene show extremely low degrees of incorporation of 1-hexene (ca. 3.5 mol%). This result is similar to that observed by Brookhart and Small [15] in the copolymerization of ethylene and 1-pentene, demonstrating that the BIP iron and cobalt catalyst systems are not able to copolymerize ethylene with higher α-olefins. A combined QM/MM study on the BIP iron complex-catalyzed ethylene/1-hexene copolymerization further confirmed this result [227]. On the other hand, the less sterically encumbered 2,6-bis(arylimino)pyridine iron catalysts, e.g., [$FeCl_2$(**L14**)], [$FeCl_2$(**L15**)], [$FeCl_2$-(**L18**)], [$FeCl_2$(**L21**)] and [$FeCl_2$(**L129**)], were found to be highly active for the oligomerization of 1-hexene, forming dimers (>80%) that consist primarily of linear products (>70%) and a small amount of methyl-branched products (ca. 30%) [85]. On the contrary, in the cobalt-catalyzed oligomerization of 1-hexene, a remarkable degree of isomerization, leading to 2-hexene and 3-hexene, was observed, along with the dimerization products, the ratio of dimerization to isomerization dramatically depending on the activator used [90]. For example, when cobalt catalysts (e.g., [$CoCl_2$(**L15**)], [$CoCl_2$(**L21**)] and [$CoCl_2$(**L23**)]) are activated with MMAO, the dimerization and isomerization of 1-hexene are competitive, whereas isomerization occurs almost exclusively when $AlEt_2Cl$ is used.

The BIP iron and cobalt complexes have also been found to be active toward the polymerization of conjugated/non-conjugated dienes such as 1,3-butadiene, isoprene and 1,6-heptadiene. Gong et al. [71] observed that iron and cobalt complexes [$FeCl_2$(**L14**)], [$FeCl_3$(**L14**)] and [$CoCl_2$(**L14**)] are highly active toward the polymerization of 1,3-butadiene upon activation with MAO, forming *trans*-1,4-/*cis*-1,4-polybutadienes along with traces of 1,2-polybutadiene. Iron complexes [$FeCl_2$(**L14**)] and [$FeCl_3$(**L14**)] primarily give *trans*-1,4-polybutadienes, while for the cobalt complex [$CoCl_2$(**L14**)], the selectivity is strongly dependent on the MAO/Co molar ratio. For instance, when a molar ratio of 50 is used, the polymer formed is essentially *trans*-1,4 (ca. 94%), while a significant increase in the *cis*-1,4 selectivity occurs (reaching 79%), along with a corresponding decrease in the *trans*-1,4 selectivity to 18%, when the MAO/Co molar ratio is increased to 75 or 100. Nakayama et al. [156] reported that several terpyridine iron complexes, such as $[Fe(\mathbf{L401})_2]^{2+}[FeCl_4]_2^{2-}$, [$FeCl_3$(**L401**)] and [$FeCl_3$(**L404**)], or the *N*-(2-pyridylmethyl)-2-hydroxy-3,5-di-*tert*-butylbenzaldimine-based iron complexes [$FeCl_2$(**L496**)] and [FeCl(**L496**)], were also highly active toward the (co)polymerization of isoprene and 1,3-butadiene, upon activation with MMAO, with approximately 99% conversion after 3 or 12 h. In the case of the homopolymerization of isoprene, the polymer obtained is mainly the 3,4-polyisoprene (50–85%), while for the homopolymerization of 1,3-butadiene, the less bulky [$FeCl_3$(**L401**)] affords *trans*-1,4 products (>90%), and the much bulkier [$FeCl_3$(**L404**)], as well as [$FeCl_2$(**L496**)] and [FeCl(**L496**)], yield almost equimolar amounts of *cis*-1,4, *trans*-1,4 and 1,2 products. All the iron complexes copolymerize isoprene and 1,3-butadiene to generate random copolymers.

Osakada and coworkers [228–231] reported that, upon activation with MMAO, the 2,6-bis(arylimino)pyridine iron and cobalt complexes can catalyze the cyclopolymerization of 1,6-heptadiene with high conversions (>60% for the iron catalysts, and >40% for the cobalt ones), forming polymers with *trans*- or *cis*-fused 1,2-cyclopentanediyl groups. The ratio of *cis*- to *trans*-fused five-membered rings in this polymer is dependent upon the steric bulk of the aryl substituents of iron and cobalt complexes. The less sterically encumbered aryl-BIP iron complexes, e.g., [$FeCl_2$(**L84**)], with *o*-methyl substituents, [$FeCl_2$(**L86**)], with *o*-ethyl substituents, [$FeCl_2$(**L108**)], with *o*-chloro substituents, [$FeCl_2$(**L115**)], with *o*-methyl and *p*-bromo substituents, and [$FeCl_2$(**L120**)], with *o*-methyl and *p*-methoxy substituents, yield polymers with higher contents of *cis*-five-membered rings than *trans* ones (>95:5), and high molecular weights (>6,000 g/mol), while the iron complexes with much bulkier *o*-substituents, such as [$FeCl_2$(**L87**)], with *o*-*iso*-propyl substituents, [$FeCl_2$(**L95**)], with *o*-cyclohexyl substituents, [$FeCl_2$(**L26**)], with a single *o*-*tert*-butyl substituent in each of aryl rings, [$FeCl_2$(**L170**)], with *o*-*iso*-propyl substituents on one of the aryl rings and a single *o*-*tert*-butyl substituent on the other, afford polymers with mixed *cis*- and *trans*-five-membered rings and lower molecular weights. The cobalt complexes, such as [$CoCl_2$(**L84**)], [$CoCl_2$(**L87**)] and [$CoCl_2$(**L97**)], give rise exclusively to polymers with *trans*-five-membered rings, regardless of the steric bulk of the *o*-substituents. These trends can also be observed in the cyclopolymerization of substituted 1,6-heptadienes, such as 4-phenyl-1,6-heptadiene, 4-tri-*iso*-propylsiloxy-1,6-heptadiene and 4-*tert*-butyldimethylsiloxy-1,6-heptadiene. The copolymerization of 1,6-heptadiene with ethylene catalyzed by iron and cobalt complexes have also been investigated, being reported that iron complexes lead only to a mixture of the corresponding homopolymers, whereas copolymers with variable incorporation of 1,6-heptadiene (3–50%) can be obtained with cobalt complexes.

Osakada and coworkers [232, 233] also found that the cobalt complex [$CoCl_2$(**L87**)] is able to catalyze the polymerization of 2-aryl-1-methylenecyclopropanes (aryl = phenyl, 4-methoxyphenyl, 4-chlorophenyl) in a living fashion, giving origin to polymers with the cyclopropyl groups intact. Additionally, it copolymerizes ethylene with 2-aryl-1-methylenecyclopropanes or 7-methylenebicyclo[4.1.0]heptane to yield alternating copolymers.

Pelascini et al. [234, 235] studied the catalytic behaviors of a series of iron and cobalt complexes based on the 2,6-bis(arylimino)pyridine ligands, such as **L26**, **L46**, **L59**, **L87**, **L103** and **L115**, in the polymerization of bicyclo[2.2.1]hept-2-ene (norbornene), upon activation with MAO. The results show that the iron complexes are completely inactive, irrespective of the reaction conditions and the ligands used, while the corresponding cobalt analogues can polymerize norbornene, via an addition mechanism, with low catalytic activities (ca. 10^3–10^4 g/(mol h)), forming polynorbornenes with high molecular weights (M_w ca. 10^5–10^6 g/mol). The inactivity of iron complexes such as [$FeCl_2$(**L26**)] and [$FeCl_2$(**L87**)], upon activation with MAO, was confirmed by Sacchi et al. [236] that also reported the cobalt complex [$CoCl_2$(**L26**)] as inactive. On the contrary, Chen et al. [37] found that iron and cobalt complexes [$FeCl_2$(**L28**)] and [$CoCl_2$(**L28**)] are both active toward the addition polymerization of norbornene, when activated with MAO, the latter being the most active one.

Considering that the iron and cobalt complexes are less oxophilic and more tolerant toward polar functionalities when compared with the early transition-metal complexes, the catalytic performance of the BIP iron and cobalt complexes on the polymerization of polar monomers such as methyl acrylate (MA), methyl methacrylate (MMA) and *tert*-butyl acrylate (*t*-BA) has also been widely investigated. Repo [61, 237, 238] and Abu-Surrah [239] reported that iron and cobalt complexes bearing *N*-aryl or *N*-alkyl groups, e.g., [$FeCl_2$(**L14**)], with phenyl rings, [$FeCl_2$(**L84**)], with 2,6-dimethylphenyl rings, [$FeCl_2$(**L245**)], with *iso*-propyl groups, and [$CoCl_2$(**L247**)], with cyclohexyl groups, are active toward the polymerization of *t*-BA, forming syndiotactic-enriched polymers. The catalytic activity is strongly dependent on both the metal center and the ligand structure. Generally, the iron complexes are more active than the corresponding cobalt ones, e.g., [$FeCl_2$(**L247**)] (2.5×10^5 g/(mol h)) vs. [$CoCl_2$(**L247**)] (1.3×10^5 g/(mol h)) [237], and the activities of the alkyl-substituted complexes are higher than those of the aryl-substituted ones. For instance, [$FeCl_2$(**L245**)] and [$FeCl_2$(**L247**)] give higher conversions than [$FeCl_2$(**L14**)] and [$FeCl_2$(**L84**)], under the same conditions [61]. In the case of the aryl-substituted complexes, an increase in the bulkiness of the aryl rings *o*-substituents leads to a reduction both in activity and in polymer molecular weight. Since the addition of free radical scavengers such as tetrakis[methylene(3,5-di-*tert*-butyl-4-hydroxyhydrocinnamate)]methane (Irganox 1010) and 2,6-di-*tert*-butyl-α-(3,5-di-*tert*-butyl-4-oxo-2,5-cyclohexadiene-1-ylidene)-*p*-tolyloxy (Galvinoxyl) does not inhibit the polymerization, a free radical mechanism is excluded. However, despite the detailed polymerization studies presented, the actual polymerization mechanism remains unclear. The [$FeCl_2$(**L85**)]/MAO system was demonstrated by Fullana et al. [240] to be highly active toward the polymerization of MA, forming atactic polymers with high molecular weights. Attempts to copolymerize MA and ethylene using the [$FeCl_2$(**L85**)]/MAO catalyst system resulted in a blend of the two homopolymers. Li and coworkers [241] studied the catalytic behaviors of the BIP iron complexes [$FeCl_2$(**L84**)] and [$FeCl_2$(**L87**)] on the polymerization of MMA, *n*-butyl methacrylate (*n*-BMA) and *tert*-butyl methacrylate (*t*-BMA), and found that [$FeCl_2$(**L84**)] and [$FeCl_2$(**L87**)] are both highly active toward the polymerization of MMA, in the presence of MMAO or a trialkylaluminum (e.g., $AlEt_3$ and Al^iBu_3), forming syndiotactic-enriched polymers, while MMAO alone displays a high catalytic activity for the polymerization of *n*-BMA and *t*-BMA. Kim et al. [242] tested the polymerization of vinyl ethers, such as ethyl vinyl ether (EVE), *n*-butyl vinyl ether (*n*-BVE) and *iso*-butyl vinyl ether (*i*-BVE), catalyzed by the BIP iron and cobalt complex/MAO systems and reported that iron and cobalt complexes based on ligands **L14**, **L84**, **L86** and **L87** are able to polymerize these vinyl ethers with moderate activities, affording the corresponding atactic polymers.

3.9 Dual Catalyst Systems

A dual catalyst system can be literally regarded as a system in which two different catalysts, being placed in a single reactor, operate independently or cooperatively to yield a product with some peculiar properties, which are not accessible by each of the catalysts independently. Certainly, if necessary, more than two catalysts can be mixed in a single reactor to generate multiple catalyst systems, but these are expected to be complicated cases. Therefore, the great majority of the multiple catalyst systems reported so far are in fact dual catalyst systems. According to the difference in the operating mechanism, the dual catalyst systems can be classified into two types: reactor blending and tandem catalysis [243, 244]. In the former method, two catalysts operate independently in a single reactor to yield a mixture of two different homopolymers, each of them being produced by an individual catalyst and possessing its original properties. This method allows a control on the molecular weight and the molecular weight distribution of the resulting polymer by an in situ blending. Since the molecular weight and the molecular weight distribution are responsible for the ultimate mechanical and rheological properties of polymers, respectively, this method is effective in the simultaneous improvement of the physical properties and processability of the polymers. In the case of the latter method, one of the two catalysts operates firstly to produce oligomers with controllable molecular weight distributions, which are further copolymerized (as comonomers) with α-olefins by the second catalyst. Finally, a copolymer with controllable branching is formed. It has been demonstrated that this method is very powerful to produce linear low-density polyethylene (LLDPE) [244].

Considering the unique abilities of the BIP iron and cobalt catalysts to produce strictly linear polyethylenes or α-olefins with high selectivities, the use of the BIP iron and cobalt catalysts, both in reactor blending and in tandem catalysis, has attracted considerable attention [245–250]. Mecking [245], using a combination of the iron complex [$FeCl_2$(**L87**)] with the α-diimine nickel(II) complex [$NiBr_2$(ArN=C(Me)-C(Me)=NAr)] (where Ar = 2,6-$Me_2C_6H_3$) or the zirconocene [Zr(^{n}BuCp)$_2$Cl$_2$] (where ^{n}BuCp = η-*n*-butylcyclopentadienyl), made the first attempt to prepare blends with a BIP iron complex in reactor blending. Since the nickel complex and the zirconocene polymerize ethylene to yield highly branched and ultra-high molecular weight polyethylenes, respectively, the combination of [$FeCl_2$(**L87**)] with the nickel complex leads to a reactor blend of linear and branched polyethylenes, while the combination with the zirconocene gives a blend of linear polyethylenes with different molecular weights or crystallinities. The differences in density, molecular weight and overall branching between the reactor blend and the original polymers formed by each catalyst are indicative of both catalysts being active and compatible. An ideal polyethylene blend in an industrial application consists of a linear fraction, with a relatively low molecular weight that provides rigidity, and a branched fraction, with high molecular weight that provides flexibility, totally showing a bimodal/broad molecular weight distribution. Considering the fact that the BIP cobalt complexes yield linear polyethylenes with relatively lower molecular weights compared with their iron analogues, under the same conditions, the cobalt complex [$CoCl_2$(**L87**)] was selected by Li and coworkers [246, 247] to polymerize ethylene, together with the nickel complexes [$NiBr_2$(ArN=C(R)-C(R)=NAr)] (where Ar = 2,6-$^iPr_2C_6H_3$, R = Me, acenaphthene or cyclohexane backbones) in a single reactor. As a result, a perfect blend was obtained with a bimodal molecular weight distribution, in which the linear low molecular weight fraction is produced by the cobalt complex and the highly branched fraction with high molecular weight is produced by the nickel complex. The linear correlation between the overall catalyst activity and percentage of nickel complex suggests that both catalysts operate independently from each other in the catalytic system. The final blends were characterized by using a combination of GPC/SEC, DSC, wide-angle X-ray diffraction (WAXD) and small-angle X-ray scattering (SAXS), and it was found that the molar ratio of the catalysts has a remarkable effect on the physical properties of the resulting blends, such as molecular weight, molecular weight distribution, crystallization rate, melting temperature (T_m), crystallization temperature (T_c) and degree of crystallinity.

Chadwick and coworkers [251, 252] reported an immobilized dual catalyst system, which is produced by co-impregnation of an iron complex ([$FeCl_2$(**L109**)] or [$FeCl_2$(**L85**)]) and a chromium complex [$CrCl_2${1-(8-quinolyl)Ind}] on

$MgCl_2/AlEt_n(OEt)_{3-n}$. This system is able to polymerize ethylene, leading to reactor blending of polyethylenes with low and high molecular weights, respectively. As a result, a bimodal molecular weight distribution of polyethylene was obtained. The proportions of high and low molecular weight fractions can be changed simply by adjusting the relative loadings of the two catalysts on the support. Similarly, a SiO_2-supported dual late transition-metal catalyst system was reported by Ivanchev et al. [253], in which the iron complex [$FeCl_2$(**L97**)] and the nickel complex [$NiBr_2$(ArN=C(An)-C(An)=NAr)] (where Ar = 2-cyclohexyl-4,6-$Me_2C_6H_2$, An = acenaphthene backbone) are concurrently or successively immobilized on silica gel, in the presence of MAO. This supported catalyst system shows high catalytic activities toward ethylene polymerization in the presence or absence of Al^iBu_3 as cocatalyst, affording a blend of linear and branched polyethylenes.

The use of tandem catalysis with the participation of BIP iron and cobalt complexes is essentially based on the high selectivities exhibited by these precatalysts to produce oligomers. At the same time, a second catalyst with good ability to copolymerize ethylene or propylene with higher α-olefins is required. Quijada et al. [248] prepared branched polyethylene, from a monomer feed consisting only of ethylene, using tandem catalysis with the participation of one of such iron complexes, as schematically represented in Scheme 3.72. In the presence of MAO, the less sterically encumbered iron complex [$FeCl_2$(**L21**)] was used to convert ethylene into oligomers first, and then a simultaneous copolymerization of ethylene with the in situ formed oligomers was performed by using a homogeneous zirconocene complex ([Zr(*rac*-$EtInd_2$)Cl_2] or [Zr(*rac*-Me_2SiInd_2)Cl_2], where Ind = indenyl) as second catalyst, finally leading to polyethylenes with degrees of branching varying in the range of 1–4 branches per 100 units of ethylene. It is noteworthy that the polymers obtained by the [$FeCl_2$(**L21**)]/[(Zr(*rac*-Me_2SiInd_2)Cl_2] tandem catalyst system have characteristic bimodal/broad molecular weight distributions, and the low molecular weight fraction becomes more pronounced as the Fe/Zr molar ratio increases, giving compositions similar to those obtained with [$FeCl_2$(**L21**)] alone. This finding indicates that a significant fraction of the oligomers produced by this iron complex is not incorporated into the growing polymer chain governed by the zirconocene sites. On the contrary, the polymers obtained by the [$FeCl_2$(**L21**)]/[Zr(*rac*-$EtInd_2$)Cl_2] system show a monomodal molecular weight distribution, and the degree of branching increases with the Fe/Zr molar ratio, indicating an efficient incorporation of the in situ formed oligomers into the polymer backbone. Similarly, a tandem catalysis system consisting of a cobalt complex ([$CoCl_2$(**L18**)] or [$CoCl_2$(**L73**)]) and the zirconocene [Zr(*rac*-$EtInd_2$)Cl_2] was reported by Hu and coworkers [254], which allow the formation of LLDPE with various branching degrees. Bianchini [249] and Frediani [250] groups reported a new homogeneous tandem catalysis system, which consists of a cobalt complex ([$CoCl_2$(**L204**)], [$CoCl_2$(**L214**)] or [$CoCl_2$(**L215**)]) and the constrained geometry catalyst [Ti{η^5,κN-C_5Me_4Si-$(Me)_2$(N^tBu)}Cl_2], to yield branched polyethylenes. It was found that both the extent of α-olefin incorporation and the number and type of branches are

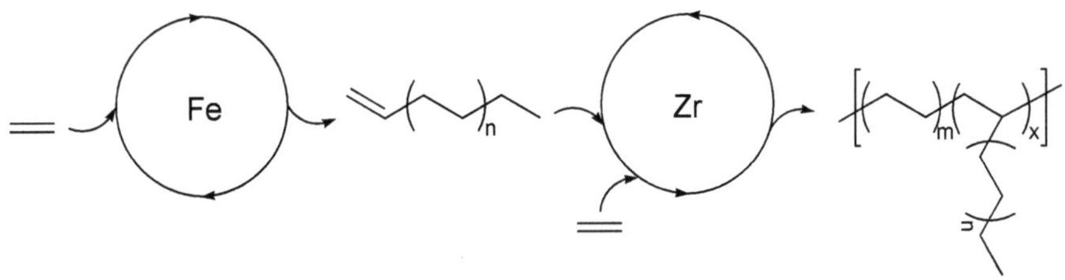

Scheme 3.72 Tandem catalysis

proportional to the molar fraction of cobalt, ranging from 0 to 0.75. An increase in these values from 0.05 to 0.75 produces an increase in the α-olefin incorporation between 5 and 50 mol%, leading to an increase in the number of branches from 27 to 254 per 1000 carbon atoms, and the polymer morphology correspondingly changes gradually from rigid solids, tough flexible white solids, white rubbery materials, to clear colorless rubbery materials. Additionally, the distribution of the branches (ethyl 82–86%, butyl 17–12%) is in excellent agreement with the molar distribution of the α-olefins produced by [$CoCl_2$(**L215**)] in the oligomerization of ethylene (butenes 82%, hexenes 15%).

By analogy with tandem catalysis employing ethylene as the single monomer (see Scheme 3.72), a comb-branched polypropylene with isotactic backbone and atactic side chains was prepared from propylene, by using a combination of [$FeCl_2$(**L84**)] and the zirconocene [Zr{*rac*-Me_2Si(2-MeBenz[*e*]Ind)$_2$}Cl_2] precatalysts [255]. Upon activation with MMAO, [$FeCl_2$(**L84**)] converts propylene into macromonomers with low molecular weights (ca. 3000 g/mol) and atactic nature, which are in situ copolymerized with propylene by the [Zr{*rac*-Me_2Si(2-MeBenz[*e*]Ind)$_2$}Cl_2] system to afford branched isotactic polypropylene. It was revealed that the catalyst addition method has a substantial effect on the nature of the resulting polymer. Simultaneous addition of both catalysts results in a mixture of atactic polypropylene and isotactic polypropylene, produced by [$FeCl_2$(**L84**)] and [Zr{*rac*-Me_2Si(2-MeBenz[*e*]Ind)$_2$}Cl_2], respectively. In contrast, a consecutive addition of [$FeCl_2$(**L84**)] and [Zr{*rac*-Me_2Si(2-MeBenz[*e*]Ind)$_2$}Cl_2] in a given time interval (30, 90, or 120 min) leads to the formation of copolymers. The degree of branching of the final polymer is dependent on the time interval and molar ratio of both catalysts. Increasing the time interval and molar ratio of iron to zirconium causes an increase in the degree of branching.

Tandem catalysis can also be performed by combining a homogeneous single-site catalyst with a heterogeneous Ziegler–Natta catalyst. Hu and coworkers [256, 257] combined the iron complex [$FeCl_2$(**L53**)] with the Ziegler–Natta catalyst $TiCl_4/MgCl_2$ to in situ polymerize ethylene in the presence of MAO or MAO/$AlEt_3$ as cocatalysts, affording branched polyethylene with total branching varying from 10.5/1000 C to 59.0/1000 C, including ethyl (ca. 50%), butyl (ca. 30%) and longer branches (ca. 20%). The result shows that the amount of cocatalyst has a marked effect on the final properties of the polymers. Recently, Xu et al. [258] reported an immobilized tandem catalyst system, in which the iron complex

$[FeCl_2(\mathbf{L104})]$ and the zirconocene $[Zr(\mathit{rac}\text{-}EtInd_2)Cl_2]$ are co-impregnated on a layered zeolite-like calcosilicate (CAS-1), to polymerize ethylene, yielding LLDPE. By comparison with the homogeneous tandem catalyst systems, supported tandem catalyst systems show improved temperature stability and enhanced molecular weights and degrees of branching of the resulting LLDPE.

3.10 Supported Iron and Cobalt Catalysts

Catalyst immobilization is a key step to industrialize the homogeneous single-site catalysts, which not only can overcome the reactor fouling resulting from the employment of homogeneous catalysts, but also improves the morphology and bulk density of polymer particles. Additionally, it can also substantially affect the properties of the resulting polymers, such as molecular weight, molecular weight distribution, branching or crystallinity, as well as the temperature stability of the catalyst and polymerization kinetics. Parallel to the modification of the BIP iron and cobalt catalysts, many efforts have been devoted to investigating their immobilization [259–267].

3.10.1 Self-immobilization

The immobilization of catalysts can be simply divided into two types: self-immobilization and immobilization on inorganic or polymeric materials. The self-immobilization implies that a catalyst is itself in situ incorporated into the growing polymer chain during the course of an olefin polymerization reaction, and the formed polymers, bearing the catalyst, agglomerate to generate a seed polymer particle. The subsequent polymerization occurs on the seed polymer particle, and not in the solution, the morphological characteristics of the seed polymer particle being replicated by the growing polymer. In order to achieve the self-immobilization of a catalyst, an active functional group is required to be attached to the catalyst itself, which can then be polymerized or copolymerized with the monomer, under olefin polymerization conditions. The presence of terminal vinyl functionalities in the molecular structure of the precatalyst is preferred because of its similarity to the ethylene monomer. Therefore, a great number of catalyst functionalizations so far reported have been directed toward the introduction of terminal vinyl groups into the precatalyst structure [78, 103, 261, 268–270].

The introduction of terminal vinyl groups at the imino carbons of the BIP ligands can be readily achieved by deprotonation of the methyl protons of the bis(arylacetimino)pyridine ligands with lithium diisopropylamide (LDA) or sodium bis(trimethylsilyl)amide, followed by treatment with ω-alkenyl bromides, leading to the terminal vinyl-functionalized ligands **L512–L520** (Scheme 3.73) [103, 268]. Using a similar synthetic protocol, Herrmann and coworkers [78] prepared the ligand **L521** (Scheme 3.73). The reaction of ligands **L512–L521** with

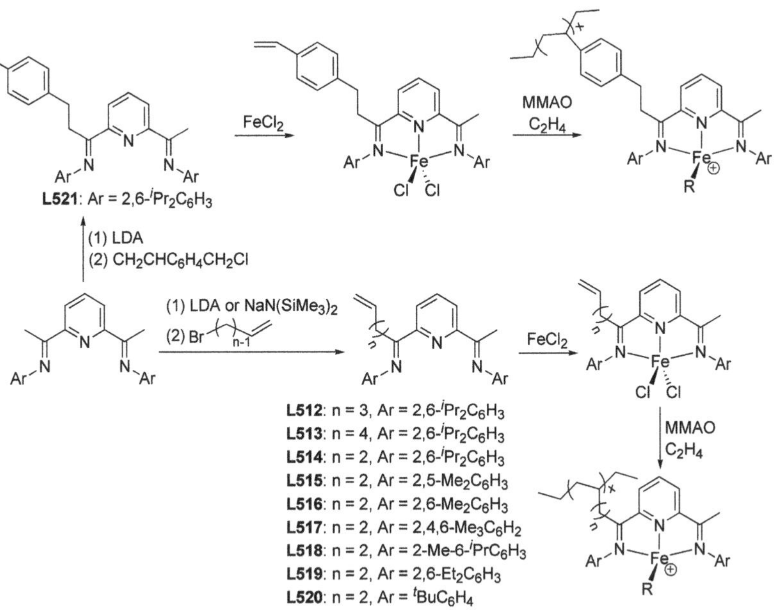

Scheme 3.73 Immobilizations derived from the introduction of terminal vinyl groups at the substituents of the BIP ligand imino carbons

iron dichloride, in tetrahydrofuran, leads to the corresponding iron complexes [$FeCl_2$(**L512**)]–[$FeCl_2$(**L521**)] bearing alkyl substituents at the imino carbons that contain terminal vinyl groups. Upon activation with MAO or MMAO, these iron complexes exhibit high activities toward ethylene polymerization. Kaul et al. [268] reported that, during the course of ethylene polymerization with [$FeCl_2$(**L512**)]–[$FeCl_2$(**L514**)], in the presence of MMAO, the terminal vinyl groups of the complexes are copolymerized with ethylene, affording in situ self-immobilized catalytically active particles, which work as supported catalysts (Scheme 3.73). These self-immobilized polymeric catalysts can be isolated from a solution of MMAO-activated iron complex, after bubbling ethylene for a short period (e.g., 5 min), and subsequently stored and reused to catalyze ethylene polymerization with high activities. In contrast, no self-immobilization was observed in the cases of [$FeCl_2$(**L515**)]–[$FeCl_2$(**L520**)]$FeCl_2$/MAO systems [103].

Jin and coworkers [269, 270] prepared a symmetrical allyl-substituted BIP ligand **L522** by direct condensation of 2,6-diacetylpyridine with 4-allyl-2,6-di-*iso*-propylaniline (Scheme 3.74). Using a similar reaction protocol, unsymmetrical allyl-substituted BIP ligands 2-(ArN=C(Me))-6-(Ar′N=C(Me))C_5H_3N (Ar = 4-allyl-2,6-$^iPr_2C_6H_2$, Ar′ = 2,6-$^iPr_2C_6H_3$ (**L523**) or 2,6-$Me_2C_6H_3$ (**L524**)) were synthesized by stepwise condensation of 2,6-diacetylpyridine with 4-allyl-2,6-di-*iso*-propylaniline and 2,6-di-*iso*-propylaniline or 2,6-dimethylaniline [270, 271]. Their corresponding iron complexes [$FeCl_2$(**L522**)]–[$FeCl_2$(**L524**)] are easily prepared by treatment of the ligands with iron dichloride. Upon activation with

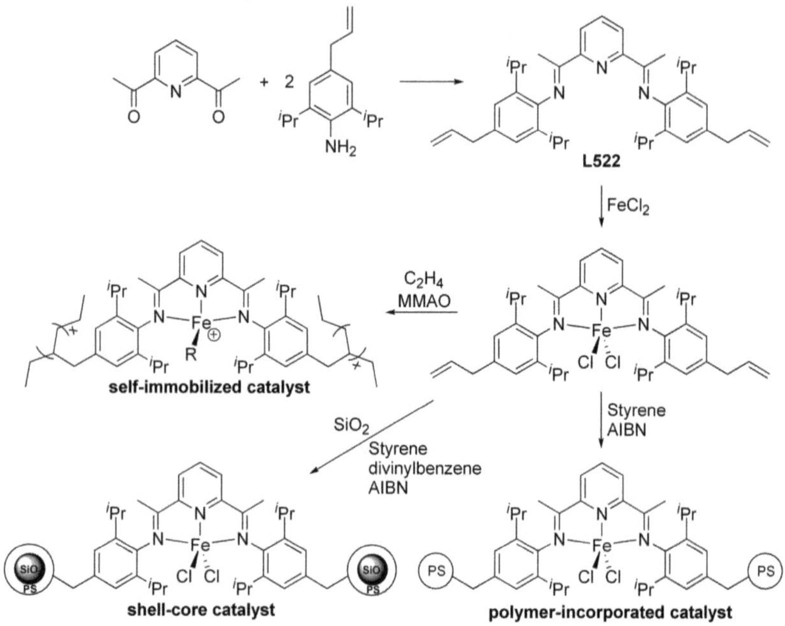

Scheme 3.74 Immobilization methods derived from the introduction of allyl groups at the *p*-positions of the BIP ligand aryl rings

MMAO, these complexes are active toward ethylene polymerization, and [$FeCl_2$(**L522**)], containing two allyl groups, shows a relatively lower activity (1.9×10^6 g/(mol h bar)) than that of the corresponding [$FeCl_2$(**L523**)], containing a single allyl group (4.02×10^6 g/(mol h bar)). In all cases, the self-immobilization phenomenon is observed, i.e., the catalyst molecules are incorporated into the growing polymer through self-catalyzed copolymerization of the allyl groups with ethylene (Scheme 3.74). However, the self-immobilization cannot lead to greatly improved morphologies of the resulting polymers since iron catalysts have extremely weak abilities to copolymerize ethylene with α-olefins. In order to improve the morphology of these polymers, two kinds of new supported catalysts were prepared, as represented in Scheme 3.74. Iron complexes containing allyl groups (e.g., [$FeCl_2$(**L522**)] and [$FeCl_2$(**L523**)]) are copolymerized with styrene in the presence of a radical initiator (azobis-*iso*-butyronitrile, AIBN), giving polymer-incorporated catalysts, whereas if iron complexes, styrene and divinylbenzene are copolymerized in the presence of AIBN and silica (SiO_2) particles, shell-core catalysts are obtained. Upon activation with MMAO, these polymer-incorporated catalysts show high catalytic activities toward ethylene polymerization, similar to those obtained with the original homogeneous iron catalysts, whereas the activities of the shell-core catalysts are substantially higher, being the molecular weights of the resulting polyethylenes greatly increased. The morphologies obtained with the polymer-incorporated catalysts are still ill-defined,

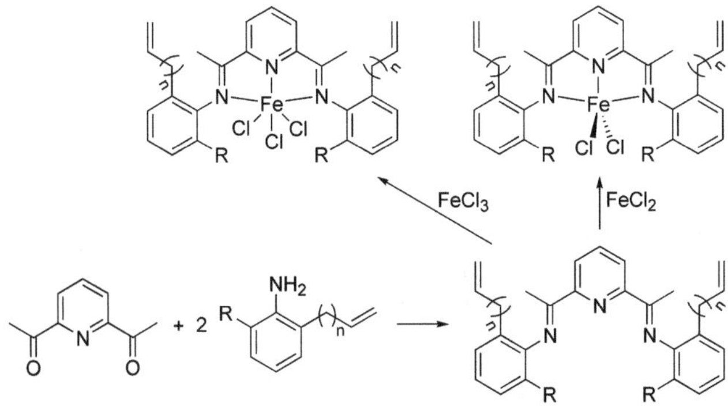

Scheme 3.75 Introduction of terminal vinyl groups at the *o*-positions of the BIP ligand aryl rings

as in the case of the self-immobilized catalysts, but the shell-core ones afford well-defined spherical polymer particles.

The introduction of terminal vinyl groups into the *o*-positions of the BIP ligand aryl rings can be easily achieved by condensation of 2,6-diacetylpyridine with 2-(ω-alkenyl)-substituted anilines. Reaction of these ligands with iron di- or trichlorides affords iron complexes [$FeCl_2$(**L525**)]–[$FeCl_2$(**L529**)] and [$FeCl_3$-(**L525**)]–[$FeCl_3$(**L529**)] (Scheme 3.75) [103]. When activated with MAO, all these complexes exhibit high catalytic activities toward ethylene polymerization, forming a mixture of oligomers and polyethylenes. It is worth to mention that, compared with their corresponding saturated iron complexes (e.g., [$FeCl_2$(**L21**)], [$FeCl_2$(**L22**)] and [$FeCl_2$(**L25**)]), the terminal vinyl-substituted iron complexes form higher contents of polymers, which corresponds to a lower rate of β-hydrogen elimination (chain transfer). For example, [$FeCl_2$(**L527**)] exclusively yields a polymer fraction, whereas its corresponding saturated iron complex [$FeCl_2$(**L25**)] gives a product with over 60 wt% of oligomers. In all cases, no self-immobilization was observed.

Alt et al. [261] and Kim et al. [272] introduced terminal vinyl groups at the 4-position of the BIP ligand pyridine ring by condensation of 4-(ω-alkenyl)-substituted 2,6-diacetylpyridine with the appropriate anilines (Scheme 3.76). Subsequent reaction with iron dichloride affords the terminal vinyl-functionalized iron complexes [$FeCl_2$(**L530**)]–[$FeCl_2$(**L532**)]. For a comparison, the saturated iron complex [$FeCl_2$(**L533**)] (**L533** = 4-^{n}BuO-2,6-(ArN=C(Me))$_2$$C_5H_2N$; Ar = 2,6-^{i}$Pr_2C_6H_3$) was synthesized using the same reaction procedure. Upon activation with MAO, all the iron complexes showed moderate catalytic activities toward ethylene polymerization. Compared with the saturated iron complex [$FeCl_2$(**L533**)]), which leads to pronounced reactor fouling, the iron complexes bearing terminal vinyl groups do not give rise to adhesion of polymer on the

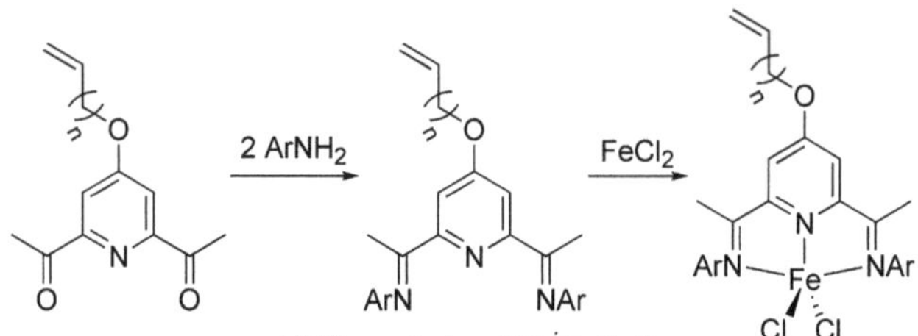

Scheme 3.76 Introduction of terminal vinyl groups at the 4-position of the BIP ligand pyridine ring

reactor walls and stirrer, indicating that self-immobilization occurred. Additionally, it was demonstrated that the self-immobilized catalysts greatly suppress the chain transfer to aluminum, leading to polyethylene with a monomodal molecular weight distribution.

3.10.2 Immobilization on Inorganic or Polymeric Supports

Silica (SiO_2) in the form of amorphous silica gel with irregularly formed particles, which can be described as agglomerates of smaller particles with large surface area (e.g., commercially available Davison SiO_2, $a = 200–300\ m^2/g$ [253, 273–275]), is the most widely used inorganic support. Immobilization of catalysts on SiO_2 can be in principle divided into three methods: (a) direct immobilization of the catalyst precursor on SiO_2 (denoted as [MCl_2L]/SiO_2, where L = the BIP ligand, M = Fe or Co); (b) immobilization of catalyst precursor on SiO_2 pretreated with MAO or trialkylaluminum (denoted as [MCl_2L]/MAO (or AlR_3)/SiO_2, where R = Me, Et or ^{i}Bu); (c) chemical tethering of the catalyst precursor ligands on SiO_2.

Sivaram [273] and Semikolenova [201] groups studied in detail the interactions of the iron precatalysts (e.g., [$FeCl_2$(**L84**)] and [$FeCl_2$(**L87**)]) with the SiO_2 support in supported catalysts obtained by the former two methods. They found that direct immobilization gives low loading of catalyst precursor, e.g., 0.56–0.63 wt% (Fe) for [$FeCl_2$(**L87**)]/SiO_2, obtained by Sivaram and coworkers, and 0.10–0.13 wt% (Fe) for [$FeCl_2$(**L84**)]/SiO_2, by Semikolenova et al., regardless of the amount and nature of the hydroxyl groups present on SiO_2. Since the loading of iron precatalyst is substantially lower than that of the terminal hydroxyl groups on SiO_2, Semikolenova et al. considered that fixation of the iron precatalyst occurs by multiple bonding of the latter to the SiO_2 surface, via the interaction of both pyridyl and phenyl groups of the ligand with the surface hydroxyl groups. However, based on the examination of the binding energy of iron by X-ray photoelectron spectroscopy (XPS), Sivaram and coworkers found that the iron precatalyst has no strong chemical interaction with the SiO_2 surface, except some weak secondary interactions between Si–OH and Fe–Cl. At the same time, the binding energy of the iron centers in the supported catalyst [$FeCl_2$(**L87**)]/MAO/SiO_2, made by method (b), revealed to be higher than those of the [$FeCl_2$(**L87**)]/SiO_2 catalyst and of the homogeneous [$FeCl_2$(**L87**)], indicating the presence of cationic iron centers. A similar increase in the binding energy of the Fe_{2p} center for [$FeCl_2$(**L87**)]/MAO/SiO_2 was observed by Ma et al. [276]. Upon activation with MAO or trialkylaluminum, directly immobilized iron catalysts, e.g., [$FeCl_2$(**L84**)]/SiO_2 [201] and [$FeCl_2$(**L87**)]/SiO_2 [273], exhibit high catalytic activities toward ethylene polymerization. Compared with the parent homogeneous catalysts, both [$FeCl_2$(**L84**)]/SiO_2 and [$FeCl_2$(**L87**)]/SiO_2 show significantly improved temperature stability and polymerization kinetics. The homogeneous iron catalysts [$FeCl_2$(**L84**)] and [$FeCl_2$(**L87**)] generally show very high initial activities, which sharply fall down in the course of the polymerization, and a temperature increase up to 70 °C causes catalyst deactivation. Conversely, the supported catalysts [$FeCl_2$(**L84**)]/SiO_2 and [$FeCl_2$(**L87**)]/SiO_2 give steady-state polymerization kinetics and exhibit high activities even at 70 °C. It is noteworthy that Sivaram and coworkers found that the catalytic activity increases with the calcination temperature of SiO_2, while Semikolenova et al. observed a reverse trend. Improved molecular weights and morphologies of the resulting polymers were observed for both supported catalysts.

Barabanov et al. [274, 275] studied the kinetics of ethylene polymerization catalyzed by the supported iron catalyst [$FeCl_2$(**L84**)]/SiO_2. The number of active species and the propagation rate constant were measured by inhibiting the polymerization reaction with radioactive ^{14}CO. It was found that the supported catalyst is highly stable, and the catalytic activity, as well as the number of active species and the propagation rate constant, are practically constant during the polymerization time (up to 60 min). The same results were also obtained for Al_2O_3- or $MgCl_2$-supported catalysts. In addition, an investigation of the effect of temperature on the number of active sites and propagation rate constant, with or without the presence of hydrogen, revealed that the variation of the number of active centers results from a reversed equilibrium of the active and "dormant" sites. Broad molecular weight distributions were obtained for all the supported catalysts, indicating a set of active sites with different reactivities, and it was demonstrated that, compared with the parent homogeneous catalysts, the supported ones possess a lower number of active sites producing low molecular weight polymer and a higher number of active sites generating the high molecular weight fraction. Using the same method, the same authors [277] investigated the supported cobalt catalyst [$CoCl_2$(**L84**)]/SiO_2, and similar results were obtained.

A new strategy to immobilize iron complexes on SiO_2 was reported by Kaul et al. [268], as schematically represented in Scheme 3.77. The iron complexes [$FeCl_2$(**L512**)]–[$FeCl_2$(**L514**)], bearing terminal vinyl groups at the BIP ligands imino carbons, were anchored on SiO_2 through the hydrosilylation reaction of the Si–H groups of a tetramethyldisilazane-modified silica surface with the terminal

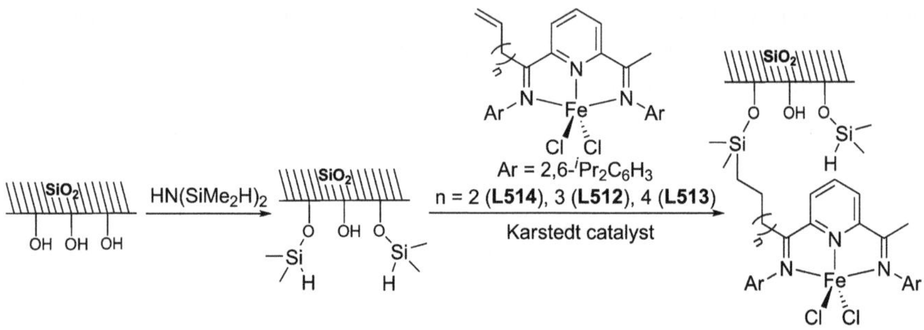

Scheme 3.77 Direct tethering of an iron complex on a tetramethyldisilazane-modified SiO_2 surface

vinyl groups of the iron complexes, in the presence of the Karstedt catalyst. This immobilization method leads to high loadings of iron catalyst (0.99–1.15 wt% Fe). Upon activation with MMAO, these supported catalysts are highly active toward ethylene polymerization, leading to polyethylenes showing bimodal molecular weight distributions, the high molecular weight fraction being predominant.

The supported iron catalysts of the type $[FeCl_2L]/MAO$ (or AlR_3)/SiO_2, made by method (b), can be described as iron catalyst precursors anchored on the SiO_2 surface through a direct interaction with MAO, that will result in the in situ formation of the active species. The cationic metal-alkyl centers thus formed become trapped and stabilized by the aluminum activator, which can be identified by XPS spectroscopy [259, 276]. Similar to the directly immobilized iron catalysts $[FeCl_2L]/SiO_2$, $[FeCl_2L]/MAO$ (or AlR_3)/SiO_2 supported catalysts also show a steady-state kinetics for ethylene polymerization, but the catalytic activities are remarkably lower than those of the parent homogeneous catalysts [253, 259]. Conversely, unlike $[FeCl_2L]/SiO_2$, a substantial increase in the polymer molecular weight is observed for the $[FeCl_2L]/MAO$ (or AlR_3)/SiO_2 systems [259, 273], indicating that β-hydrogen transfer is markedly suppressed.

Wang et al. [217] studied the effect of the activator on the bimodal molecular weight distribution produced by supported iron catalysts $[FeCl_2($**L84**$)]/TEAO/SiO_2$ or $[FeCl_2($**L87**$)]/TEAO/SiO_2$ (TEAO = tetraethylaluminoxane). It was revealed that the immobilization of the iron catalyst can significantly improve the M_n of the high molecular weight fraction of the bimodal polyethylene, but it has an insignificant effect on the low molecular weight fraction. Such bimodal molecular weight distribution can be easily adjusted by the modification of the polymerization conditions, such as the ethylene pressure and molar ratio of aluminum to iron.

Ivanchev et al. [253] prepared a series of supported single or dual catalyst systems, consisting of the iron complex $[FeCl_2($**L97**$)]$ and the nickel complex $[NiBr_2(ArN{=}C(An){-}C(An){=}NAr)]$ (where Ar = 2-cyclohexyl-4,6-$Me_2C_6H_2$, An = acenaphthene backbone), by varying the order of immobilization, e.g., Fe/MAO/SiO_2 (**I**), Ni/MAO/SiO_2 (**II**), (Fe + Ni)/MAO/SiO_2 (**III**), Ni/Fe/MAO/SiO_2 (**IV**), Fe/Ni/MAO/SiO_2 (**V**), and **I** + **II** (**VI**). All these supported catalysts show high activities toward ethylene polymerization, in the presence or absence of $Al^{i}Bu_3$. The dual catalyst systems **III–VI** afford a blend of linear and branched polyethylenes, the degree of branching notably varying in each of these catalyst systems.

A novel layered zeolite-like calcosilicate (CAS-1) was employed by Xu et al. [278] to immobilize the iron precatalyst [$FeCl_2$(**L87**)], leading to a supported catalyst [$FeCl_2$(**L87**)]/MAO/CAS-1, which exhibits high catalytic activities toward ethylene polymerization, upon activation with MAO or $AlEt_3$, giving rise to polyethylenes not only with higher molecular weights but also unique morphologies.

Recently, Alt and coworkers [260, 279] described a new method to immobilize iron precatalysts on SiO_2. First, trimethylaluminum ($AlMe_3$) is partially hydrolyzed by the addition of water as steam, in the presence of SiO_2, to yield a SiO_2-supported cocatalyst, which is denoted as PHT. The latter is then treated with iron complex, leading to the supported catalysts, e.g., [$FeCl_2$(**L57**)]/PHT and [$FeCl_2$(**L84**)]/PHT. Using the same method, binary or ternary supported catalysts based on [$FeCl_2$(**L57**)], [$ZrCp_2Cl_2$] and [$TiCp_2Cl_2$], were prepared. Upon activation with $Al^{i}Bu_3$, these supported catalysts are active toward ethylene polymerization. For [$FeCl_2$(**L57**)]/PHT, the highest activity (5.34×10^6 g/(mol h bar)) is obtained when a water/aluminum molar ratio of 1:1 is applied, being approximately 10 times higher than that of the parent homogeneous catalyst (6.30×10^5 g/(mol h bar)). On the contrary, the ternary supported catalyst (Fe + Zr + Ti)/PHT gives a surprisingly low activity ($0.50–2.34 \times 10^5$ g/(mol h bar)), probably due to unfavorable interactions of the different catalyst species. The same immobilization protocol was also used on the other supports (e.g., MCM-41) by the same authors [279].

Chemical tethering of the precatalyst ligands on SiO_2 for the preparation of supported catalysts (method (c)) has recently attracted considerable attention [271, 272, 280, 281]. Han et al. [280, 281] reported a strategy to anchor the ligand of an iron precatalyst on a flat SiO_2 model (based on silicon (100) single crystals covered with a thin layer of silica (20 nm) with a surface roughness below 1 nm), as displayed in Scheme 3.78. The terminal vinyl group of the alkyl substituent at the BIP ligand **L514** imino carbon was hydrosilylated with dichloromethylsilane, in the presence of the Karstedt catalyst, and subsequently reacted with the hydroxyl groups of SiO_2, leading to a SiO_2-supported ligand. The latter was further reacted with iron dichloride yielding the desired supported iron catalyst. Its binding energy of Fe_{2p}, determined by XPS spectroscopy, is similar to that of the corresponding parent homogeneous iron complex, suggesting a very close chemical environment for iron in both catalysts. Upon activation with $Al^{i}Bu_3$, the supported iron catalyst shows a moderate activity ($2.5–5.3 \times 10^5$ g/(mol h bar)) toward ethylene polymerization. Polyethylene is formed as a thin film on the silica surface, and no polymer formation is observed in the solution. This is because the iron catalyst is covalently anchored to the surface, and the polymer grows in a single direction, forming a film with a constant height on the flat SiO_2 surface. Using the same preparation strategy, Li [271] and Kim [272] groups anchored the

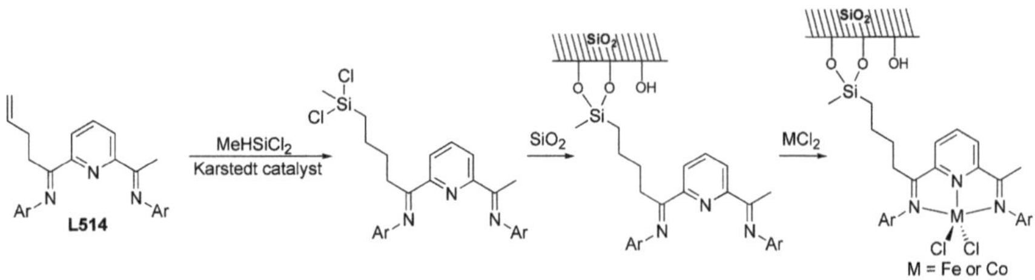

Scheme 3.78 Chemical tethering of the BIP ligand on SiO_2 via the functionalization of the terminal vinyl groups at the imino carbons

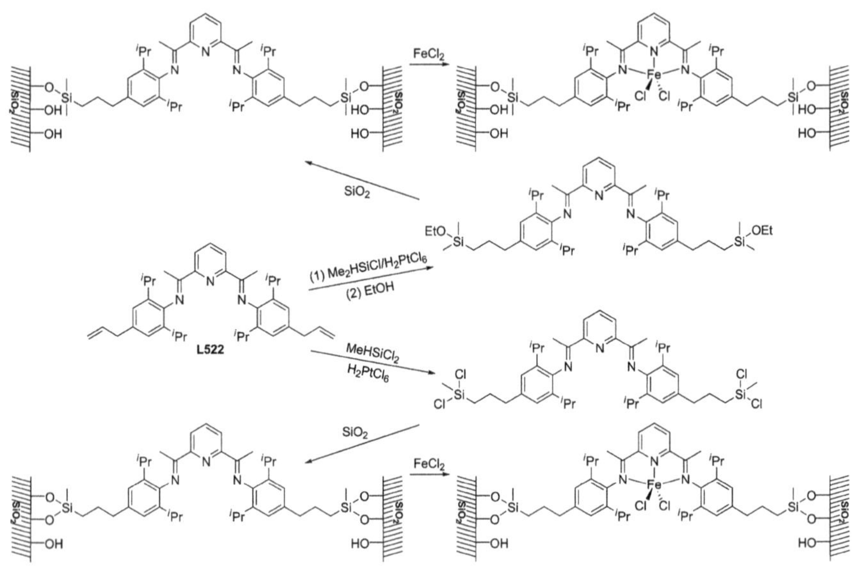

Scheme 3.79 Chemical tethering of the BIP ligand **L522** on SiO_2 via the functionalization of the terminal vinyl groups at the aryl rings *p*-positions

ligands **L522–L524** and **L531** on SiO_2, respectively, and prepared the corresponding supported iron and cobalt catalysts (Schemes 3.79 and 3.80). Upon activation with MAO or MMAO, these supported catalysts show higher activities toward ethylene polymerization and yield polyethylenes with higher molecular weights in comparison with those of the corresponding parent homogeneous catalysts. However, the **L531**-based supported iron and cobalt catalysts show activities about 100 times lower than those of their corresponding homogeneous catalysts.

MCM-41 and SBA-15 are mesoporous zeolites, and have large surface areas, narrow pore size distributions, large pore volumes and a hexagonal arrangement of uniform cylindrical pores (approximately 1.5 to 10 nm in diameter). They also have been used to immobilize the BIP iron and cobalt precatalysts [279, 282–284]. By impregnating the iron complexes [$FeCl_2$(**L18**)] and [$FeCl_2$(**L52**)] on

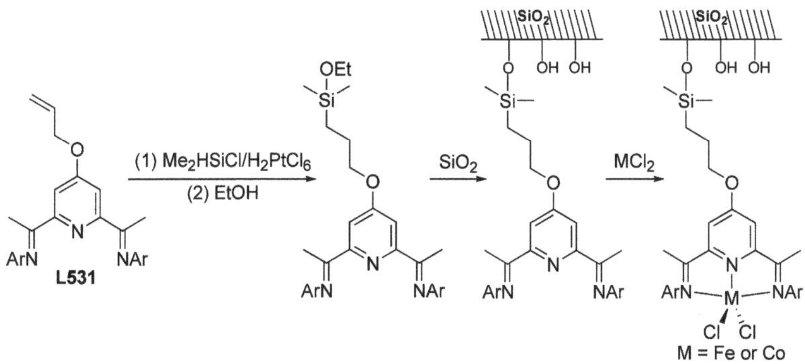

Scheme 3.80 Chemical tethering of the BIP ligand on SiO_2 via the functionalization of the terminal vinyl groups at the 4-position of the pyridine ring

MAO-pretreated MCM-41 or SBA-15, Guo et al. [282] prepared supported iron catalysts [$FeCl_2$(**L18**)]/MAO/MCM-41 (or SBA-15) and [$FeCl_2$(**L52**)]/MAO/MCM-41 (or SBA-15). Upon activation with MAO, all the supported catalysts exhibit high catalytic activities toward ethylene oligomerization, forming linear α-olefins. The MCM-41 zeolite, with relatively narrow pore size, leads to much lower molecular weight α-olefins. Similarly, Zhang et al. [283] prepared the supported iron catalyst [$FeCl_2$(**L87**)]/MAO/MCM-41, which is active toward ethylene polymerization, when activated with MAO or $AlEt_3$. Although its catalytic activity is much lower than that of the corresponding parent homogeneous catalyst, the resulting polyethylenes have molecular weights substantially higher than those obtained with the corresponding homogeneous catalyst. Interestingly, when the supported catalyst [$FeCl_2$(**L87**)]/MAO/SBA-15 was used to polymerize ethylene, a polyethylene with a fibrous morphology was obtained, since the two-dimensional ordered channel structure of SBA-15 controls the growth direction of the polymer chains [284]. This phenomenon is termed as extrusion ethylene polymerization. Paulino and Schuchardt [264] prepared supported catalysts [$FeCl_2$(**L87**)]/MCM-41 and [$FeCl_2$(**L115**)]/MCM-41 by direct immobilization. These two supported catalysts are active toward ethylene polymerization, upon activation with MAO. The activities are much lower than those of the corresponding homogeneous catalysts, but the molecular weights of the polyethylenes obtained are substantially higher than those produced by the corresponding homogeneous catalysts.

Mao [263, 285, 286] and Chadwick [252, 262, 267] prepared a series of $MgCl_2$ supports with relatively high surface area and high porosity by partial dealcoholization of spherical $MgCl_2$/ethanol adducts (e.g., $MgCl_2{\cdot}2.56C_2H_5OH$ [263, 285] and $MgCl_2{\cdot}1.1C_2H_5OH$ [262, 267]), which can be achieved through a thermal treatment. The morphology, crystallinity and mechanical strength of the formed support are substantially dependent on the temperature of the thermal treatment. When treated at 170 °C, for 4 h, an ideal $MgCl_2$ support $MgCl_2{\cdot}0.047C_2H_5OH$ is obtained, being characterized by a surface area of ca.

95 m^2/g and pore volume of 0.56 cm^3/g, as well as good spherical morphology and mechanical strength. A further increase in the pretreatment temperature (>200 °C) gives rise to fragmentation of the support. Additionally, dealcoholization can also be performed by reaction with $AlEt_3$ to afford a new $MgCl_2$ support $AlEt_n(OEt)_{3-n}/MgCl_2$. The immobilization of iron precatalysts on these two $MgCl_2$ supports leads to the corresponding supported iron catalysts, e.g., [$FeCl_2$(**L85**)]/$MgCl_2{\cdot}nC_2H_5OH$ [263, 286], [$FeCl_2$(**L87**)]/$MgCl_2$ [263], [$FeCl_2$(**L85**)]/$AlEt_n(OEt)_{3-n}/MgCl_2$ [263, 285], [$FeCl_2$(**L87**)]/$AlEt_n(OEt)_{3-n}/MgCl_2$ [263]. Upon activation with $AlEt_3$, Al^iBu_3 or MAO, all the supported iron catalysts are active toward ethylene polymerization. The studies show that the pretreatment temperature has a significant effect on the catalytic activity, polymerization kinetics, and morphology of the resulting polyethylenes. For the supported iron catalysts [$FeCl_2$L]/$MgCl_2{\cdot}nC_2H_5OH$ (where L = BIP ligand), an increase of the pretreatment temperature results in marked increases both in catalytic activity and bulk density of the resulting polyethylene, and a considerably steadier polymerization kinetics. In the case of the supported catalysts [$FeCl_2$L]/$AlEt_n(OEt)_{3-n}/MgCl_2$, on increasing the pretreatment temperature, the catalytic activity first rises, and then drops down, reaching its maximum value at 100 °C, while the molecular weight and bulk density of the resulting polyethylenes continuously increase. Huang et al. [267] studied the effect of hydrogen on the oligomerization and polymerization of ethylene with the supported catalysts [$FeCl_2$L]/$AlEt_n(OEt)_{3-n}/MgCl_2$. It was found that the hydrogen has a significant activating effect on the supported catalysts bearing sterically encumbered BIP ligands, e.g., [$FeCl_2$(**L87**)]/$AlEt_n(OEt)_{3-n}/MgCl_2$, but similar observation is not applicable in the ethylene oligomerization with a supported catalyst containing a less bulky BIP ligand, e.g., [$FeCl_2$(**L85**)]/$AlEt_n(OEt)_{3-n}/MgCl_2$. The presence of hydrogen in the latter case leads to a decrease in the activity and an overall increase in the product molecular weight, indicating deactivation of the active species responsible for the oligomers and low molecular weight polymer. Additionally, Huang et al. [252] prepared a supported binary catalyst system by co-immobilizing [$FeCl_2$(**L85**)] and [$NiBr_2$(ArN=C(Me)-C(Me)=NAr)] (where Ar = 2,6-$^iPr_2C_6H_3$) complexes on the support $AlEt_n(OEt)_{3-n}/MgCl_2$. In comparison with the single-component iron catalyst [$FeCl_2$(**L85**)]/$AlEt_n(OEt)_{3-n}/MgCl_2$, this binary catalyst system leads to a remarkably improved catalytic activity toward ethylene polymerization, even if a tiny amount of nickel complex is incorporated. This synergetic effect of the nickel complex in the binary catalyst system is attributed to the branched polyethylenes produced by the nickel catalyst during the early stages of the polymerization, which reduce the ethylene monomer diffusion limitation inherent in ethylene polymerization, thereby increasing the activity of the main catalyst component. Mikenas et al. [287] studied the kinetics of the polymerization of ethylene catalyzed by the supported iron catalyst [$FeCl_2$(**L85**)]/$MgCl_2$, in the presence of Al^iBu_3, and found that the corresponding activation energy (11.9 kcal/mol) is close to those of the supported Ziegler–Natta catalysts, the polymerization reaction being first order with respect to monomer, at ethylene concentrations higher than 0.2 mol/L.

Several layered montmorillonite clay-immobilized iron catalysts, prepared in situ by treatment of the corresponding homogeneous iron precatalysts (e.g., [$FeCl_2$(**L85**)] and [$FeCl_2$(**L87**)]) with MAO- or MMAO-modified clay, were used by Tritto [266] and Sivaram [288] groups, respectively, to synthesize hybrid exfoliated polyethylene-clay nanocomposites. These clay-immobilized iron catalysts display a longer polymerization lifetime and turn ethylene polymerization more efficient than the corresponding parent homogeneous systems, the polyethylenes obtained having also higher molecular weights. The studies show that a treatment of the clay with MAO or MMAO gives rise to an increase in the *d*-spacing of the clay galleries, but no further increase in this parameter is observed when the iron precatalyst is immobilized. In the course of ethylene polymerization, the active species of the iron catalyst are intercalated into the layered clay galleries and initiate the polymer growth directly from the filler lamellae interlayer, thus promoting deagglomeration of the clay and its effective dispersion in the polymer. The exfoliation of the clay inside the polymer matrix is dependent on various parameters such as the clay content, catalyst amount, and Al/Fe molar ratio.

Armspach et al. [265] described a new protocol to immobilize iron precatalysts on methylated α- and β-cyclodextrin derivatives, as shown in Scheme 3.81. Amination of α- and β-cyclodextrin dimesylate gives rise to the α- and β-cyclodextrin diamine derivatives, which, followed by condensation with 2,6-pyridinedicarboxaldehyde, in dilute acetonitrile, afford the BIP ligand-capped α- and β-cyclodextrin derivatives. Complexation of these two α- and β-cyclodextrin-supported BIP ligands to iron dichloride leads to the desired supported iron precatalysts. Upon activation with MAO, the α-cyclodextrin-supported iron system is almost inactive toward ethylene polymerization, whereas the β-cyclodextrin-supported one shows an appreciable but still very low catalytic activity (56–583 g/

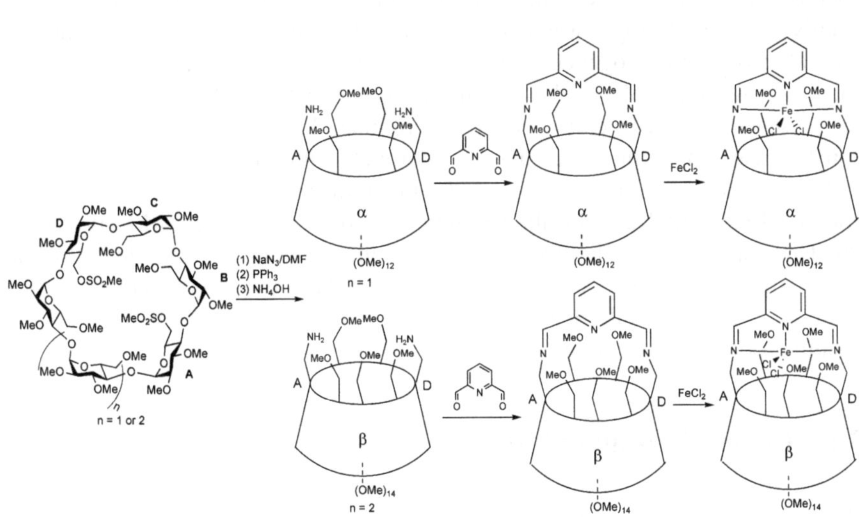

Scheme 3.81 Immobilization of an iron precatalyst on methylated α- and β-cyclodextrin derivatives

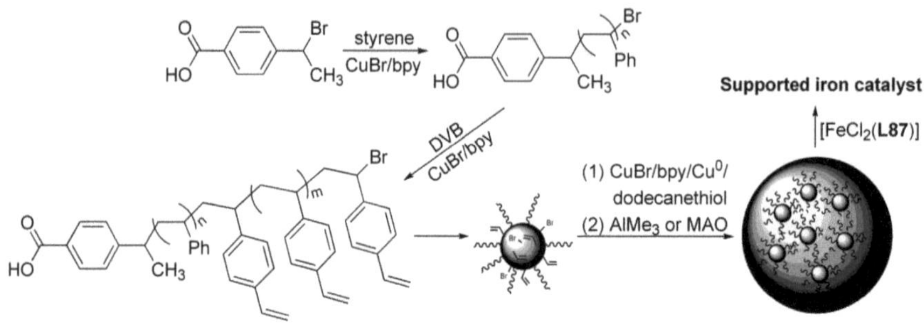

Scheme 3.82 Immobilization of an iron precatalyst on polystyrene microgels

(mol h bar)). The resulting polyethylenes have moderate molecular weights ($M_w = 1.28–2.73 \times 10^4$ g/mol), indicating that the β-cyclodextrin support provides adequate steric protection of the active site for the polymer chain to reach an appreciable length, although at the expense of activity.

Cramail et al. [289, 290] recently described a straightforward and novel route to the preparation of star-like polystyrene (PS)-supported iron catalysts, as displayed in Scheme 3.82. Using 4-(1-bromoethyl)benzoic acid as initiator, styrene is polymerized by atom transfer radical polymerization (ATRP), in the presence of copper(I) bromide and 2,2′-bipyridine (bpy), yielding benzoic acid-functionalized polystyrene macroinitiators, which are cross-linked using divinylbenzene, as coupling agent, in combination with copper(I) bromide and 2,2′-bipyridine, affording PS stars composed of microgel cores and benzoic acid end-capped arms ($M_w = 5 \times 10^4$ g/mol, with an average number of 10 arms per molecule). Using the same synthetic route, PS microgels with arms end-capped by either benzophenone or ethylene oxide units bearing methoxy or hydroxyl end groups can be readily prepared. Since it was experimentally demonstrated that the presence of bromide atoms in microgel cores would deactivate the polymerization of ethylene catalyzed by iron catalysts, these atoms were eliminated by hydrogen transfer in the presence of dodecanethiol, copper(I) bromide, 2,2′-bipyridine (bpy) and copper(0). Treatment of these star-like PS microgels with $AlMe_3$ (the corresponding hydrodynamic radii vary from 6 to 300 nm), followed by reaction with the iron complex [$FeCl_2$(**L87**)], yields the PS-supported iron catalyst [$FeCl_2$(**L87**)]/$AlMe_3$/PS. Upon activation with MAO, it shows high catalytic activities for ethylene polymerization, which are comparable to those of the corresponding homogeneous catalyst. Increased bulk density and improved spherical morphology of the resulting polyethylene are observed.

3.11 Conclusions and Outlook

Originated from the sterically encumbered 2,6-bis(arylimino)pyridine iron and cobalt complexes, discovered by Brookhart and Gibson groups, a vast number of 2,6-bis(imino)pyridine and related ligands have been designed and synthesized so far in order to mimic the original BIP model and to develop the potential of this kind of ligands. In this chapter, these ligands were collected, assembled and classified according to the relevance to each other, and their structural features and reactivities of complexation to iron and cobalt metal salts were discussed. Particular attention was paid to the description of the catalytic performance of their corresponding iron and cobalt catalysts in oligo- or polymerization of ethylene and of other olefins, including the related mechanisms of oligo- or polymerization.

The experimental facts demonstrate that the BIP ligands are virtually amazing chemicals with wealthy and versatile chemistries, which not only is reflected by their excellent complexation to various transition metals (e.g., iron, cobalt, nickel, titanium, vanadium, chromium, zirconium, hafnium, molybdenum, etc.) yielding catalytically active precatalysts for the polymerization of olefins, but also can be looked from the perspective of their ligand-participating chemical reactivities, including the formation of ligand radical anion-stabilized metal complexes, ligand alkylation, ligand deprotonation, and ligand dimerization. In fact, the abundance of chemistries of the BIP ligands seems to be the main reason for the challenging identification of the active species operating in the BIP-based metal catalysts.

The BIP-based iron and cobalt catalysts demonstrated to be highly active catalyst systems toward the ethylene oligo- or polymerization, yielding oligomers with high selectivities or polymers with strictly linear microstructures. The productivities, molecular weights and molecular weight distributions of the products are readily tuned by the modifications of the BIP-type ligands on the different positions of the ligand backbones, such as varying the aryl substituents, employing different aryl rings (e.g., phenyl, naphthyl, anthracenyl and pyrenyl), replacing the central pyridine ring by other heterocyclic rings (e.g., triazine, pyrrole, furan, thiophene and carbazole rings) and tailoring the imino arms.

The high catalytic activities can be obtained by activation with diverse cocatalysts such as MAO, MMAO, AlR_3 (R = Me, Et, ^{i}Bu, ^{n}Hex or Oct), tetraalkylaluminoxanes ($R^1R^2AlOAlR^3R^4$, where R^1, R^2, R^3, R^4 = Et or ^{i}Bu), $AlMe_3$/$B(C_6F_5)_3$ and Al^iBu_3/$[Ph_3C][B(C_6F_5)_4]$.

This kind of iron and cobalt catalysts exhibit low abilities for the homo- or copolymerization of propylene, 1-butene, 1-hexene or other higher α-olefins, but show good capabilities in the oligomerization of these olefin monomers with high selectivities (e.g., dimerization). Additionally, these catalysts also display catalytic activities in the polymerization of conjugated or non-conjugated olefin monomers, such as 1,3-butadiene, isoprene and 1,6-heptadiene. In view of the lower oxophilicities and higher tolerance of the iron and cobalt metal centers toward polar functionalities, the BIP-based iron and cobalt catalysts have also been used to polymerize polar monomers, e.g., MA, MMA, *t*-BA, EVE and *n*-BVE.

Immobilization of the iron and cobalt precatalysts on inorganic or polymeric supports, such as SiO_2, $MgCl_2$, clays and polystyrene, were widely investigated, and the resulting supported catalysts were used to polymerize ethylene with improved polymerization kinetics and polymer morphology. The presence of vinyl groups in the molecular structures of the precatalysts enables them to copolymerize with the ethylene monomer, at the early stages of polymerization, generating self-immobilized supported catalyst systems that function as real supported catalysts and efficiently affects the catalytic performance of the systems and properties of the resulting polymers.

The studies of the polymerization mechanism showed that the formation of the iron and cobalt active species involves metal-to-ligand electron transfer, variation of the central metal oxidation and spin states, and change of ligand charge distribution, which leads to ambiguity in the recognition of the nature of the active species, but meanwhile it provides insight into the understanding of the polymerization mechanism. Nevertheless, a large amount of experimental and theoretical findings pointed to the conclusion that iron- or cobalt-alkyl intermediates virtually are the active species to initiate the oligo- or polymerization, being followed by chain propagation via a migratory insertion mechanism, and chain transfer reactions, such as β-hydrogen transfer (or elimination), chain transfer to aluminum and intramolecular isomerization, which are present during the course of the polymerization, generally leading to broad molecular weight distributions.

The success of the BIP-based iron and cobalt catalysts in olefin oligo- and polymerization highlights the significance of the ligand on determining the performance of the catalyst, and proves the feasibility of ligand-oriented catalyst design ideology. At the same time, the abundance and diversity of this type of ligands provides chemists with a reservoir of selectivities to approach the ideal catalyst system. As the saying goes, all roads lead to Rome. One can believe that, in the future, the more efficient catalyst systems can be achieved by tailoring the ligands, affording polymers with excellent properties.

Acknowledgment We thank the Fundação para Ciência e Tecnologia for financial support (Projects PPCDT/QUI/59025/2004, PTDC/QUI/65474/2006 and PTDC/EQU-EQU/110313/2009, co-financed by FEDER) and a postdoctoral fellowship to L.L. (SFRH/BPD/30733/2006).

References

1. Sinn H, Kaminsky W, Vollmer HJ, Woldt R (1980) Living polymers on polymerization with extremely productive Ziegler catalysts. Angew Chem Int Ed 19:390–392
2. Sinn H, Kaminsky W (1980) Ziegler-Natta catalysis. Adv Organomet Chem 18:99–149
3. Farina M, Silvestro GD, Sozzani P (1993) Hemiisotactic polypropylene: a key point in the elucidation of the polymerization mechanism with metallocene catalysts. Macromolecules 26:946–950
4. Coates GW, Waymouth RM (1995) Oscillating stereocontrol: a strategy for the synthesis of thermoplastic elastomeric polypropylene. Science 267:217–219
5. Hauptman E, Waymouth RM, Ziller JW (1995) Stereoblock polypropylene: ligand effects on the stereospecificity of 2-arylindene zirconocene catalysts. J Am Chem Soc 117:11586–11587
6. Smith JA, Brintzinger HH (1981) Ansa-metallocene derivatives 3. Influence of an interannular ethylene bridge on the reactivity of titanocene derivatives. J Organomet Chem 218:159–167

7. Wild FRWP, Zsolnal L, Huttner G, Brintzinger HH (1982) Ansa-metallocene derivatives 4. Synthesis and molecular structures of chiral ansa-titanocene derivatives with bridge tetrahydroindenyl ligands. J Organomet Chem 232:233–247
8. Ewen JA (1984) Mechanisms of stereochemical control in propylene polymerizations with soluble Group 4B metallocene/methylalumoxane cata lysts. J Am Chem Soc 106:6355–6364
9. Ewen JA, Jones RL, Razavi A, Ferrara JD (1988) Syndiospecific propylene polymerizations with Group 4 metallocenes. J Am Chem Soc 110:6255–6256
10. Canich JAM (Exxon Chemical Patents, Inc.) (1991) Process for producing crystalline poly-alpha-olefins with a monocyclopentadienyl transition metal catalyst system. US Patent 5026798
11. Canich JAM, Licciardi GF (Exxon Chemical Patents, Inc.) (1991) Mono-Cp heteroatom containing group IVB transition metal complexes with MAO: supported catalyst for olefin polymerization. US Patent 5057475
12. Canich JAM (Exxon Chemical Patents, Inc.) (1991) Olefin polymerization catalysts. Eur. Patent 0420436
13. Johnson LK, Killian CM, Brookhart M (1995) New Pd(II)- and Ni(II)-based catalysts for polymerization of ethylene and α-olefins. J Am Chem Soc 117:6414–6415
14. Small BL, Brookhart M, Bennett AMA (1998) Highly active iron and cobalt catalysts for the polymerization of ethylene. J Am Chem Soc 120:4049–4050
15. Small BL, Brookhart M (1998) Iron-based catalysts with exceptionally high activities and selectivities for oligomerization of ethylene to linear α-olefins. J Am Chem Soc 120:7143–7144
16. Britovsek GJP, Gibson VC, Kimberley BS, Maddox PJ, McTavish SJ, Solan GA, White AJP, Williams DJ (1998) Novel olefin polymerization catalysts based on iron and cobalt. Chem Commun 849–850
17. Fujita T, Tohi Y, Mitani M, Matsui S, Saito J, Nitabaru M, Sugi K, Makio H, Tsutsui T (Mitsui Chemicals, Inc.) (1998) Olefin polymerization catalysts, transition metal compounds, processes for olefin polymerization, and alpha-olefin/conjugated diene copolymers. Eur Patent 0874005
18. Wang C, Friedrich S, Younkin TR, Li RT, Grubbs RH, Bansleben DA, Day MW (1998) Neutral nickel(II)-based catalysts for ethylene polymerization. Organometallics 17:3149–3151
19. Mitani M, Mohri J, Yoshida Y, Saito J, Ishii S, Tsuru K, Matsui S, Furuyama R, Nakano T, Tanaka H, Kojoh S, Matsugi T, Kashiwa N, Fujita T (2002) Living polymerization of ethylene catalyzed by titanium complexes having fluorine-containing phenoxy-imine chelate ligands. J Am Chem Soc 124:3327–3336
20. Mitani M, Furuyama R, Mohri J, Saito J, Ishii S, Terao H, Nakano T, Tanaka H, Fujita T (2003) Syndiospecific living propylene polymerization catalyzed by titanium complexes having fluorine-containing phenoxy-imine chelate ligands. J Am Chem Soc 125:4293–4305
21. Mitani M, Nakano T, Fujita T (2003) Unprecedented living olefin polymerization derived from an attractive interaction between a ligand and a growing polymer chain. Chem Eur J 9:2396–2403
22. Bianchini C, Giambastiani G, Rios IG, Mantovani G, Meli A, Segarra AM (2006) Ethylene oligomerization, homopolymerization and copolymerization by iron and cobalt catalysts with 2, 6-(bis-organylimino)pyridyl ligands. Coord Chem Rev 250:1391–1418
23. Gibson VC, Redshaw C, Solan GA (2007) Bis(imino)pyridines: surprisingly reactive ligands and a gateway to new families of catalysts. Chem Rev 107:1745–1776
24. Bianchini C, Giambastiani G, Luconi L, Meli A (2010) Olefin oligomerization, homopolymerization and copolymerization by late transition metals supported by (imino)pyridine ligands. Coord Chem Rev 254:431–455
25. Calderazzo F, Englert U, Pampaloni G, Santi R, Sommazzi A, Zinna M (2005) Bis(arylimino)pyridine derivatives of Group 4 metals: preparation, characterization and activity in ethylene polymerization. Dalton Trans 914–922

26. Reardon D, Conan F, Gambarotta S, Yap G, Wang Q (1999) Life and death of an active ethylene polymerization catalyst. Ligand involvement in catalyst activation and deactivation. Isolation and characterization of two unprecedented neutral and anionic vanadium(I) alkyls. J Am Chem Soc 121:9318–9325
27. Colamarco E, Milione S, Cuomo C, Grassi A (2004) Homo- and copolymerization of butadiene catalyzed by an bis(imino)pyridyl vanadium complex. Macromol Rapid Commun 25:450–454
28. Schmidt R, Welch MB, Knudsen RD, Gottfried S, Alt HG (2004) N,N,N-Tridentate iron(II) and vanadium(III) complexes Part I. Synthesis and characterization. J Mol Catal A: Chem 222:9–15
29. Schmidt R, Welch MB, Knudsen RD, Gottfried S, Alt HG (2004) N,N,N-Tridentate iron(II) and vanadium(III) complexes Part II: catalytic behavior for the oligomerization and polymerization of ethene and characterization of the resulting products. J Mol Catal A Chem 222:17–25
30. Schmidt R, Das PK, Welch MR, Knudsen RD (2004) N, N, N-Tridentate iron(II) and vanadium(III) complexes Part III. UV–vis spectroscopic studies of reactions of ethene-oligomerization and polymerization catalysts with methyl aluminoxane cocatalyst. J Mol Catal A Chem 222:27–45
31. Hanton MJ, Tenza K (2008) Bis(imino)pyridine complexes of the first-row transition metals: alternative methods of activation. Organometallics 27:5712–5716
32. Semikolenova NV, Zakharov VA, Echevskaja LG, Matsko MA, Bryliakov KP, Talsi EP (2009) Homogeneous catalysts for ethylene polymerization based on bis(imino)pyridine complexes of iron, cobalt, vanadium and chromium. Catal Today 144:334–340
33. Romero J, Carrillo-Hermosilla F, Antiñolo A, Otero A (2009) Homogeneous and supported bis(imino)pyridyl vanadium(III) catalysts. J Mol Catal A Chem 304:180–186
34. Esteruelas MA, López AM, Méndez L, Oliván M, Oñate E (2003) Preparation, structure, and ethylene polymerization behavior of bis(imino)pyridyl chromium(III) complexes. Organometallics 22:395–406
35. Small BL, Carney MJ, Holman DM, O'Rourke CE, Halfen JA (2004) New chromium complexes for ethylene oligomerization: Extended use of tridentate ligands in metal-catalyzed olefin polymerization. Macromolecules 37:4375–4386
36. Nakayama Y, Sogo K, Yasuda H, Shiono T (2005) Unique catalytic behavior of chromium complexes having halogenated bis(imino)pyridine ligands for ethylene polymerization. J Polym Sci Part A: Polym Chem 43:3368–3375
37. Chen J, Huang Y, Li Z, Zhang Z, Wei C, Lan T, Zhang W (2006) Syntheses of iron, cobalt, chromium, copper and zinc complexes with bulky bis(imino)pyridyl ligands and their catalytic behaviors in ethylene polymerization and vinyl polymerization of norbornene. J Mol Catal A: Chem 259:133–141
38. Vidyaratne I, Scott J, Gambarotta S, Duchateau R (2007) Reactivity of chromium complexes of a bis(imino)pyridine ligand: highly active ethylene polymerization catalysts carrying the metal in a formally low oxidation state. Organometallics 26:3201–3211
39. Hiya K, Nakayama Y, Yasuda H (2003) Facile syntheses of bis[1-(arylimino)ethyl]pyridine-$MoCl_3$/MMAO catalytic systems and their dual catalytic functions for ROMP of norbornene and linear polymerization of ethylene. Macromolecules 36:7916–7922
40. Suzuki H, Matsumura S, Satoh Y, Sogoh K, Yasuda H (2004) Random copolymerizations of norbornene with other monomers catalyzed by novel Ni compounds involving N- or O-donated ligands. React Funct Polym 58:77–91
41. Suzuki H, Matsumura S, Satoh Y, Sogoh K, Yasuda H (2004) Random and block copolymerizations of norbornene with conjugated 1, 3-dienes catalyzed by novel Ni compounds involving N- or O-donated ligands. React Funct Polym 59:253–266
42. Su B, Zhao J, Zhang Q, Qin W (2009) Cobalt(II) and nickel(II) complexes bearing mono(imino)pyridyl and bis(imino)pyridyl ligands: preparation, structure and ethylene polymerization/oligomerization behaviors. Polym Int 58:1051–1057

43. Knijnenburg Q, Gambarotta S, Budzelaar PHM (2006) Ligand-centred reactivity in diiminepyridine complexes. Dalton Trans 5442–5448
44. Mecking S (2001) Olefin polymerization by late transition metal complexes-A root of Ziegler catalysts gains new ground. Angew Chem Int Ed 40:534–540
45. Keim W, Kowaldt FH, Goddard R, Kruger C (1978) Novel coordination of (benzoylmethylene)triphenylphosphorane in a nickel oligomerization catalyst. Angew Chem Int Ed 17:466–467
46. Peuckert M, Keim W (1983) A new nickel-complex for the oligomerization of ethylene. Organometallics 2:594–597
47. Lions F, Martin KV (1957) Tridentate chelate compounds. I. J Am Chem Soc 79:2733–2738
48. Figgins PE, Busch DH (1960) Complexes of iron(II), cobalt(II) and nickel(II) with biacetyl-bis-methylimine, 2-pyridinal-methylimine and 2, 6-pyridindial-bis-methylimine. J Am Chem Soc 82:820–824
49. Figgins PE, Busch DH (1961) Infrared spectra of octahedral complexes of iron(II), cobalt(II) and nickel (II) with biacetyl-bis-methylimine and pyridinal methylamines. J Phys Chem 65:2236–2240
50. Restivo RJ, Ferguson G (1976) Structural characterization of metal-complexes of 2,6-diacetylpyridine-bis(imines)-crystal and molecular-structure of dinitrato[2,6-bis[1-(phenylimino)ethyl]pyridine]copper(II). Dalton Trans 518–521
51. Small BL, Brookhart M (1999) Polymerization of propylene by a new generation of iron catalysts: mechanisms of chain initiation, propagation, and termination. Macromolecules 32:2120–2130
52. Bristovsek GJP, Mastroianni S, Solan GA, Baugh SPD, Redshaw C, Gibson VC, White AJP, Williams DJ, Elsegood MRJ (2000) Oligomerisation of ethylene by bis(imino)pyridyliron and -cobalt complexes. Chem Eur J 6:2221–2231
53. Britovsek GJP, Bruce M, Gibson VC, Kimberley BS, Maddox PJ, Mastroianni S, McTavish SJ, Redshaw C, Solan GA, Strömberg S, White AJP, Williams DJ (1999) Iron and cobalt ethylene polymerization catalysts bearing 2, 6-bis(imino)pyridyl ligands: synthesis, structures, and polymerization studies. J Am Chem Soc 121:8728–8740
54. Wang J, Li W, Jiang B, Yang Y (2009) $Fe(acac)_n$ and $Co(acac)_n$ bearing different bis(imino)pyridine ligands for ethylene polymerization and oligomerization. J Appl Polym Sci 113:2378–2391
55. Bluhm ME, Folli C, Döring M (2004) New iron-based bis(imino)pyridine and acetyliminopyridine complexes as single-site catalysts for the oligomerization of ethylene. J Mol Catal A Chem 212:13–18
56. Cámpora J, Cartes MÁ, Rodríguez-Delgado A, Naz AM, Palma P, Pérez CM (2009) Studies on the atropisomerism of Fe(II) 2, 6-bis(N-arylimino)pyridine complexes. Inorg Chem 48:3679–3691
57. Görl C, Alt HG (2007) Influence of the para-substitution in bis(arylimino)pyridine iron complexes on the catalytic oligomerization and polymerization of ethylene. J Organomet Chem 692:4580–4592
58. Abu-Surrah AS, Lappalainen K, Piironen U, Lehmus P, Repo T, Leskelä M (2002) New bis(imino)pyridine-iron(II)- and cobalt(II)-based catalysts: synthesis, characterization and activity toward polymerization of ethylene. J Organomet Chem 648:55–61
59. Schmidt R, Welch MB, Palackal SJ, Alt HG (2002) Heterogenized iron(II) complexes as highly active ethene polymerization catalysts. J Mol Catal A Chem 179:155–173
60. Chen Y, Chen R, Qian C, Dong X, Sun J (2003) Halogen-substituted 2, 6-bis(imino)pyridyl iron and cobalt complexes: highly active catalysts for polymerization and oligomerization of ethylene. Organometallics 22:4312–4321
61. Castro PM, Lappalainen K, Ahalgrén M, Leskelä M, Repo T (2003) Iron-based catalysts bearing bis(imido)-pyridine ligands for the polymerization of *tert*-butyl acrylate. J Polym Sci Part A Polym Chem 41:1380–1389

62. Long Z, Wua B, Yang P, Li G, Liu Y, Yang X-J (2009) Synthesis and characterization of para-nitro substituted 2, 6-bis(phenylimino) pyridyl Fe(II) and Co(II) complexes and their ethylene polymerization properties. J Organomet Chem 694:3793–3799
63. Chen Y, Qian C, Sun J (2003) Fluoro-substituted 2, 6-bis(imino)pyridyl iron and cobalt complexes: high-activity ethylene oligomerization catalysts. Organometallics 22:1231–1236
64. Kleigrewe N, Steffen W, Blömker T, Kehr G, Fröhlich R, Wibbeling B, Erker G, Wasilke J-C, Wu G, Bazan GC (2005) Chelate bis(imino)pyridine cobalt complexes: synthesis, reduction, and evidence for the generation of ethane polymerization catalysts by Li^+ cation activation. J Am Chem Soc 127:13955–13968
65. Jurca T, Dawson K, Mallov I, Burchell T, Yap GPA, Richeson DS (2010) Disproportionation and radical formation in the coordination of “GaI” with bis(imino)pyridines. Dalton Trans 39:1266–1272
66. McTavish S, Britovsek GJP, Smit TM, Gibson VC, White AJP, Williams DJ (2007) Iron-based ethylene polymerization catalysts supported by bis(imino)pyridine ligands: derivatization via deprotonation/alkylation at the ketimine methyl position. J Mol Catal A Chem 261:293–300
67. Steves JE, Kennedy MD, Chiang KP, Kassel WS, Dougherty WG, Dudley TJ, Zubris DL (2009) Synthesis of a bulky bis(imino)pyridine compound: a methodology for systematic variation of steric bulk and energetic implications for metalation. Dalton Trans 1214–1222
68. Smit TM, Tomov AK, Gibson VC, White AJP, Williams DJ (2004) Dramatic effect of heteroatom backbone substituents on the ethylene polymerization behavior of bis(imino)pyridine iron catalysts. Inorg Chem 43:6511–6512
69. Ionkin AS, Marshall WJ, Adelman DJ, Fones BB, Fish BM, Schiffhauer MF (2006) High-temperature catalysts for the production of α-olefins based on iron(II) and iron(III) tridentate bis(imino)pyridine complexes with double pattern of substitution: ortho-methyl plus meta-aryl. Organometallics 25:2978–2992
70. Gibson VC, Long NJ, Oxford PJ, White AJP, Williams DJ (2006) Ferrocene-substituted bis(imino)pyridine iron and cobalt complexes: toward redox-active catalysts for the polymerization of ethylene. Organometallics 25:1932–1939
71. Gong D, Wang B, Bai C, Bi J, Wang F, Dong W, Zhang X, Jiang L (2009) Metal dependent control of cis-/trans-1, 4 regioselectivity in 1, 3-butadiene polymerization catalyzed by transition metal complexes supported by 2, 6-bis[1-(iminophenyl)ethyl]pyridine. Polymer 50:6259–6264
72. Zhang Z, Zou J, Cui N, Ke Y, Hu Y (2004) Ethylene oligomerization catalyzed by a novel iron complex containing fluoro and methyl substituents. J Mol Catal A Chem 219:249–254
73. Archer AM, Bouwkamp MW, Cortez M, Lobkovsky E, Chirik PJ (2006) Arene coordination in bis(imino)pyridine iron complexes: identification of catalyst deactivation pathways in iron-catalyzed hydrogenation and hydrosilation. Organometallics 25:4269–4278
74. Tellmann KP, Gibson VC, White AJP, Williams DJ (2005) Selective dimerization/oligomerization of α-olefins by cobalt bis(imino)pyridine catalysts stabilized by trifluoromethyl substituents: Group 9 metal catalysts with productivities matching those of iron systems. Organometallics 24:280–286
75. Ionkin AS, Marshall WJ, Adelman DJ, Fones BB, Fish BM, Scheffhauer MF, Soper PD, Waterland RL, Spence RE, Xie T (2008) High-temperature catalysts for the production of α-olefins based on iron(II) and iron(III) tridentate bis(imino)pyridine complexes modified by nitrilo group. J Polym Sci Part A Polym Chem 46:585–611
76. Ionkin AS, Marshall WJ, Adelman DJ, Shoe AL, Spence RE, Xie T (2006) Nitro-substituted iron(II) tridentate bis(imino)pyridine complexes as high-temperature catalysts for the production of α-olefins. J Polym Sci Part A Polym Chem 44:2615–2635
77. Ionkin AS, Marshall WJ, Adelman DJ, Fones BB, Fish BM, Schiffhauer MF (2008) High-temperature catalysts for the production of α-olefins based on iron(II) and cobalt(II)

tridentate bis(imino)pyridine complexes with a double pattern of substitution: o-methyl plus o-fluorine in the same imino arm. Organometallics 27:1147–1156
78. Kaul FAR, Puchta GT, Frey GD, Herdtweck E, Herrmann WA (2007) Iminopyridine complexes of 3d metals for ethylene polymerization: comparative structural studies and ligand size controlled chain termination. Organometallics 26:988–999
79. Bryliakov KP, Semikolenova NV, Zudin VN, Zakharov VA, Talsi EP (2004) Ferrous rather than ferric species are the active sites in bis(imino)pyridine iron ethylene polymerization catalysts. Catal Commun 5:45–48
80. Britovsek GJP, Clentsmith GKB, Gibson VC, Goodgame DML, McTavish SJ, Pankhurst QA (2002) The nature of the active site in bis(imino)pyridine iron ethylene polymerisation catalysts. Catal Commun 3:207–211
81. Martínez J, Cruz V, Ramos J, Gutiérrez-Oliva S, Martínez-Salazar J, Toro-Labbe A (2008) On the nature of the active site in bis(imino)pyridyl iron, a catalyst for olefin polymerization. J Phys Chem C 112:5023–5028
82. Cruz VL, Ramos J, Martínez-Salazar J, Gutiérrez-Oliva S, Toro-Labbé A (2009) Theoretical study on a multicenter model based on different metal oxidation states for the bis(imino)pyridine iron catalysts in ethylene polymerization. Organometallics 28:5889–5895
83. Raucoules R, Bruin T, Raybaud P, Adamo C (2008) Evidence for the iron(III) oxidation state in bis(imino)pyridine catalysts. A density functional theory study. Organometallics 27:3368–3377
84. Souane R, Isel F, Peruch F, Lutz PJ (2002) Pyridine bis(imine) cobalt or iron complexes for ethylene and 1-hexene (co)polymerization. C R Chim 5:43–48
85. Small BL, Marcucci AJ (2001) Iron catalysts for the head-to-head dimerization of α-olefins and mechanistic implications for the production of linear α-olefins. Organometallics 20:5738–5744
86. Babik ST, Fink G (2002) Propylene polymerization with a bisiminepyridine iron complex: activation with $Ph_3C[B(C_6F_5)_4]$ and AlR_3; iron hydride species in the catalytic cycle. J Mol Catal A Chem 188:245–253
87. Babik ST, Fink G (2003) Propylene bulk phase oligomerization with bisiminepyridine iron complexes in a calorimeter: mechanistic investigation of 1,2 versus 2,1 propylene insertion. J Organomet Chem 683:209–219
88. Pellecchia C, Mazzeo M, Pappalardo D (1998) Isotactic-specific polymerization of propene with an iron-based catalyst: polymer end groups and regiochemistry of propagation. Macromol Rapid Commun 19:651–655
89. Raucoules R, Bruin T, Raybaud P, Adamo C (2009) Theoretical unraveling of selective 1-butene oligomerization catalyzed by iron-bis(arylimino)pyridine. Organometallics 28:5358–5367
90. Small BL (2003) Tridentate cobalt catalysts for linear dimerization and isomerization of α-olefins. Organometallics 22:3178–3183
91. Pelascini F, Peruch F, Lutz PJ, Wesolek M, Kress J (2005) Pyridine bis(imino) iron and cobalt complexes for ethylene polymerization: influence of the aryl imino substituents. Eur Polym J 41:1288–1295
92. Liu J-Y, Zheng Y, Li Y-G, Pan L, Li Y-S, Hu N-H (2005) Fe(II) and Co(II) pyridinebisimine complexes bearing different substituents on *ortho*- and *para*-position of imines: synthesis, characterization and behavior of ethylene polymerization. J Organomet Chem 690:1233–1239
93. Zhang Y, Sun W-H, Li T, Yang X (2004) Influence of electronic effect on catalytic activity of bis(imino)pyridyl Fe(II) and bis(imino)pyrimidyl Fe(II) complexes. J Mol Catal A Chem 218:119–124
94. Ivanchev SS, Tolstikov GA, Badaev VK, Oleinik II, Ivancheva NI, Rogozin DG, Oleinik IV, Myakin SV (2004) New bis(arylimino)pyridyl complexes as components of catalysts for ethylene polymerization. Kinet Catal 45:176–182

95. Kissin YV, Qian C, Xie G, Chen Y (2006) Multi-center nature of ethylene polymerization catalysts based on 2, 6-bis(imino)pyridyl complexes of iron and cobalt. J Polym Sci Part A Polym Chem 44:6159–6170
96. Saito J, Mitani M, Mohri J, Yoshida Y, Matsui S, Ishii S, Kojoh S, Kashiwa N, Fujita T (2001) Living polymerization of ethylene with a titanium complex containing two phenoxy-imine chelate ligands. Angew Chem Int Ed 40:2918–2920
97. Saito J, Mitani M, Onda M, Mohri J, Ishii S, Yoshida Y, Nakano T, Tanaka H, Matsugi T, Kojoh S, Kashiwa N, Fujita T (2001) Microstructure of highly syndiotactic living poly(propylene)s produced from a titanium complex with chelating fluorine-containing phenoxyimine ligands (an FI Catalyst). Macromol Rapid Commun 22:1072–1075
98. Qiu J, Li Y, Hu Y (2000) A new iron-based catalyst for ethylene polymerization. Polym Int 49:5–7
99. Ma Z, Wang H, Qiu J, Xu D, Hu Y (2001) A novel highly active iron/2, 6-bis(imino)pyridyl catalyst for ethylene polymerization. Macromol Rapid Commun 22:1280–1283
100. Ma Z, Sun W-H, Li Z-L, Shao C-X, Hu Y-L, Li X-H (2002) Ethylene polymerization by iron complexes with symmetrical and unsymmetrical ligands. Polym Int 51:994–997
101. Bianchini C, Giambastiani G, Guerrero IR, Meli A, Passaglia E, Gragnoli T (2004) Simultaneous polymerization and Schulz-Flory oligomerization of ethylene made possible by activation with MAO of a C_1-symmetric [2, 6-bis(arylimino)pyridyl]iron dichloride precursor. Organometallics 23:6087–6089
102. Kim I, Han BH, Ha Y-S, Ha C-S, Park D-W (2004) Effect of substituent position on the ethylene polymerization by Fe(II) and Co(II) pyridyl bis-imine catalysts. Catal Today 93–95:281–285
103. Görl C, Alt HG (2007) Iron complexes with ω-alkenyl substituted bis(arylimino)pyridine ligands as catalyst precursors for the oligomerization and polymerization of ethylene. J Mol Catal A Chem 273:118–132
104. Zhang Z, Chen S, Zhang X, Li H, Ke Y, Lu Y, Hu Y (2005) A series of novel 2, 6-bis(imino)pyridyl iron catalysts: synthesis, characterization and ethylene oligomerization. J Mol Catal A Chem 230:1–8
105. Bianchini C, Giambastiani G, Mantovani G, Meli A, Mimeau D (2004) Oligomerisation of ethylene to linear α-olefins by tetrahedral cobalt(II) precursors stabilised by benzo[*b*]thiophen-2-yl-substituted (imino)pyridine ligands. J Organomet Chem 689:1356–1361
106. Bianchini C, Gatteschi D, Giambastiani G, Rios IG, Ienco A, Laschi F, Mealli C, Meli A, Sorace L, Toti A, Vizza F (2007) Electronic influence of the thienyl sulfur atom on the oligomerization of ethylene by cobalt(II) 6-(thienyl)-2-(imino)pyridine catalysis. Organometallics 26:726–739
107. Bianchini C, Mantovani G, Meli A, Migliacci F, Laschi F (2003) Selective oligomerization of ethylene to linear α-olefins by tetrahedral cobalt(II) complexes with 6-(organyl)-2-(imino)pyridyl ligands: influence of the heteroatom in the organyl group on the catalytic activity. Organometallics 22:2545–2547
108. Champouret YDM, Maréchal J, Chaggar RK, Fawcett J, Singh K, Maseras F, Solan GA (2007) Factors affecting imine coordination in (iminoterpyridine)MX_2 (M = Fe, Co, Ni, Zn): synthesis, structures, DFT calculations and ethylene oligomerisation studies. New J Chem 31:75–85
109. Sun W-H, Tang X, Gao T, Wu B, Zhang W, Ma H (2004) Synthesis, characterization, and ethylene oligomerization and polymerization of ferrous and cobaltous 2-(ethylcarboxylato)-6-iminopyridyl complexes. Organometallics 23:5037–5047
110. Fernandes S, Bellabarba RM, Ribeiro AFG, Gomes PT, Ascenso J, Mano JF, Dias AR, Marques MM (2002) Polymerisation of ethylene catalysed by mono-imine-2, 6-diacetylpyridine iron/methylaluminoxane (MAO) catalyst system: effect of the ligand on polymer microstructure. Polym Int 51:1301–1303

111. Sun W-H, Hao P, Li G, Zhang S, Wang W, Yi J, Asma M, Tang N (2007) Synthesis and characterization of iron and cobalt dichloride bearing 2-quinoxalinyl-6-iminopyridines and their catalytic behavior toward ethylene reactivity. J Organomet Chem 692:4506–4518
112. Su B-Y, Zhao J-S (2006) Preparation, structure and ethylene polymerization/oligomerization behavior of 2-carbethoxy-6-iminopyridine-based transition metal complexes. Polyhedron 25:3289–3298
113. Su B-Y, Zhao J-S, Gao Q-C (2007) Crystal structure and ethylene oligomerization catalytic activity of {2-carbethoxy-6-[1-[(2, 6-diethylphenyl) imino]ethyl]pyridine}$CoCl_2$. Chin J Chem 25:121–124
114. Gibson VC, Redshaw C, Solan GA, White AJP, Williams DJ (2007) Aluminum alkyl-mediated route to novel N, N, O-chelates for five-coordinate iron(II) chloride complexes: synthesis, structures, and ethylene polymerization studies. Organometallics 26:5119–5123
115. Wang K, Wedeking K, Zuo W, Zhang D, Sun W-H (2008) Iron(II) and cobalt(II) complexes bearing N-((pyridin-2-yl)methylene)-quinolin-8-amine derivatives: synthesis and application to ethylene oligomerization. J Organomet Chem 693:1073–1080
116. Lappalainen K, Yliheikkilä K, Abu-Surrah AS, Polamo M, Leskelä M, Repo T (2005) Iron(II)- and cobalt(II) complexes with tridentate bis(imino)pyridine nitrogen ligands bearing chiral bulky aliphatic and aromatic substituents: crystal structure of [$CoCl_2${2, 6-bis[R-(+)-(bornylimino)methyl]pyridine}]. Z Anorg Allg Chem 631:763–768
117. Davis RN, Tanski JM, Adrian JC Jr, Tyler LA (2007) Variations in the coordination environment of Co^{2+}, Cu^{2+} and Zn^{2+} complexes prepared from a tridentate (imino)pyridine ligand and their structural comparisons. Inorg Chim Acta 360:3061–3068
118. Bianchini C, Mantovani G, Meli A, Migliacci F, Zanobini F, Laschi F, Sommazzi A (2003) Oligomerisation of ethylene to linear α-olefins by new C_s- and C_1-symmetric [2,6-bis(imino)pyridyl]iron and -cobalt dichloride complexes. Eur J Inorg Chem 1620–1631
119. Wallenhorst C, Kehr G, Luftmann H, Fröhlich R, Erker G (2008) Bis(iminoethyl)pyridine systems with a pendant alkenyl group. Part A: cobalt and iron complexes and their catalytic behavior. Organometallics 27:6547–6556
120. Pelascini F, Wesolek M, Peruch F, Lutz PJ (2006) Modified pyridine-bis(imine) iron and cobalt complexes: synthesis, structure, and ethylene polymerization study. Eur J Inorg Chem 4309–4316
121. Cámpora J, Pérez CM, Rodríguez-Delgado A, Naz AM, Palma P, Alvarez E (2007) Selective alkylation of 2, 6-diiminopyridine ligands by dialkylmanganese reagents: a "one-pot" synthetic methodology. Organometallics 26:1104–1107
122. Cámpora J, Naz AM, Palma P, Rodríguez-Delgado A, Álvarez E, Tritto I, Boggioni L (2008) Iron and cobalt complexes of 4-alkyl-2,6-diiminopyridine ligands: synthesis and ethylene polymerization catalysis. Eur J Inorg Chem 1871–1879
123. Nückel S, Burger P (2001) Transition metal complexes with sterically demanding ligands, 3. Synthetic access to square-planar terdentate pyridine-diimine rhodium(I) and iridium(I) methyl complexes: successful detour via reactive triflate and methoxide complexes. Organometallics 20:4345–4359
124. Ionkin AS, Johnson LK (2008) Catalysts for olefin polymerization or oligomerization. US7442819
125. Beaufort L, Benvenuti F, Noels AF (2006) Iron(II)–ethylene polymerization catalysts bearing 2, 6-bis(imino)pyrazine ligands Part I. Synthesis and characterization. J Mol Catal A Chem 260:210–214
126. Beaufort L, Benvenuti F, Noels AF (2006) Iron(II)–ethylene polymerization catalysts bearing 2, 6-bis(imino)pyrazine ligands Part II. Catalytic behaviour, homogeneous and heterogeneous insights. J Mol Catal A Chem 260:215–220
127. Britovsek GJP, Gibson VC, Hoarau OD, Spitzmesser SK, White AJP, Williams DJ (2003) Iron and cobalt ethylene polymerization catalysts: variations on the central donor. Inorg Chem 42:3454–3465

128. Tenza K, Hanton MJ, Slawin AMZ (2009) Ethylene oligomerization using first-row transition metal complexes featuring heterocyclic variants of bis(imino)pyridine ligands. Organometallics 28:4852–4867
129. Dawson DM, Walker DA, Thornton-Pett M, Bochmann M (2000) Synthesis and reactivity of sterically hindered iminopyrrolato complexes of zirconium, iron, cobalt and nickel. Dalton Trans 459–466
130. Gibson VC, Spitzmesser SK, White AJP, Williams DJ (2003) Synthesis and reactivity of 1,8-bis(imino)carbazolide complexes of iron, cobalt and manganese. Dalton Trans 2718–2727
131. Wang L, Sun W-H, Han L, Li Z, Hu Y, He C, Yan C (2002) Cobalt and nickel complexes bearing 2, 6-bis (imino) phenoxy ligands: syntheses, structures and oligomerization studies. J Organomet Chem 650:59–64
132. Han L, Du J, Yang H, Wang H, Leng X, Galstyan A, Zarić S, Sun W (2003) Crystal structure and modeling calculation of the columnar helix 2, 6-bis(imino)phenoxy iron(III) chloride. Inorg Chem Comm 6:5–9
133. Zhang S, Sun W-H, Xiao T, Hao X (2010) Ferrous and cobaltous chlorides bearing 2, 8-bis(imino)quinolines: highly active catalysts for ethylene polymerization at high temperature. Organometallics 29:1168–1173
134. Britovsek GJP, Baugh SPD, Hoarau O, Gibson VC, Wass DF, White AJP, Williams DJ (2003) The role of bulky substituents in the polymerization of ethylene using late transition metal catalysts: a comparative study of nickel and iron catalyst systems. Inorg Chim Acta 345:279–291
135. Wang L, Sun W-H, Han L, Yang H, Hu Y, Jin X (2002) Late transition metal complexes bearing 2, 9-bis(imino)-1, 10-phenanthrolinyl ligands: synthesis, characterization and their ethylene activity. J Organomet Chem 658:62–70
136. Overett MJ, Meijboom R, Moss JR (2005) New iron bis(imino)pyridyl complexes containing dendritic wedges for alkene oligomerisation. Dalton Trans 551–555
137. Amort C, Malaun M, Krajete A, Kopacka H, Wurst K, Christ M, Lilge D, Kristen MO, Bildstein B (2002) N-pyrrolyl-[N, N, N]-bis(imino)pyridyl iron(II) and cobalt(II) olefin polymerization catalysts. Appl Organomet Chem 16:506–516
138. Britovsek GJP, Gibson VC, Mastroianni S, Oakes DCH, Redshaw C, Solan GA, White AJP, Williams DJ (2001) Imine versus amine donors in iron-based ethylene polymerisation catalysts. Eur J Inorg Chem 431–437
139. Bruce M, Gibson VC, Redshaw C, Solan GA, White AJP, Williams DJ (1998) Cationic alkyl aluminium ethylene polymerization catalysts based on monoanionic N,N,N-pyridyliminoamide ligands. Chem Commun 2523–2524
140. Guérin F, McConville DH, Vittal JJ (1996) Conformationally rigid diamide complexes of zirconium: electron deficient analogues of Cp_2Zr. Organometallics 15:5586–5590
141. Britovsek GJP, England J, Spitzmesser SK, White AJP, Williams DJ (2005) Synthesis of iron(II), manganese(II), cobalt(II) and ruthenium(II) complexes containing tridentate nitrogen ligands and their application in the catalytic oxidation of alkanes. Dalton Trans 945–955
142. Vedder C, Schaper F, Brintzinger H, Kettunen M, Babik S, Fink G (2005) Chiral iron(II) and cobalt(II) complexes with biphenyl-bridged bis(pyridylimine) ligands—syntheses, structures and reactivities. Eur J Inorg Chem 1071–1080
143. Gao R, Wang K, Li Y, Wang F, Sun W-H, Redshaw C, Bochmann M (2009) 2-Benzoxazolyl-6-(1-(arylimino)ethyl)pyridyl cobalt (II) chlorides: a temperature switch catalyst in oligomerization and polymerization of ethylene. J Mol Catal A Chem 309:166–171
144. Gao R, Li Y, Wang F, Sun W-H, Bochmann M (2009) 2-Benzoxazolyl-6-[1-(arylimino) ethyl]pyridyliron(II) chlorides as ethylene oligomerization catalysts. Eur J Inorg Chem 4149–4156

145. Gao R, Zhang M, Liang T, Wang F, Sun W (2008) Nickel(II) complexes chelated by 2-arylimino-6-benzoxazolylpyridine: syntheses, characterization, and ethylene oligomerization. Organometallics 27:5641–5648
146. Hao P, Chen Y, Xiao T, Sun W-H (2010) Iron(III) complexes bearing 2-(benzimidazole)-6-(1-aryliminoethyl)pyridines: synthesis, characterization and their catalytic behaviors toward ethylene oligomerization and polymerization. J Organomet Chem 695:90–95
147. Chen Y, Hao P, Zuo W, Gao K, Sun W-H (2008) 2-(1-Isopropyl-2-benzimidazolyl)-6-(1-aryliminoethyl)pyridyl transition metal (Fe, Co, and Ni) dichlorides: syntheses, characterizations and their catalytic behaviors toward ethylene reactivity. J Organomet Chem 693:1829–1840
148. Xiao L, Gao R, Zhang M, Li Y, Cao X, Sun W-H (2009) 2-(1*H*–2-benzimidazolyl)-6-(1-(arylimino)ethyl)pyridyl iron(II) and cobalt(II) dichlorides: syntheses, characterizations, and catalytic behaviors toward ethylene reactivity. Organometallics 28:2225–2233
149. Sun W-H, Hao P, Zhang S, Shi Q, Zuo W, Tang X, Lu X (2007) Iron(II) and cobalt(II) 2-(benzimidazolyl)-6-(1-(arylimino)ethyl)pyridyl complexes as catalysts for ethylene oligomerization and polymerization. Organometallics 26:2720–2734
150. Wang K, Hao P, Zhang D, Sun W-H (2008) Tridentate N^N^N iron(II) and cobalt(II) complexes of ion-paired structures: synthesis, characterization and magnetism. J Mol Struct 890:95–100
151. Karam AR, Catarí EL, López-Linares F, Agrifoglio G, Albano CL, Díaz-Barrios A, Lehmann TE, Pekerar SV, Albornoz LA, Atencio R, González T, Ortega HB, Joskowics P (2005) Synthesis, characterization and olefin polymerization studies of iron(II) and cobalt(II) catalysts bearing 2, 6-bis(pyrazol-1-yl)pyridines and 2, 6-bis(pyrazol-1-ylmethyl)pyridines ligands. Appl Catal A Gen 280:165–173
152. Calderazzo F, Englert U, Hub C, Marchetti F, Pampaloni G, Passarelli V, Romano A, Santi R (2003) Synthesis and structural characterization of iron(II) derivatives of heterocyclic tridentate amines. Inorg Chim Acta 344:197–206
153. Zabel D, Schubert A, Wolmershäuser G, Jones Jr RL, Thiel WR (2008) Iron and cobalt complexes of tridentate N-donor ligands in ethylene polymerization: efficient shielding of the active sites by simple phenyl groups. Eur J Inorg Chem 3648–3654
154. Imanishi Y, Nomura K (2000) Olefin polymerization and copolymerization with soluble transition-metal complex catalysts. J Polym Sci Part A Polym Chem 38:4613–4626
155. Tondreau AM, Darmon LM, Wile BM, Floyd SK, Lobkovsky E, Chirik PJ (2009) Enantiopure pyridine bis(oxazoline) "Pybox" and bis(oxazoline) "Box" iron dialkyl complexes: comparison to bis(imino)pyridine compounds and application to catalytic hydrosilylation of ketones. Organometallics 28:3928–3940
156. Nakayama Y, Baba Y, Yasuda H, Kawakita K, Ueyama N (2003) Stereospecific polymerizations of conjugated dienes by single site iron complexes having chelating N, N, N-donor ligands. Macromolecules 36:7953–7958
157. Danopoulos AA, Tsoureas N, Wright JA, Light ME (2004) N-Heterocyclic pincer dicarbene complexes of iron(II): C-2 and C-5 metalated carbenes on the same metal center. Organometallics 23:166–168
158. McGuinness DS, Gibson VC, Steed JW (2004) Bis(carbene)pyridine complexes of the early to middle transition metals: survey of ethylene oligomerization and polymerization capability. Organometallics 23:6288–6292
159. Danopoulos AA, Wright JA, Motherwell WB, Ellwood S (2004) N-Heterocyclic "Pincer" dicarbene complexes of cobalt(I), cobalt(II), and cobalt(III). Organometallics 23:4807–4810
160. Al-Benna S, Sarsfield MJ, Thornton-Pett M, Ormsby DL, Maddox PJ, Brès P, Bochmann M (2000) Sterically hindered iminophosphorane complexes of vanadium, iron, cobalt and nickel: a synthetic, structural and catalytic study. Dalton Trans 4247–4257
161. Müllera G, Klinga M, Leskelä M, Rieger B (2002) Iron and cobalt complexes of a series of tridentate P, N, P ligands synthesis, characterization, and application in ethene polymerization reactions. Z Anorg Allg Chem 628:2839–2846

162. Sun W-H, Jie S, Zhang S, Zhang W, Song Y, Ma H, Chen J, Wedeking K, Fröhlich R (2006) Iron complexes bearing 2-imino-1, 10-phenanthrolinyl ligands as highly active catalysts for ethylene oligomerization. Organometallics 25:666–677
163. Jie S, Zhang S, Wedeking K, Zhang W, Ma H, Lu X, Deng Y, Sun W-H (2006) Cobalt(II) complexes bearing 2-imino-1, 10-phenanthroline ligands: synthesis, characterization and ethylene oligomerization. C R Chim 9:1500–1509
164. Jie S, Zhang S, Sun W-H, Kuang X, Liu T, Guo J (2007) Iron(II) complexes ligated by 2-imino-1, 10-phenanthrolines: preparation and catalytic behavior toward ethylene oligomerization. J Mol Catal A Chem 269:85–96
165. Zhang M, Zhang W, Xiao T, Xiang J-F, Hao X, Sun W-H (2010) 2-Ethyl-ketimino-1, 10-phenanthroline iron(II) complexes as highly active catalysts for ethylene oligomerization. J Mol Catal A Chem 320:92–96
166. Jie S, Zhang S, Sun W-H (2007) 2-Arylimino-9-phenyl-1,10-phenanthrolinyl-iron, -cobalt and -nickel complexes: synthesis, characterization and ethylene oligomerization behavior. Eur J Inorg Chem 5584–5598
167. Pelletier JDA, Champouret YDM, Cadarso J, Clowes L, Gañete M, Singh K, Thanarajasingham V, Solan GA (2006) Electronically variable imino-phenanthrolinyl-cobalt complexes; synthesis, structures and ethylene oligomerisation studies. J Organomet Chem 691:4114–4123
168. Zhang M, Hao P, Zuo W, Jie S, Sun W-H (2008) 2-(Benzimidazol-2-yl)-1, 10-phenanthrolyl metal (Fe and Co) complexes and their catalytic behaviors toward ethylene oligomerization. J Organomet Chem 693:483–491
169. Zhang M, Gao R, Hao X, Sun W-H (2008) 2-Oxazoline/benzoxazole-1, 10-phenanthrolinylmetal (iron, cobalt or nickel) dichloride: synthesis, characterization and their catalytic reactivity for the ethylene oligomerization. J Organomet Chem 693:3867–3877
170. Wang L, Ren H, Sun J (2008) Fe(acac)$_3$-bis(imino)pyridine/MAO: a new catalytic system for ethylene polymerization. J Appl Polym Sci 108:167–173
171. Wang L, Sun J (2008) Methylene bridged binuclear bis(imino)pyridyl iron(II) complexes and their use as catalysts together with Al(*i*-Bu)$_3$ for ethylene polymerization. Inorg Chim Acta 361:1843–1849
172. Liu J, Li Y, Liu J, Li Z (2005) Ethylene polymerization with a highly active and long-lifetime macrocycle trinuclear 2, 6-bis(imino)pyridyliron. Macromolecules 38:2559–2563
173. Tolstikov GA, Ivanchev SS, Oleinik II, Ivancheva NI, Oleinik IV (2005) New multifunctional bis(imino)pyridine-iron chloride complexes and ethylene polymerization catalysts on their basis. Doklady Phys Chem 404:182–185
174. Zheng Z-J, Chen J, Li Y-S (2004) The synthesis and catalytic activity of poly(bis(imino)pyridyl) iron(II) metallodendrimer. J Organomet Chem 689:3040–3045
175. Khamker Q, Champouret YDM, Singh K, Solan GA (2009) Bis(imino)quaterpyridine-bearing multimetallic late transition metal complexes as ethylene oligomerisation catalysts. Dalton Trans 8935–8944
176. Barbaro P, Bianchini C, Giambastiani G, Rios IG, Meli A, Oberhauser W, Segarra AM, Sorace L, Toti A (2007) Synthesis of new polydentate nitrogen ligands and their use in ethylene polymerization in conjunction with iron(II) and cobalt(II) bis-halides and methylaluminoxane. Organometallics 26:4639–4651
177. Bianchini C, Giambastiani G, Rios IG, Meli A, Oberhauser W, Sorace L, Toti A (2007) Synthesis of a new polydentate ligand obtained by coupling 2, 6-bis(imino)pyridine and (imino)pyridine moieties and its use in ethylene oligomerization in conjunction with iron(II) and cobalt(II) bis-halides. Organometallics 26:5066–5078
178. Zhang S, Vystorop I, Tang Z, Sun W-H (2007) Bimetallic (iron or cobalt) complexes bearing 2-methyl-2, 4-bis(6-iminopyridin-2-yl)-1*H*–1, 5-benzodiazepines for ethylene reactivity. Organometallics 26:2456–2460
179. Talsi EP, Babushkin DE, Semikolenova NV, Zudin VN, Panchenko VN, Zakharov VA (2001) Polymerization of ethylene catalyzed by iron complex bearing 2, 6-bis(imine)pyridyl

ligand: ^{1}H and ^{2}H NMR monitoring of ferrous species formed via catalyst activation with $AlMe_3$, MAO, $AlMe_3/B(C_6F_5)_3$ and $AlMe_3/CPh_3B(C_6F_5)_4$. Macromol Chem Phys 202: 2046–2051
180. Castro PM, Lahtinen P, Axenov K, Viidanoja J, Kotiaho T, Leskelä M, Repo T (2005) Activation of 2, 6-bis(imino)pyridine iron(II) chloride complexes by methylaluminoxane: an electrospray ionization tandem mass spectrometry investigation. Organometallics 24:3664–3670
181. Bryliakov KP, Talsi EP, Semikolenova NV, Zakharov VA (2009) Formation and nature of the active sites in bis(imino)pyridine iron-based polymerization catalysts. Organometallics 28:3225–3232
182. Deng L, Margl P, Ziegler T (1999) Mechanistic aspects of ethylene polymerization by iron(II)-bisimine pyridine catalysts: a combined density functional theory and molecular mechanics study. J Am Chem Soc 121:6479–6487
183. Griffiths EAH, Britovsek GJP, Gibson VC, Gould IR (1999) Highly active ethylene polymerisation catalysts based on iron: an ab initio study. Chem Commun 1333–1334
184. Khoroshun DV, Musaev DG, Vreven T, Morokuma K (2001) Theoretical study on bis(imino)pyridyl-Fe(II) olefin poly- and oligomerization catalysts. Dominance of different spin states in propagation and β-hydride transfer pathways. Organometallics 20:2007–2026
185. Bryliakov KP, Semikolenova NV, Zakharov VA, Talsi EP (2004) Active intermediates of ethylene polymerization over 2, 6-bis(imino)pyridyl iron complex activated with aluminum trialkyls and methylaluminoxane. Organometallics 23:5375–5378
186. Bouwkamp MW, Bart SC, Hawrelak EJ, Trovitch RJ, Lobkovsky E, Chirik PJ (2005) Square planar bis(imino)pyridine iron halide and alkyl complexe. Chem Commun 3406–3408
187. Fernández I, Trovitch RJ, Lobkovsky E, Chirik PJ (2008) Synthesis of bis(imino)pyridine iron di- and monoalkyl complexes: stability differences between $FeCH_2SiMe_3$ and $FeCH_2CMe_3$ derivatives. Organometallics 27:109–118
188. Scott J, Gambarotta S, Korobkov I, Budzelaar PHM (2005) Reduction of (diiminopyridine)iron: evidence for a noncationic polymerization pathway? Organometallics 24:6298–6300
189. Cámpora J, Naz AM, Palma P, Álvarez E, Reyes ML (2005) 2, 6-Diiminopyridine iron(II) dialkyl complexes. Interaction with aluminum alkyls and ethylene polymerization catalysis. Organometallics 24:4878–4881
190. Scott J, Gambarotta S, Korobkov I, Budzelaar PHM (2005) Metal versus ligand alkylation in the reactivity of the (bis-iminopyridinato)Fe catalyst. J Am Chem Soc 127:13019–13029
191. Bouwkamp MW, Lobkovsky E, Chirik PJ (2005) Bis(imino)pyridine iron(II) alkyl cations for olefin polymerization. J Am Chem Soc 127:9660–9661
192. Trovitch RJ, Lobkovsky E, Chirik PJ (2008) Bis(imino)pyridine iron alkyls containing β-hydrogens: synthesis, evaluation of kinetic stability, and decomposition pathways involving chelate participation. J Am Chem Soc 130:11631–11640
193. Bart SC, Chłopek K, Bill E, Bouwkamp MW, Lobkovsky E, Neese F, Wieghardt K, Chirik PJ (2006) Electronic structure of bis(imino)pyridine iron dichloride, monochloride, and neutral ligand complexes: a combined structural, spectroscopic, and computational study. J Am Chem Soc 128:13901–13912
194. Bart SC, Lobkovsky E, Chirik PJ (2004) Preparation and molecular and electronic structures of iron(0) dinitrogen and silane complexes and their application to catalytic hydrogenation and hydrosilation. J Am Chem Soc 126:13794–13807
195. Margl P, Deng L, Ziegler T (1999) Cobalt(II) imino pyridine assisted ethylene polymerization: a quantum-mechanical/molecular-mechanical density functional theory investigation. Organometallics 18:5701–5708
196. Gibson VC, Humphries MJ, Tellmann KP, Wass DF, White AJP, Williams DJ (2001) The nature of the active species in bis(imino)pyridyl cobalt ethylene polymerisation catalysts. Chem Commun 2252–2253

197. Kooistra TM, Knijnenburg Q, Smits JMM, Horton AD, Budzelaar PHM, Gal AW (2001) Olefin polymerization with [{bis(imino)pyridyl}$Co^{II}Cl_2$]: generation of the active species involves CoI. Angew Chem Int Ed 40:4719–4722
198. Steffen W, Blömker T, Kleigrewe N, Kehr G, Fröhlich R, Erker G (2004) Generation of bis(iminoalkyl)pyridine cobalt(I) cations under metal alkyl free conditions that polymerize ethane. Chem Commun 1188–1189
199. Gibson VC, Tellmann KP, Humphries MJ, Wass DF (2002) Bis(imino)pyridine cobalt alkyl complexes and their reactivity toward ethylene: a model system for β-hydrogen chain transfer. Chem Commun 2316–2317
200. Tellmann KP, Humphries MJ, Rzepa HS, Gibson VC (2004) Experimental and computational study of β-H transfer between cobalt(I) alkyl complexes and 1-alkenes. Organometallics 23:5503–5513
201. Semikolenova NV, Zakharov VA, Talsi EP, Babushkin DE, Sobolev AP, Echevskaya LG, Khysniyarov MM (2002) Study of the ethylene polymerization over homogeneous and supported catalysts based on 2, 6-bis(imino)pyridyl complexes of Fe(II) and Co(II). J Mol Catal A Chem 182–183:283–294
202. Soshnikov IE, Semikolenova NV, Bushmelev AN, Bryliakov KP, Lyakin OY, Redshaw C, Zakharov VA, Talsi EP (2009) Investigating the nature of the active species in bis(imino)pyridine cobalt ethylene polymerization catalysts. Organometallics 28: 6003–6013
203. Humphries MJ, Tellmann KP, Gibson VC, White AJP, Williams DJ (2005) Investigations into the mechanism of activation and initiation of ethylene polymerization by bis(imino)pyridine cobalt catalysts: synthesis, structures, and deuterium labeling studies. Organometallics 24:2039–2050
204. Knijnenburg Q, Hetterscheid D, Kooistra TM, Budzelaar PHM (2004) The electronic structure of (diiminopyridine)cobalt(I) complexes. Eur J Inorg Chem 1204–1211
205. Bowman AC, Milsmann C, Atienza CCH, Lobkovsky E, Wieghardt K, Chirik PJ (2010) Synthesis and molecular and electronic structures of reduced bis(imino)pyridine cobalt dinitrogen complexes: ligand versus metal reduction. J Am Chem Soc 132:1676–1684
206. Tomov AK, Gibson VC, Britovsek GJP, Long RJ, Meurs M, Jones DJ, Tellmann KP, Chirinos JJ (2009) Distinguishing chain growth mechanisms in metal-catalyzed olefin oligomerization and polymerization systems: C_2H_4/C_2D_4 co-oligomerization/polymerization experiments using chromium, iron, and cobalt catalysts. Organometallics 28:7033–7040
207. Barabanov AA, Bukatov GD, Zakharov VA, Semikolenova NV, Echevskaja LG, Matsko MA (2005) Kinetic peculiarities of ethylene polymerization over homogeneous bis(imino)pyridine Fe(II) catalysts with different activators. Macromol Chem Phys 206:2292–2298
208. Wang S, Liu D, Mao B (2009) Characterization of compositional heterogeneity in the polyethylene prepared with bis(imino)pyridyl iron(II) precatalyst and triethylaluminum. J Appl Polym Sci 111:1151–1155
209. Wang S, Liu D, Huang R, Zhang Y, Mao B (2006) Studies on the activation and polymerization mechanism of ethylene polymerization catalyzed by bis(imino)pyridyl iron(II) precatalyst with alkylaluminum. J Mol Catal A Chem 245:122–131
210. Britovsek GJP, Cohen SA, Gibson VC, Maddox PJ, Meurs M (2002) Iron-catalyzed polyethylene chain growth on zinc: linear α-olefins with a Poisson distribution. Angew Chem Int Ed 41:489–491
211. Britovsek GJP, Cohen SA, Gibson VC, Meurs M (2004) Iron catalyzed polyethylene chain growth on zinc: a study of the factors delineating chain transfer versus catalyzed chain growth in zinc and related metal alkyl systems. J Am Chem Soc 126:10701–10712
212. Wang Q, Li L, Fan Z (2005) Effects of tetraalkylaluminoxane activators on ethylene polymerization catalyzed by iron-based complexes. Eur Polym J 41:1170–1176

213. Li L, Wang Q, Fan ZQ (2004) Synthesis of polyethylene with bimodal molecular weight distribution by iron-based late transition metal catalyst/various aluminoxanes system. Chem J Chin Univ 25:1777–1779
214. Wang Q, Li L (2004) Effect of aluminoxane on molecular weight and molecular weight distribution of polyethylene prepared by an iron-based catalyst. Polym Int 53:1473–1478
215. Wang Q, Li L, Fan Z (2004) Polyethylene with bimodal molecular weight distribution synthesized by 2, 6-bis(imino)pyridyl complexes of Fe(II) activated with various activators. Eur Polym J 40:1881–1886
216. Wang Q, Li L, Fan Z (2005) Effect of alkylaluminum on ethylene polymerization catalyzed by 2, 6-bis(imino)pyridyl complexes of Fe(II). J Polym Sci Part A Polym Chem 43:1599–1606
217. Li L, Wang Q (2004) Synthesis of polyethylene with bimodal molecular weight distribution by supported iron-based catalyst. J Polym Sci Part A Polym Chem 42:5662–5669
218. Radhakrishnan K, Cramail H, Deffieux A, François P, Momtaz A (2003) Influence of alkylaluminium activators and mixtures thereof on ethylene polymerization with a tridentate bis(imino)pyridinyliron complex. Macromol Rapid Commun 24:251–254
219. Kumar KR, Sivaram S (2000) Ethylene polymerization using iron(II)bis(imino) pyridyl and nickel (diimine) catalysts: effect of cocatalysts and reaction parameters. Macromol Chem Phys 201:1513–1520
220. Wang Q, Yang H, Fan Z (2002) Efficient activators for an iron catalyst in the polymerization of ethylene. Macromol Rapid Commun 23:639–642
221. Knijnenburg Q, Smits JMM, Budzelaar PHM (2006) Reaction of the diimine pyridine ligand with aluminum alkyls: an unexpectedly complex reaction. Organometallics 25:1036–1046
222. Knijnenburg Q, Smits JMM, Budzelaar PHM (2004) Reaction of $AlEt_2Cl$ with the diiminepyridine ligand: an unexpected product. C R Chim 7:865–869
223. Scott J, Gambarotta S, Korobkov I, Knijnenburg Q, Bruin B, Budzelaar PHM (2005) Formation of a paramagnetic Al complex and extrusion of Fe during the reaction of (diiminepyridine)Fe with AlR_3 (R = Me, Et). J Am Chem Soc 127:17204–17206
224. Yang H, Wang Q, Fan Z, Xu H (2004) Improved high-temperature performance of iron (II) complexes for ethylene polymerization by variation of aluminoxanes. Polym Int 53:37–40
225. Wang Q, Yang H, Fan Z, Xu H (2004) Performance of various activators in ethylene polymerization based on an iron(II) catalyst system. J Polym Sci Part A Polym Chem 42:1093–1099
226. Tudella J, Ribeiro MR, Cramail H, Deffieux A (2007) Novel organic activators for single site iron catalysts. Macromol Chem Phys 208:815–822
227. Ramos J, Cruz V, Muñoz-Escalona A, Martínez-Salazar J (2002) A computational study of iron-based Gibson-Brookhart catalysts for the copolymerization of ethylene and 1-hexene. Polymer 43:3635–3645
228. Takeuchi D, Matsuura R, Park S, Osakada K (2007) Cyclopolymerization of 1, 6-heptadienes catalyzed by iron and cobalt complexes: synthesis of polymers with trans- or cis-fused 1, 2-cyclopentanediyl groups depending on the catalyst. J Am Chem Soc 129:7002–7003
229. Takeuchi D, Matsuura R, Fukuda Y, Osakada K (2009) Selective cyclopolymerization of α,ω-dienes and copolymerization with ethylene catalyzed by Fe and Co complexes. Dalton Trans 8955–8962
230. Takeuchi D, Osakada K (2008) New polymerization of dienes and related monomers catalyzed by late transition metal complexes. Polymer 49:4911–4924
231. Takeuchi D, Matsuura R, Osakada K (2008) Copolymerization of hepta-1, 6-diene with ethylene catalyzed by cobalt complexes. Macromol Rapid Commun 29:1932–1936
232. Takeuchi D, Anada K, Osakada K (2004) Cobalt-complex-catalyzed copolymerization of ethylene with 2-aryl-1-methylenecyclopropanes. Angew Chem Int Ed 43:1233–1235
233. Takeuchi D, Osakada K (2005) Alternating copolymerization of ethylene with 7-methylenebicyclo[4.1.0]heptane promoted by the cobalt complex. Highly regulated

structure and thermal rearrangement of the obtained copolymer. Macromolecules 38:1528–1530
234. Pelascini F, Peruch F, Lutz PJ, Wesolek M, Kress J (2003) Polymerization of norbornene with $CoCl_2$ and pyridine bisimine cobalt(II) complexes activated with MAO. Macromol Rapid Commun 24:768–771
235. Pelascini F, Peruch F, Lutz PJ, Wesolek M, Kress J (2004) Polymerization of norbornene with Co(II) complexes. Macromol Symp 213:265–274
236. Sacchi MC, Sonzogni M, Losio S, Forlini F, Locatelli P, Tritto I, Licchelli M (2001) Vinylic polymerization of norbornene by late transition metal-based catalysis. Macromol Chem Phys 202:2052–2058
237. Yliheikkilä K, Lappalainen K, Castro PM, Ibrahim K, Abu-Surrah A, Leskelä M, Repo T (2006) Polymerization of acrylate monomers with MAO activated iron(II) and cobalt(II) complexes bearing tri- and tetradentate nitrogen ligands. Eur Polym J 42:92–100
238. Castro PM, Lankinen MP, Uusitalo A, Leskelä M, Repo T (2004) Polymerization of acrylate monomers by iron(II) complexes bearing bis(imino)pyridyl or phosphine ligands. Macromol Symp 213:199–208
239. Abu-Surrah AS, Qaroush AK (2007) Polymerization of vinyl monomers via MAO activated iron(II) dichloro complexes bearing bis(imino)pyridine-, quinolinaldimine- and thiophenaldimine-based tridentate nitrogen ligands. Eur Polym J 43:2967–2974
240. Fullana MJ, Miri MJ, Vadhavkar SS, Kolhatkar N, Delis AC (2008) Polymerization of methyl acrylate and as comonomer with ethylene using a bis(imino)pyridine iron(II) chloride/methylaluminoxane catalyst. J Polym Sci Part A Polym Chem 46:5542–5558
241. Liu J, Zheng Y, Li Y (2004) Polymerization of methyl methacrylate by iron(II) pyridinebisimine complexes. Polymer 45:2297–2301
242. Kim I, Ha YS, Ha C-S (2004) Polymerization of vinyl ethers by iron(II) and cobalt(II) pyridyl bis(imine) complexes in the presence of methylaluminoxane. Macromol Rapid Commun 25:1069–1072
243. de Souza RF, Casagrande OL Jr (2001) Recent advances in olefin polymerization using binary catalyst systems. Macromol Rapid Commun 22:1293–1301
244. Komon ZJA, Bazan GC (2001) Synthesis of branched polyethylene by tandem catalysis. Macromol Rapid Commun 22:467–478
245. Mecking S (1999) Reactor blending with early/late transition metal catalyst combinations in ethylene polymerization. Macromol Rapid Commun 20:139–143
246. Long Y-Y, Liu S-R, Cui L, Li Y-S (2010) Influence of branching on the thermal and crystallization behavior of bimodal polyethylenes synthesized with binary late-transition-metal catalyst combinations. J Appl Polym Sci 115:3045–3055
247. Pan L, Zhang KY, Li YG, Bo SQ, Li YS (2007) Thermal and crystallization behaviors of polyethylene blends synthesized by binary late transition metal catalysts combinations. J Appl Polym Sci 104:4188–4198
248. Quijada R, Rojas R, Bazan G, Komon ZJA, Mauler RS, Galland GB (2001) Synthesis of branched polyethylene from ethylene by tandem action of iron and zirconium single site catalysts. Macromolecules 34:2411–2417
249. Bianchini C, Frediani M, Giambastiani G, Kaminsky W, Meli A, Passaglia E (2005) Amorphous polyethylene by tandem action of cobalt and titanium single-site catalysts. Macromol Rapid Commun 26:1218–1223
250. Frediani M, Piel C, Kaminsky W, Bianchini C, Rosi L (2006) Tandem copolymerization: an effective control of the level of branching and molecular weight distribution. Macromol Symp 236:124–133
251. Kukalyekar N, Balzano L, Peters GWM, Rastogi S, Chadwick JC (2009) Characteristics of bimodal polyethylene prepared via co-immobilization of chromium and iron catalysts on an $MgCl_2$-based support. Macromol React Eng 3:448–454
252. Huang R, Koning C, Chadwick JC (2007) Synergetic effect of a nickel diimine in ethylene polymerization with immobilized Fe-, Cr-, and Ti-based catalysts on $MgCl_2$ supports. Macromolecules 40:3021–3029

253. Ivanchev SS, Ivancheva NI, Khaikin SY, Sviridova EV, Rogozin DG (2006) Polymerization of ethylene by SiO_2-supported two-component catalytic systems containing bis(imino)pyridine and bis(imine) ligands. Polym Sci Ser A 48:251–256
254. Wang H, Ma Z, Ke Y, Hu Y (2003) Synthesis of linear low density polyethylene (LLDPE) by in situ copolymerization with novel cobalt and zirconium catalysts. Polym Int 52:1546–1552
255. Ye Z, Zhu S (2003) Synthesis of branched polypropylene with isotactic backbone and atactic side chains by binary iron and zirconium single-site catalysts. J Polym Sci Part A Polym Chem 41:1152–1159
256. Zhang Z, Lua Z, Chen S, Li H, Zhang X, Lu Y, Hu Y (2005) Synthesis of branched polyethylene from ethylene stock by an interference-free tandem catalysis of $TiCl_4/MgCl_2$ and iron catalyst. J Mol Catal A Chem 236:87–93
257. Lu Z, Zhang Z, Li Y, Wu C, Hu Y (2006) Synthesis of branched polyethylene by in situ polymerization of ethylene with combined iron catalyst and Ziegler–Natta catalyst. J Appl Polym Sci 99:2898–2903
258. Xu H, Guo C-Y, Zhang M, Yang H-J, Dong J, Yuan G (2007) In situ copolymerization of ethylene to linear low-density polyethylene (LLDPE) with calcosilicate (CAS-1) supported dual-functional catalytic system. Catal Commun 8:2143–2149
259. Ma Z, Ke Y, Wang H, Guo C, Zhang M, Sun W-H, Hu Y (2003) Ethylene polymerization with a silica-supported iron-based diimine catalyst. J Appl Polym Sci 88:466–469
260. Helldörfer M, Alt HG, Ebenhoch J (2002) Microgels as new support materials for heterogeneous cocatalysts in ethylene polymerization reactions. J Appl Polym Sci 86:3021–3029
261. Seitz M, Milius W, Alt HG (2007) Iron(II) coordination compounds with ω-alkenyl substituted bis(imino)pyridine ligands: self-immobilizing catalysts for the polymerization of ethylene. J Mol Catal A Chem 261:246–253
262. Huanga R, Kukalyekar N, Koning CE, Chadwick JC (2006) Immobilization and activation of 2,6-bis(imino)pyridyl Fe, Cr and V precatalysts using a $MgCl_2/AlR_n(OEt)_{3-n}$ support: effects on polyethylene molecular weight and molecular weight distribution. J Mol Catal A Chem 260:135–143
263. Huanga R, Liu D, Wang S, Mao B (2005) Preparation of spherical $MgCl_2$ supported bis(imino)pyridyl iron(II) precatalyst for ethylene polymerization. J Mol Catal A Chem 233:91–97
264. Paulino IS, Schuchardt U (2004) Ethylene polymerization using iron catalysts heterogenized in MCM-41. Catal Commun 5:5–7
265. Armspach D, Matt D, Peruch F, Lutz P (2003) Cyclodextrin-encapsulated iron catalysts for the polymerization of ethylene. Eur J Inorg Chem 805–809
266. Leone G, Bertini F, Canetti M, Boggioni L, Conzatti L, Tritto I (2009) Long-lived layered silicates-immobilized 2, 6-bis(imino)pyridyl iron(II) catalysts for hybrid polyethylene nanocomposites by in situ polymerization: effect of aryl ligand and silicate modification. J Polym Sci Part A Polym Chem 47:548–564
267. Huang R, Koning CE, Chadwick JC (2007) Effects of hydrogen in ethylene polymerization and oligomerization with magnesium chloride-supported bis(imino)pyridyl iron catalysts. J Polym Sci Part A Polym Chem 45:4054–4061
268. Kaul FAR, Puchta GT, Schneider H, Bielert F, Mihalios D, Herrmann WA (2002) Immobilization of bis(imino)pyridyliron(II) complexes on silica. Organometallics 21:74–82
269. Jin G, Zhang D (2004) Micron-granula polyolefin with self-immobilized nickel and iron diimine catalysts bearing one or two allyl groups. J Polym Sci Part A Polym Chem 42:1018–1024
270. Liu C, Jin G (2002) Polymer-incorporated iron catalysts for ethylene polymerization—a new approach to immobilize iron olefin catalysts on polystyrene chains. New J Chem 26:1485–1489
271. Zheng Z, Liu J, Li Y (2005) Ethylene polymerization with silica-supported bis(imino)-pyridyl iron(II) catalysts. J Catal 234:101–110

272. Kim I, Han BH, Ha C-S, Kim J-K, Suh H (2003) Preparation of silica-supported bis(imino)pyridyl iron(II) and cobalt(II) catalysts for ethylene polymerization. Macromolecules 36:6689–6691
273. Ray S, Sivaram S (2006) Silica-supported bis(imino)pyridyl iron(II) catalyst: nature of the support–catalyst interactions. Polym Int 55:854–861
274. Barabanov AA, Bukatov GD, Zakharov VA (2008) Effect of temperature on the number of active sites and propagation rate constant at ethylene polymerization over supported bis(imino)pyridine iron catalysts. J Polym Sci Part A Polym Chem 46:6621–6629
275. Barabanov AA, Bukatov GD, Zakharov VA, Semikolenova NV, Mikenas TB, Echevskaja LG, Matsko MA (2006) Kinetic study of ethylene polymerization over supported bis(imino)pyridine iron(II) catalysts. Macromol Chem Phys 207:1368–1375
276. Ma Z, Sun W-H, Zhu N, Li Z, Shao C, Hu Y (2002) Preparation of silica-supported late transition metal catalyst and ethylene polymerization. Polym Int 51:349–352
277. Barabanov AA, Bukatov GD, Zakharov VA, Semikolenova NV, Echevskaja LG, Matsko MA (2008) Ethylene polymerization over homogeneous and supported catalysts based on bis(imino)pyridine Co(II) complex: data on the number of active centers and propagation rate constant. Macromol Chem Phys 209:2510–2515
278. Xu H, Guo C, Xue C, Ma Z, Zhang M, Dong J, Wang J, Hu Y (2006) Novel layered calcosilicate-immobilized iron-based diimine catalyst for ethylene polymerization. Eur Polym J 42:203–208
279. Schilling M, Bal R, Görl C, Alt HG (2007) Heterogeneous catalyst mixtures for the polymerization of ethylene. Polymer 48:7461–7475
280. Han W, Müller C, Vogt D, Niemantsverdriet H, Thüne PC (2006) Introducing a flat model of the silica-supported bis(imino)pyridyl iron(II) polyolefin catalyst. Macromol Rapid Commun 27:279–283
281. Han W, Niemantsverdriet H, Thüne PC (2007) A flat model approach to tethered bis(imino)pyridyl iron ethylene polymerization catalysts. Macromol Symp 260:147–153
282. Guo C-Y, Xu H, Zhang M, Zhang X, Yan F, Yuan G (2009) Immobilization of bis(imino)pyridine iron complexes onto mesoporous molecular sieves and their catalytic performance in ethylene oligomerization. Catal Commun 10:1467–1471
283. Zhang M, Xu H, Guo C, Ma Z, Dong J, Ke Y, Hu Y (2005) Ethylene polymerization with iron-based diimine catalyst supported on MCM-41. Polym Int 54:274–278
284. Guo C, Jin G-X, Wang F (2004) Preparation and characterization of SBA-15 supported iron(II)-bisimine pyridine catalyst for ethylene polymerization. J Polym Sci Part A Polym Chem 42:4830–4837
285. Huang R, Liu D, Wang S, Mao B (2004) Spherical $MgCl_2$ supported iron catalyst for ethylene polymerization: effect of the preparation procedure on catalyst activity and the morphology of polyethylene particles. Macromol Chem Phys 205:966–972
286. Xu R, Liu D, Wang S, Mao B (2006) Preparation of spherical $MgCl_2$-supported late-transition metal catalysts for ethylene polymerization. Macromol Chem Phys 207:779–786
287. Mikenas TB, Zakharov VA, Echevskaya LG, Matsko MA (2007) Kinetic features of ethylene polymerization over supported catalysts [2, 6-bis(imino)pyridyl iron dichloride/magnesium dichloride] with AlR_3 as an activator. J Polym Sci Part A Polym Chem 45:5057–5066
288. Ray S, Galgali G, Lele A, Sivaram S (2005) In situ polymerization of ethylene with bis(imino)pyridine iron(II) catalysts supported on clay: the synthesis and characterization of polyethylene–clay nanocomposites. J Polym Sci Part A Polym Chem 43:304–318
289. Bouilhac C, Cloutet E, Deffieux A, Taton D, Cramail H (2007) Functional star-like polystyrenes as organic supports of MeDIP(2, 6-iPrPh)$_2$FeCl$_2$ catalyst toward ethylene polymerization. Macromol Chem Phys 208:1349–1361
290. Bouilhac C, Cloutet E, Cramail H, Deffieux A, Taton D (2005) Functionalized star-like polystyrenes as organic supports of a tridentate bis(imino)pyridinyliron/aluminic derivative catalytic system for ethylene polymerization. Macromol Rapid Commun 26:1619–1625

291. Laren MW, Duin MA, Klerk C, Naglia M, Rogolino D, Pelagatti P, Bacchi A, Pelizzi C, Elsevier CJ (2002) Palladium(0) complexes with unsymmetric bidentate nitrogen ligands for the stereoselective hydrogenation of 1-phenyl-1-propyne to (Z)-1-phenyl-1-propene. Organometallics 21:1546–1553
292. Arnáiz A, Cuevas JV, García-Herbosa G, Carbayo A, Casares JA, Gutierrez-Puebla E (2002) Revealing the diastereomeric nature of pincer terdentate nitrogen ligands 2,6-(arylaminomethyl)pyridine through coordination to palladium. Dalton Trans 2581–2586
293. Balamurugan R, Palaniandavar M, Halcrow MA (2006) Copper(II) complexes of sterically hindered Schiff base ligands: synthesis, structure, spectra and electrochemistry. Polyhedron 25:1077–1088
294. Holland JM, Liu X, Zhao JP, Mabbs FE, Kilner CA, Thornton-Pett M, Halcrow MA (2000) Copper complexes of 2,6-bis(iminomethyl)pyridine derivatives and of 1,3-bis(pyridine-2-yl)pyrazole. Effects of ligand bulk and conformational strain on the ground state of a six-coordinate copper(II) ion. Dalton Trans 3316–3324
295. Granifo J, Bird SJ, Orrell KG, Osborne AG, Šik V (1999) Synthesis and dynamic NMR studies of stereochemical non-rigidity in rhenium(I) complexes of 2, 6-bis[1-(phenylimino)ethyl]pyridine derivatives. Inorg Chim Acta 295:56–63
296. Çetinkaya B, Çetinkaya E, Brookhart M, White PS (1999) Ruthenium II complexes with 2, 6-pyridyl-diimine ligands: synthesis, characterization and catalytic activity in epoxidation reactions. J Mol Catal A Chem 142:101–112
297. Schmidt R, Hammon U, Gottfried S, Welch MB, Alt HG (2003) Iron(II)-based catalysts for ethene oligomerization. J Appl Polym Sci 88:476–482
298. Oleinik II, Oleinik IV, Abdrakhmanov IB, Ivanchev SS, Tolstikov GA (2004) Design of Schiff base-like postmetallocene catalytic systems for polymerization of olefins: II. Synthesis of 2, 6-bis(aryliminoalkyl)pyridines with cycloalkyl substituents. Russ J Gen Chem 74:1575–1578
299. Ionkin AS, Marshall WJ, Adelman DJ, Fones BB, Fish BM, Schiffhauer MF (2008) Modification of iron(II) tridentate bis(imino)pyridine complexes by a boryl group for the production of α-olefins at high temperature. Organometallics 27:1902–1911
300. Liu J-Y, Zheng Y, Hu N-H, Li Y-S (2006) Electron effect of *p*-substituent on iron(II) and cobalt(II) pyridinebisimine catalyst for ethylene polymerization. Chin J Chem 24:1447–1451
301. Paulino IS, Schuchardt U (2004) A new iron catalyst for ethylene polymerization. J Mol Catal A Chem 211:55–58
302. Zhilovskii GS, Oleinik II, Ivanchev SS, Tolstikov GA (2009) Design of postmetallocene Schiff base-like catalytic systems for polymerization of olefins: XI. Synthesis of Schiff bases containing cycloalkyl substituents from 2-acetyl-6-bromopyridine. Russ J Org Chem 45:44–47
303. Toti A, Giambastiani G, Bianchini C, Meli A, Luconi L (2008) New functionalized vinyl monomers by ethylene-(functionalized)-norbornene hetero-trimerization catalyzed by cobalt(II)-(imino)pyridine complexes. Adv Synth Catal 350:1855–1866
304. Kuwabara J, Takeuchi D, Osakada K (2009) Structures of Co, Pd and Ni complexes with iminopyridine ligands having an hydroxymethyl or acrylate pendant group. Polyhedron 28:2459–2465

Chapter 4
Challenges and Breakthroughs in Transition Metal Catalyzed Copolymerization of Polar and Non-Polar Olefins

Juan Cámpora and Mikael Brasse

Abstract This chapter describes the state of the art in copolymerization of non-polar olefins and functionalized monomers with transition metal-based catalysts operating by coordination-insertion mechanisms. A critical approach has been adopted in order to analyze what are the difficulties of making compatible olefin polymerization catalysts and functionalized monomers, and the strategies that have been proposed to overcome such difficulties. Special attention is devoted to mechanistic and computational insights that led to the development of our present ideas and to the understanding of this complex topic.

List of Abbreviations

AA	Acrylamide
acac	Acetylacetonate
AN	Acrylonitrile
BAr^{f-}_4	Tetrakis(3,5-bistrifluoromethylphenyl)borate
dba	Dibenzylideneacetone
dmso	Dimethylsulfoxide
E	Ethene
EMA	Polyethylene-co-methyl acrylate copolymer
EVA	Polyethylene-co-vinyl acetate copolymer
FMO	Frontier molecular orbitals
MA	Methyl acrylate

J. Cámpora (✉) · M. Brasse
Instituto de Investigaciones Químicas, CSIC-Universidad de Sevilla,
41092 Sevilla, Spain
e-mail: campora@iiq.csic.es

M. Brasse
e-mail: mikael@iiq.csic.es

G. Giambastiani and J. Cámpora (eds.), *Olefin Upgrading Catalysis by Nitrogen-based Metal Complexes II*, Catalysis by Metal Complexes, 36,
DOI: 10.1007/978-94-007-0696-5_4,

tmda	*N,N,N′,N′*-tetramethylethylenediamine
VA^f	Vinyl trifluoroacetate
VCl	Vinyl chloride
VF	Vinyl fluoride

4.1 Introduction

Ziegler–Natta catalysts provide access to an extremely wide range of polyolefin materials. These catalysts allow a precise control over the properties of the polyolefins, expanding the range of their uses and improving their adaptation to specific applications [1]. However, the essentially non-polar nature of polyethylene or polypropylene is the origin of some important limitations of these materials. For example, the surface of these polymers has low affinity for pigments and dyes, and has poor adhesive properties. The introduction of polar functional groups helps to overcome these limitations and greatly improves some other physical and chemical properties [2, 3]. Unfortunately, Ziegler–Natta systems, including both heterogeneous and homogeneous polymerization catalysts that operate by coordination-insertion mechanisms, are in general incompatible with functionalized monomers such as methyl acrylate (MA), vinyl acetate (VA) or acrylonitrile (AN) due to the ability of polar groups to interact with the metal center, disrupting the catalytic process [4].

Polar polyolefin copolymers, such as ethylene–methyl acrylate (EMA) or ethylene–vinyl acetate (EVA), are important commercial materials. Compared to non-polar olefins, these show better surface properties, gloss, toughness and mar resistance. Ionizable functions, e.g., carboxylic or sulfonic groups, called *ionomers*, can provide extreme toughness by extensive crosslinking, or excellent adhesive and sealing properties. These materials are produced by radical polymerization methods under high temperature and pressures that have scarcely evolved in the last 50 years. The resulting polymers usually have branched structures and broad molecular weight distributions [2, 3]. In addition, the possibility of controlling the copolymer composition is very limited with these methods, which invariably afford products containing more than 50% of the polar comonomer [5].

Improving the tolerance of Ziegler–Natta catalysts to polar monomers would permit the synthesis of polar copolymers under mild conditions, reducing energy costs and allowing better control on their microstructure and molecular weight. Moreover, these catalysts would allow regulating the polar comonomer contents in the copolymers. Since polar comonomers are much more expensive than ethene or propene, this would also contribute to significantly lower the cost of polar copolymers. It is documented that even a very small proportion of polar functionalities has a dramatic influence on the polyolefin properties. For example, it is known that less than 1% of maleic anhydride incorporation in polypropylene can improve properties of polypropylene allowing direct application of compatibilizing agents on its surface [6]. Therefore, it is not strange that an intense research effort to overcome the difficulties associated with the controlled copolymerization of polar and non-polar comonomers has been sustained for many years [3, 4, 7–9]. This goal is still far from being fully accomplished, but very significant progress has been achieved in recent years. Brookhart's discovery of the first highly active catalysts for olefin polymerization based on late transition metals in 1995 [10, 11] spurred a quest for the synthesis of new catalysts that could potentially copolymerize olefins and polar monomers. When shortly after Pd catalysts were shown to copolymerize ethene with methyl acrylate and other polar olefins [12], prospects appeared to be confirmed, at least in part. It was soon appreciated, however, that such capability is rather unusual even for late transition metal catalysts, and that incorporation of polar comonomers has to pay a toll in the form of a dramatic drop of the catalytic activity.

Since the discovery of MAO as activating cocatalyst in the early 1980s, early transition metal-based complexes containing Cp-based ligands have dominated the field of molecular catalysts for olefin polymerization [13]. Sophisticated Cp ligands can shape the steric and electronic environment of the metal to achieve impressive results in terms of productivity and stereoselectivity, but at the cost of a considerable synthetic effort. In contrast, Brookhart's catalysts containing readily available nitrogen-based α-diimine ligands are quite simple and easy to prepare [12]. These have inspired the development of an ample family of catalysts containing nitrogen-based ligands. With the important exception of palladium phosphinosulfonato complexes [9, 14], most catalysts that exhibit some capability for incorporating polar comonomers belong to this family. This justifies the inclusion of this chapter in a book specifically devoted to nitrogen-based ligands.

Rather than a comprehensive description of the activity and selectivity of the different copolymerization catalysts, which has been the subject of two recent reviews [8, 9], the purpose of this chapter is to provide a vision of the nature of the difficulties met in the quest for catalysts for polar/non-polar olefin copolymerization, and the different approaches devised to overcome them.

4.2 Types of Catalysts Used for the Copolymerization of Polar and Non-polar Comonomers

Olefin polymerization catalysts containing early transition metals are essentially incompatible with functionalized monomers [2–4]. However, group 4 metallocene catalysts can be used to prepare polar copolymers if the functional groups are previously transformed to render them more compatible. As shown in Scheme 4.1, the functional groups of the monomers can be masked by treatment with Lewis acids or silylating agents [15–23] (up), or replaced by a boryl group (bottom) [24]. Recently, titanium catalysts containing Schiff-base ligands, shown in Fig. 4.1, have been used to copolymerize ethene with masked polar comonomers [25]. These

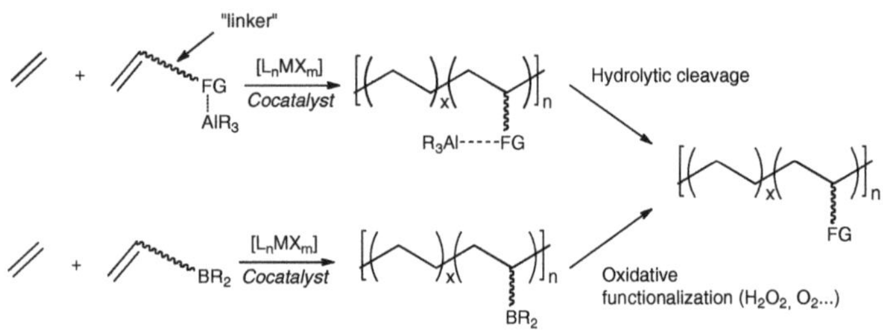

Scheme 4.1 Copolymerization of protected polar comonomers

Fig. 4.1 Efficient titanium catalysts for the copolymerization of ethene with protected polar comonomers

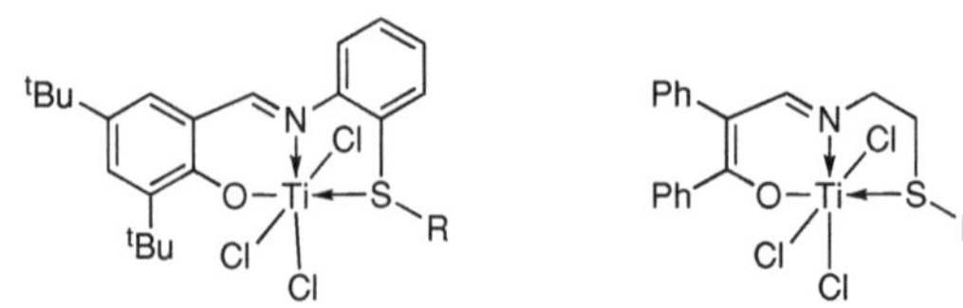

methods have a number of serious disadvantages. In any case, a post-polymerization process is necessary to produce the polar copolymer. Masking or protecting elements can be removed by hydrolysis, and boryl groups can be oxidatively cleaved generating alcohol functionalities. In addition, the success of these early transition catalysts is limited to special comonomers in which the polar functionality and the polymerizable vinyl group are separated, such as ω-functionalized α-olefins [15–25] or functionalized norbornenes [22, 23]. This constitutes a severe limitation since industrially relevant copolymers are limited to those containing *fundamental monomers* [9], which are small and cheap molecules that contain directly connected vinyl and polar groups, such as MA, MMA or VA.

Catalysts containing less oxophilic late transition elements are more tolerant to polar functionalities, and offer more opportunities for the copolymerization of polar and non-polar comonomers without previous compatibilization treatments. So far, the vast majority of the catalysts capable of such task are based on group 10 metals, Ni and Pd. Figure 4.2 shows a classification of these catalysts according to their general structures. Early work by Klabunde and Ittel in the late 1980s showed that nickel phosphinoenolato complexes (**A**) used for ethene oligomerization in the *Shell higher olefin process* (SHOP) can be turned into polymerization catalysts when the ancillary PPh_3 ligand is removed with a phosphine scavenger such as $Ni(cod)_2$ or $Rh(acac)(C_2H_4)_2$ [26]. These nickel catalysts are very tolerant to polar substances and can operate even in alcohol solution (this is how they are used in the SHOP process). Klabunde and Ittel demonstrated that such catalysts not only tolerate polar molecules, but they can directly copolymerize ethene with polar α-olefins containing ester, keto, alkylsiloxanes or fluorinated functionalities, provided that these are separated from the vinyl group by a polymethylene linker. However, simple functional

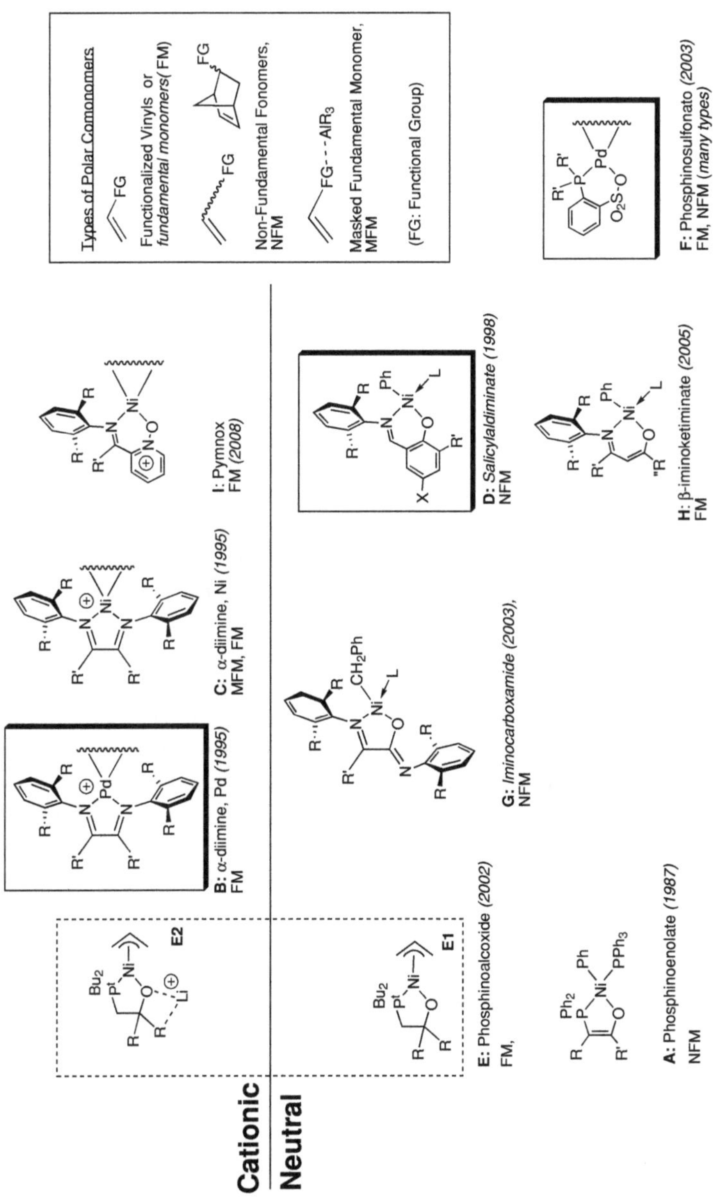

Fig. 4.2 Group 10 catalysts successfully applied to the copolymerization of non-polar and polar vinyl monomers. The *boxes* highlight those that have been subject to mechanistic investigation

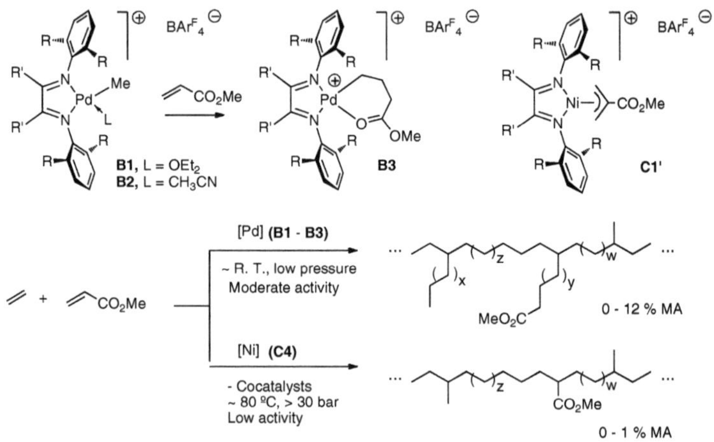

Scheme 4.2 Ethene–MA copolymerization with α-diimine catalysts

vinyls such as MA or VA could not be incorporated [27]. More recently, Claverie carried out such copolymerization reactions in aqueous emulsion, improving the incorporation of polar comonomers [28].

Widespread interest on late transition metal catalysts for the copolymerization of polar and non-polar olefins was not triggered until Brookhart's report on α-diimine complexes **B** and **C** [10]. Using complexes **B**, Brookhart demonstrated in 1996 that fundamental vinyl monomers such as alkyl acrylates or methyl vinyl ketone can be directly copolymerized with ethene or propene by a coordination-insertion mechanism [12]. The catalysts used for this purpose are ionic complexes containing the low-coordinating borate anion $[B(3,5\text{-}C_6H_3(CF_3)_2)_4]^-$ (BAr_4^{f-}) (Scheme 4.2). Initially, cationic methylpalladium derivatives stabilized with a labile diethylether (**B1**) or acetonitrile (**B2**) ligands were used. It was discovered that when these react with MA, they produce chelate complexes, **B3**, which are very stable and more convenient catalyst precursors. Copolymerization of ethene and MA with palladium α-diimine catalysts affords branched EMA copolymers with polar comonomer contents of up to 12%. The methoxycarbonyl groups are found exclusively at terminal positions of the branches [29]. This property has been exploited by Guan for the synthesis of functionalized polyethylenes displaying different types of functional groups at the branch edges (see Sect. 4.5.1) [30–32]. Compared with the palladium α-diimine derivatives, the analogous Ni catalysts **C** are much more sensitive to polar substances, failing to incorporate MA under mild conditions [7]. However, $[NiX_2(\alpha\text{-diimine})]$ complexes catalyze the copolymerization of ethene with polar comonomers if these are previously masked with aluminum alkyls [21, 33–36]. In contrast to early transition metal catalysts, these nickel catalysts can incorporate *fundamental* vinyls such as MA, VA and AN [34], and even acrylamide (AA) [33, 36], which is one of the most challenging functional comonomers. Direct copolymerization of ethene and free MA can also be achieved with Ni(α-diimine) catalysts (introduced in the form of a cationic η^3-allylnickel complex, **C1'**, Scheme 4.2) at high temperature and pressure

(80 °C, >30 bar). Under these conditions, Ni catalysts produce slightly branched copolymers with the acrylate units incorporated in main polymer chain. MA incorporation is usually below 1%, and the catalytic activities are also quite low although they are somewhat improved by the addition of large amounts of $B(C_6F_5)_3$ and $Na(BAr^f_4)$ as co-catalysts [37, 38].

Neutral salicylaldiminato complexes of nickel, (**D**) described by Grubbs in 1998, are the next milestone in this account (Fig. 4.2). The design of these complexes is clearly inspired by the phosphinoenolato catalysts, replacing the soft P donor group by a hard imine fragment bearing a bulky aryl substituent, which is the key feature of Brookhart's α-diimine model [7]. This leads to significantly higher activities and polymer molecular weights. Stabilized with L = PPh_3 these catalysts can be very active in ethene homopolymerization either alone or in the presence of ligand scavengers ($Ni(cod)_2$ or $B(C_6F_5)_3$) [39]. The catalyst design is further improved when PPh_3 is replaced by the labile ligand MeCN [40]. Just as the phosphinoenolato catalysts, salicylaldiminato complexes tolerate polar solvents such as ethers [39], ketones, or even water.[1] The two types of nickel catalysts resemble in their ability to copolymerize ethene with special polar comonomers containing separated vinyl and polar functionalities such as ω-functionalized α-olefins but, disappointingly, they are also similar in their incapacity to incorporate simple functional vinyls such as MA or VA [40, 42]. The rate of incorporation of ω-functionalized α-olefins with nickel salicylaldiminato catalysts is rather low (0.2–2 mol%), but improves for functionalized norbornenes monomers and other polycyclic alkenes (5–10%) [43, 44]. Another class of neutral nickel catalysts are iminocarboxamidato complexes, **G**, introduced by Bazan [45]. These complexes can exist in two isomeric forms that differ in the ligand coordination mode, κ-*N*, *O* or κ-*N*, *N* (Fig. 4.3). Remarkably, only the κ-*N*, *O* isomer is active, which renders the system more akin to salicylaldiminato catalysts than to α-diimines. Iminocarboxamidato and salicylaldiminato catalysts also resemble in their capability to incorporate polar norbornene derivatives. A phosphine sponge such as $Ni(cod)_2$ efficiently activates the PMe_3-containing iminocarboxamidato catalysts, which readily achieve comonomer incorporation of ca. 15 mol% and productivities in the range 50–150 kg of copolymer/mol Ni h [46–49].

Other types of nickel catalysts have also been described that represent additional variations of electrically neutral molecules with mixed-donor ligands. The DuPont group reported a series of allylnickel complexes displaying P–O ligands

[1] The problem of polymerization catalysts tolerance to polar substances is closely related to polar comonomer copolymerization. Like the nickel phosphinoenolato catalysts, their salicylaldiminato counterparts are also capable to operate in the presence of alcohols and water. Olefin polymerization in aqueous media is important because it directly affords polyolefin latexes, and has been addressed by different groups. Palladium diimines have also been applied for this purpose, but their tolerance to water is due to catalyst encapsulation into the polyolefin particle rather than to genuine tolerance. It is a noteworthy fact that, in spite of their higher performance in these polar media, nickel catalysts are usually inferior to palladium in their ability to copolymerize polar and non-polar comonomers. For some leading references, see for example Soula et al. [28] and [41].

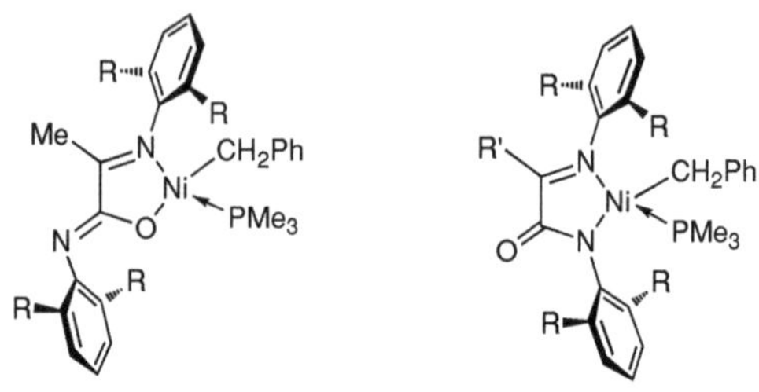

Fig. 4.3 Catalytically active and inactive forms of nickel iminocarboxamidato catalysts

including derivatives **E1** (Fig. 4.2) that copolymerize ethene and alkyl acrylates in the presence of a large excess of $B(C_6F_5)_3$ [38, 50]. The performance of these catalysts is further improved by complexation with a Li^+ cation (**E2**), but little information is available on this intriguing effect [50]. Structurally related nickel phosphinophenolato complexes have also been shown to catalyze the copolymerization of ethene with long-chain functionalized olefins [51]. The β-ketiminato complexes **H,** activated with MMAO, catalyze the copolymerization of ethene and MMA affording copolymers with up to 17 mol% comonomer incorporation, depending on the MMA concentration. ^{1}H NMR analyses of these polymers indicate that the MMA units are uniformly distributed in the polyethylene chain. These features strongly suggest that MMA is incorporated by a true coordination-insertion mechanism, which is unusual for this monomer [52]. Catalyst type **I** contains PymNox ligands, which relate to the salicylaldiminato anion in a way that the phenolato fragment has been replaced by a neutral pyridine-*N*-oxide unit. Therefore, PymNox catalysts can be regarded as cationic analogues of the salicylaldiminato family. Simple [$NiBr_2$(PymNox)] complexes activated with MMAO have been tested in MA/ethene copolymerization. Only the derivative containing an acetaldimine-based PymNox ligand (R = Me) produces EMA with 0.7 mol% MA. NMR analysis of the polymer showed that the CO_2Me groups are directly attached to the main polymer chain or form short CH_2CO_2Et branches [53].

The diversity of neutral nickel polymerization catalysts contrasts with the paucity of active catalysts based on neutral palladium complexes. Although some neutral palladium complexes containing anionic pyrrole-imine [54], N–O salicylaldiminato [55] or even simple monodentate phosphine groups [56] have been reported to copolymerize ethene and acrylates, these involve radical initiated processes rather than controlled coordination-insertion mechanisms. Thus, the discovery of a new class of palladium polymerization catalysts based on anionic phosphinosulfonato ligands (type **F** in Fig. 4.2) came as a surprise. However, phosphinosulfonato catalysts have proven to be a major breakthrough in olefin/polar monomer copolymerization [57]. In their preliminary communication in 2003, Pugh and Drent generated the active species in situ, by reacting the free phosphinosulfonic acids with palladium sources such as $Pd(OAc)_2$ or $Pd(dba)_2$ [58]. The resulting catalysts copolymerize ethene with MA and other alkyl acrylates, producing essentially linear copolymers with up to 17 mol% acrylate incorporation

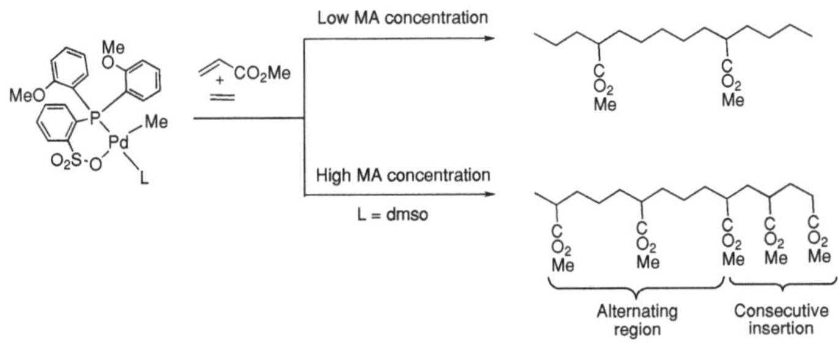

Scheme 4.3 EMA copolymerization with Pd-phosphinosulfonato catalysts

and productivities of ca 5–15 kg/mol h. In contrast with Pd α-di-imine complexes, the phosphinosulfonato catalysts incorporate MA units directly into the polymer backbone (Scheme 4.3). In recent years, phosphinosulfonato catalysts have been the subject of intense research by several groups. A number of well-defined palladium phosphinosulfonato complexes have been prepared and characterized containing different types of auxiliary ligands L such as Cl^- (anionic complexes) [59], pyridine and bulky pyridine derivatives [60–66], or amines [65] and weakly bound solvent molecules such as dmso [67, 68] methanol or water [69], as well as base-free, dimeric species [69, 70]. The syntheses of well-defined species lead to improved productivities and more efficient co-monomer incorporation. For example, using catalysts with weakly bound dmso, Mecking has achieved ethene-MA copolymers with more than 50 mol% incorporation [67, 70], bridging the gap between the composition ranges provided by copolymerization catalysts based on insertion and radical mechanisms [5]. The NMR spectra of such comonomer-rich copolymers reveal the presence of alternating E–MA-E–MA regions, together with consecutively inserted MA units, as shown in Scheme 4.3. In the absence of ethene, dmso-stabilized phosphinosulfonato complex catalyzes the homooligomerization of MA [67].

In addition to alkyl acrylates [58, 59, 62, 67, 70] and functionalized norbornene comonomers [71], palladium phosphinosulfonato catalysts demonstrated a unique capability for copolymerizing ethene with various comonomers, including industrially relevant AN [61], or VA [66]. Other challenging monomers copolymerized with these catalysts are vinyl fluoride (VF) [64], vinyl sulfones [68], *N*-vinyl pyrrolidone and acrylamide derivatives [65]. In all these cases, essentially linear copolymers containing evenly distributed comonomer units were obtained. Although phosphinosulfonato catalysts fail to copolymerize ethene with propene or other α-olefins, they efficiently incorporate norbornene [71]. Their versatility was further demonstrated when shortly after the initial report on phosphinosulfonato catalysts, Pugh and Drent communicated that the same in situ generated catalysts copolymerize ethene and CO [72]. Contrary to all the catalysts known up to that date, the copolymer was not perfectly alternating and contained ethene excess. Further investigation in this field has shown the possibility of decreasing the CO content to as low as 10 mol% [69, 73].

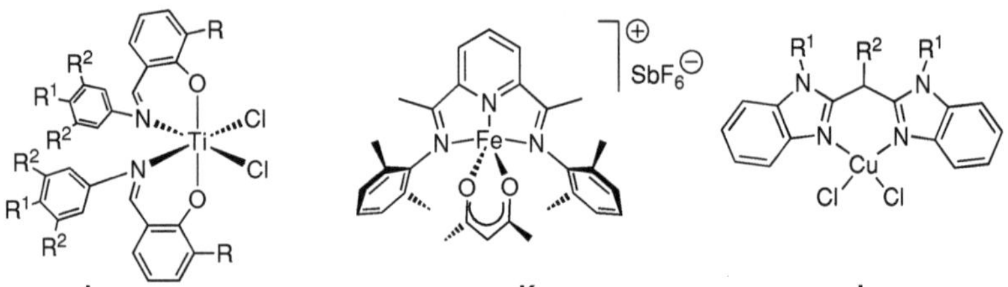

J K L

Fig. 4.4 Some catalysts based in non-group 10 elements tested in polar comonomer copolymerization

A number of well-defined nickel organometallic complexes with phosphinosulfonato ligands have also been prepared [74–77]. Although these compounds behave as very active ethene homopolymerization or oligomerization catalysts, no copolymerizations with polar monomers have been reported so far. It is worth noting the analogy with the α-diimine system with regard to the low compatibility of the Ni catalysts with polar comonomers.

In addition to Ni and Pd, other metals have also been investigated for polar-comonomer incorporation (Fig. 4.4). However, only the Ti-based phenoxiimine catalysts developed by Fujita, **J**, have shown the capability of copolymerizing polar comonomers without previous masking. This is certainly an exceptional feature for an early-transition metal-based system. Catalysts **J**, when activated with small amounts of MAO, incorporate up to 1 mol% of hexane-1-yl-acetate into polyethylene, achieving high productivities of about 500 kg/mol h [78] (see Chap. 1, Vol. 35). Complexes containing elements that are closer to Ni and Pd in the periodic table should be in principle better candidates than Ti. However, this is not the case. In spite of the structural similarity of Fe or Co 2,6-bisiminopyridine and the Ni, Pd α-diimine systems, the former have been found unable to copolymerize ethene with polar comonomers. The cationic iron derivative **K**, activated with MAO, was tested for ethene copolymerization in the presence of MMA, MA and 2-vinyl-1,3-dioxolane. Although the ethene homopolymerization activity was not entirely suppressed (especially in the case of MMA), no comonomer incorporation was detected [79]. On the other hand, 2,6-bisiminopyridine iron complexes initiate acrylate homopolymerization, presumably by a radical mechanism [80]. Thus, it is not surprising that attempts to copolymerize ethene and MMA with Fe(2,6-bisiminopyridine) catalysts afford mixtures of ethene and MMA homopolymers rather than true copolymers [81].

A number of copper complexes [82–85] such as the bis(benzoimidazolyl)-methane (BMIM) derivative **L**, described by Stibrany [82, 83] copolymerize ethene and MMA. Furthermore, several BMIM catalysts have a small activity in ethene homopolymerization. The latter is very likely the result of an insertion mechanism, therefore it was suggested that the same mechanism could also be responsible for ethene/MMA copolymerization [82, 83]. However, the EMA copolymers have high polar comonomer (>40 mol%), which strongly suggests that they are produced by a radical initiated mechanism [85].

4.3 Basic Mechanisms

Before examining the difficulties of polar and non-polar comonomer copolymerization, this section will summarize the essential aspects of polymerization mechanism by late transition metal complexes, and how these are perturbed by the presence of polar functions in the co-monomers.

Ni and Pd α-diimine complexes are by far the best understood olefin polymerization late transition catalysts [86]. Since Brookhart's initial communication in 1995 [10], an intense effort has been devoted to fully develop all their potential [7, 8]. Intensive mechanistic research and computational modeling have been combined to draw a detailed picture of this system [87–91]. The conceptual framework developed for the α-diimine catalysts is a powerful tool for the analysis of related catalyst systems, where the experimental mechanistic data available is limited.

General features of olefin polymerization mechanism by Ni and Pd α-diimine catalysts are summarized in Scheme 4.4. The upper part displays the sequence of steps for the normal propagation of the polymer chain, i.e. recurrent insertion of monomer units into the active M–C bond, not different in essence from the standard Cossee–Arlman mechanism [1]. However, the tendency of Ni and Pd catalysts to undergo rapid and reversible β-hydrogen elimination leads to the reactivity manifold depicted in the lower part of the Scheme. Chain transfer occurs by rapid associative displacement of the resulting olefin ligand by the monomer (bottom, right side). This process limits the polymerization degree [88, 91]. In α-diimine complexes, the substituents of the *N*-aryl fragments hinder this process by obstructing the access of free monomer to the axial coordination positions of the metal. This effect tends to increase the molecular weight of the polymer, in an extent that depends directly on the size of the aryl substituents. Theoretical calculations have disclosed an alternative pathway for chain transfer that does not require formation of an intermediate hydride species and involves concerted H transfer from the alkyl chain to the monomer [92]. Since both mechanisms are kinetically equivalent, it would be difficult to distinguish between them on an experimental basis.

Blocking the associative monomer exchange enables the process termed *chain walking* [89, 90] to become competitive with chain transfer, which is the cause of the branching of the polymer chain. The β-H elimination step is a reversible process, but re-insertion of the olefin back into the metal-hydride bond is not regioselective and can take place either in 1,2 or 2,1 fashion. The latter leads to the formation of a secondary alkyl species. If a new monomer unit inserts in this Pd–C bond, the result is the generation of a methyl branch in the growing chain. The combination of successive episodes of β-H elimination followed by non-selective olefin re-insertion causes the effective displacement of the metal center along the polymer chain, or *chain walking*. This mechanism gives rise to methyl, ethyl and higher branches. For Pd, the energy barrier for β-hydrogen elimination is significantly lower than for the monomer insertion, which means that *chain walking* and

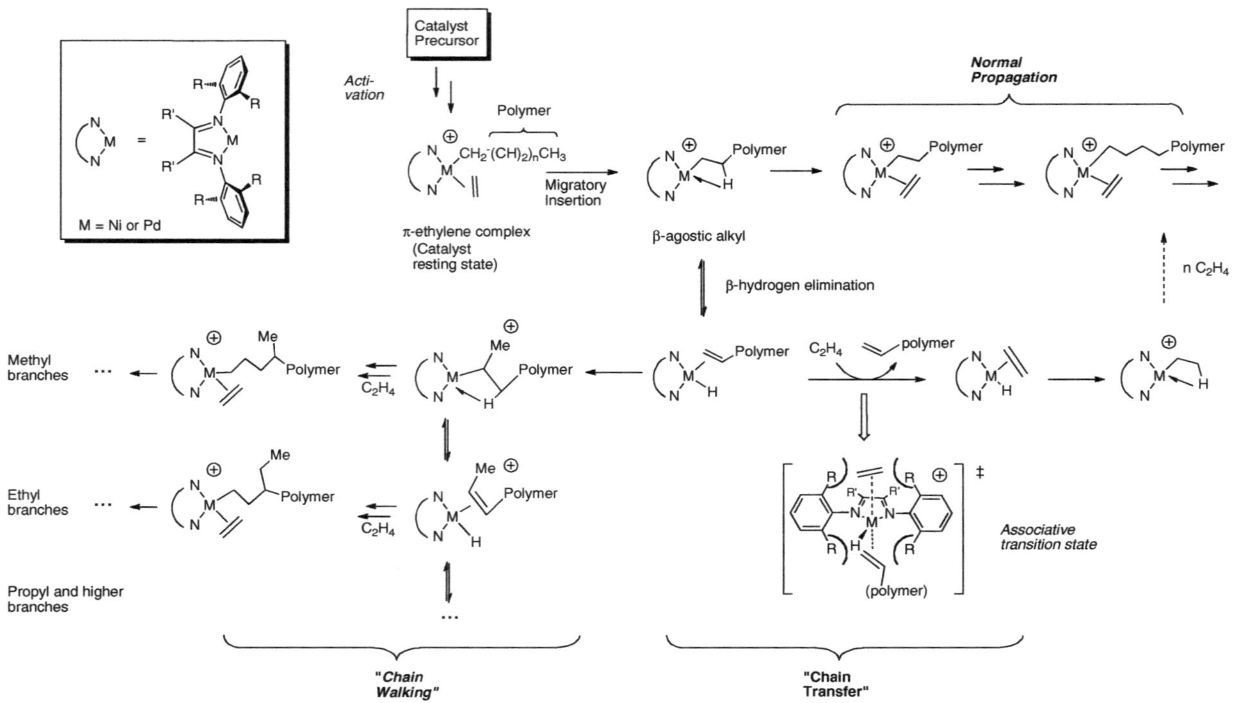

Scheme 4.4 Essential mechanistic features of olefin polymerization with Ni and Pd α-diimine catalysts

branch generation is faster than normal propagation. Since the metal center moves freely along the polymer chain prior to each insertion, the total number of branches does not depend on the monomer concentration (however, monomer concentration does affect the shape of the branches, see Sect. 4.5.4.1). In contrast, *chain walking* and monomer insertion have comparable energy barriers for Ni catalysts. As a result, polymers produced with Ni are more linear. Decreasing the monomer concentration and increasing the temperature favors the *chain walking* process in comparison to chain transfer and normal propagation, leading to an increased number of branches [87, 91, 93].

Catalysts containing ligands with two different donor atoms (e.g. salicylaldiminato or phosphinosulfonato complexes), give rise to a special mechanistic situation because each of the intermediates can exist in two possible *cis* or *trans* configurations. Ziegler has carried out DFT calculations in order to examine the mechanism of ethene polymerization by nickel salicylaldiminato catalysts [94]. The basic features of this mechanism are shown in Scheme 4.5. As in the α-diimine system, the resting state of the catalyst is a π-olefin complex. Isomer $\mathbf{D1_O}$, is more stable than $\mathbf{D1_N}$ by 3.0 kcal/mol because the stronger σ-donor groups, the alkyl and the imine, avoid mutually *trans* positions. However, the least stable isomer $\mathbf{D1_N}$ is also the most reactive one. The energy barrier for migratory insertion, 15.0 kcal/mol, is substantially smaller than that for $\mathbf{D1_O}$, 25.0 kcal/mol. Ethene insertion leads to the corresponding agostic alkyl intermediates $\mathbf{D2_O}$ and $\mathbf{D2_N}$. Note that migratory insertion inverts the relative configuration of the products, i.e., $\mathbf{D1_O}$ affords $\mathbf{D2_N}$, and $\mathbf{D1_N}$ produces $\mathbf{D2_O}$. At first sight, the propagation mechanism would alternate between the two possible insertion steps with the overall rate controlled by the most difficult of them, $\mathbf{D1_O}$ to $\mathbf{D2_N}$. Due to height of the energy barrier associated to the latter, the catalyst would be rather poor. However, the conformational lability of square-planar Ni(II) complexes avoids this problem, because the intermediates readily exchange their configuration via tetrahedral isomer. The energy barrier for the *cis–trans* isomerization of the π-ethene complexes **D1** has been calculated in 11.4 kcal/mol which is appreciably lower than that for the insertion step. Thus, the most stable π-ethene complex $\mathbf{D1_O}$ will

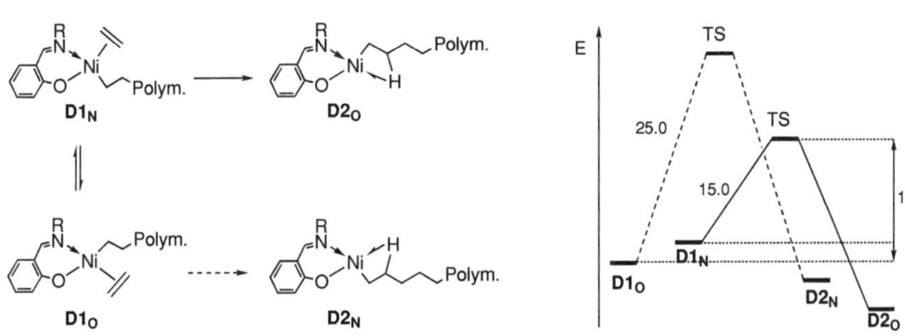

Scheme 4.5 Propagation mechanism in ethene polymerization with Ni salicylaldiminato catalysts. The diagram on the *right side* represents calculated energies in kcal/mol

isomerize to its reactive isomer **D1**$_N$ prior to undergoing migratory insertion. As shown in the diagram of Scheme 4.5, the overall energy barrier for this process is calculated as the difference between the energies of the two π-complexes plus the lowest insertion barrier, i.e. 3.0 + 15.0 = 18.0 kcal/mol. In contrast, the isomerization of the agostic species **D2** is much more difficult, with a barrier of 36.6 kcal/mol. A similar situation has been described for the structurally related PymNox catalyst (**I**, Fig. 4.2). For this cationic catalyst, the composed energy barrier for the overall insertion process is only 13.0 kcal/mol [95]. It is worth mentioning that this propagation scheme cannot be immediately extended to palladium, because formation of tetrahedral intermediates is less favorable for second and third-row than for first-row transition elements. An alternative isomerization pathway has been proposed for Pd phosphinosulfonato catalysts (see Sect. 4.5.4).

Insertion of polar and non-polar olefins into M–C bonds proceeds by qualitatively identical mechanisms. However, the presence of a substituent on the double bond leads to two possible regioisomeric products arising from 2,1 or 1,2 insertion (Scheme 4.6). The 2,1 product is usually favored for steric reasons (the alkyl group migrates to the least hindered carbon on the olefin), and electron-withdrawing groups further increase this tendency. Thus, while propene insertion in cationic α-diimine alkylpalladium complexes leads to a 8:2 mixture of the 2,1 and 1,2 insertion products [89], fully selective 2,1 insertions have been observed for vinyl monomers bearing electron-withdrawing groups such as MA [29], VA [96], AN [97, 98] or vinyl chloride (VCl) [99]. Reactions of different Pd complexes with AN [100], MA [67], VA [101] or vinyl sulfones [66] provide further examples of selective 2,1 insertions. The reaction of VA with a Ni α-diimine methyl complex also favors the 2,1-insertion mode but proceeds with lower selectivity, affording a mixture of the two insertion products [96]. Electronic effects can override the intrinsic preference for 2,1 insertion. Vinyl ethers (in which the OR group is a π-electron donor) react with cationic Pd(α-diimine) complexes affording 1,2 insertion products [102]. Selective 1,2 insertions have also been observed in the case of acrolein dimethylacetal [103] and the nitrogen-based monomers *N*, *N*-dimethylallylamine and *N*-pentenylcarbazole [104]. Insertion of MMA in Pd-α-diimine complexes poses an interesting case. Here, steric effects originating in the bulky metal–ligand unit play a dominant role because the normal 2,1 insertion mode would produce a very crowded quaternary center directly bound to Pd. In consequence, 1,2 insertion takes place selectively [105].

Theoretical models reproduce reasonably well the general trends observed in the regioselectivity of the insertion of simple functionalized vinyl monomers

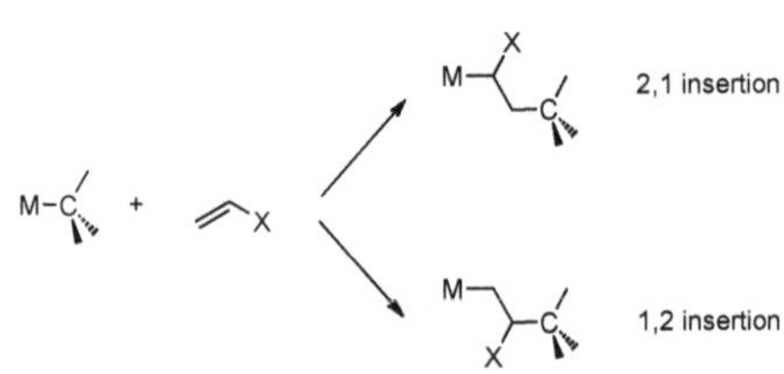

Scheme 4.6 Possible regioisomeric outcomes of the insertion of substituted vinyl monomers

[99, 106–109]. However, DFT calculations are not, in general, an easy guide for the interpretation of chemical phenomena in familiar terms such as steric or electronic influences. This requires an extra effort in the analysis of the computational data. The generic preference for 2,1 over 1,2 insertion has been rationalized in terms of the extra energy cost of distorting the monomer fragment going from the ground to the transition states in the latter case [106, 107]. Explaining the electronic influence of substituents in the olefin fragment is even more difficult, but an attempt has been made to rationalize such effects on the basis of a frontier molecular orbital analysis of the monomers [110]. In a detailed study of the copolymerization of MA and ethene and other α–olefins, Brookhart has shown that α-functionalized alkyls formed after the insertion of the polar monomer rearrange to the 6-membered chelates **B3′**. This process, shown in Scheme 4.7, involves two consecutive *chain walking* steps [12, 29]. The cyclic species **B3′** is very stable (in fact, the parent complex **B3**, prepared as shown in Scheme 4.2, is readily isolable). The subsequent insertion step requires the displacement of the cyclic structure of **B3′** by an incoming unit of ethene, a thermodynamically unfavorable process that raises the global energy barrier for insertion and controls the overall copolymerization rate. Opening the chelate ring by other olefins different from ethene, such as propene or 1-hexene, is even more difficult. Although no experimental data is available for the displacement of the chelate by a second MA unit, theoretical calculations show that this should be energetically prohibitive [106], preventing consecutive insertions of acrylate to take place. It is worth comparing this behavior with that of the analogous chelate in palladium catalyzed ethene/CO copolymerization. This very stable 5-membered chelate allows CO insertion but cannot be displaced by ethene, leading to the strictly alternating copolymerization observed with most Pd catalysts. The structure of chelate **B3′** also ensures that the carboxylate groups will end-up in the terminal position of branches that will have at

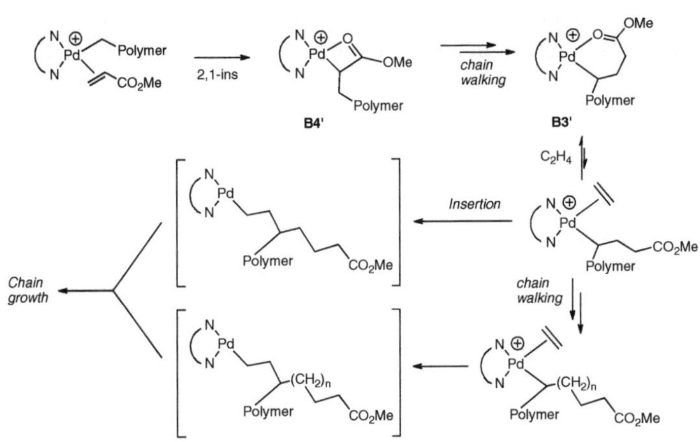

Scheme 4.7 Chain propagation and branch building in ethene–MA copolymerization by Pd-α-diimine catalysts

least three carbon units, although the *chain walking* mechanism also allows building longer carboxylate-capped branches, as depicted in Scheme 4.7.

Once the chelate is opened and a new nonpolar monomer unit is inserted, the propagating species becomes essentially the same as in the homopolymerization process, and both types of monomers will compete for the active site. Measurements of the insertion rates in complexes of type $[Pd(Me)(CH_2{=}CHX)(N{-}N)]^+$ have shown that the rate of insertion is accelerated by electronegative substituents directly attached to the double bond. For example, for N–N = α-diimine, the rates decrease in the order X = CO_2Me > Br > H > Me [111]. The same trend has been observed for the insertion of *p*-substituted styrene derivatives (with N–N = bipyridiyl), and has been attributed to the destabilization of the π-complexes of the electron-poorer olefins relative to the transition state (a ground-state effect) [112].

When two monomers are copolymerized, their relative reactivity determine the ratio at which they are incorporated into the polymer. Two factors determine the relative reactivities of the monomers: their capability to bind to the metal center, and the energy barriers for migratory insertion. Mechanistic studies on the copolymerization of ethene and MA in the Pd α-diimine system have shown that the exchange between free and coordinated monomers is rapid, and that the corresponding π-complexes are in equilibrium prior to undergoing migratory insertion [29]. This leads to the typical Curtin–Hammett situation drawn in Scheme 4.8, where the relative incorporation rates of monomer and comonomer are defined by the equilibrium constant K_{eq}, the individual insertion rates of the monomers (k_E and k_{MA}) and the comonomer concentration ratio, [MA]/[ethene]. Due to its small size and better electron-donor properties, Pd binds ethene much more strongly than MA. However, the advantage of the non-polar monomer is partially offset by the faster insertion rate of MA. Feeding the experimentally determined values of equilibrium constant and reaction rates into the equation shown in Scheme 4.8 gives an MA incorporation ratio of ca. 5 mol%, which is in good agreement with the actual comonomer content of the copolymers obtained with this system. Another successful prediction of this model is that the comonomer incorporation will increase proportionally with the comonomer ratio in the feed. However,

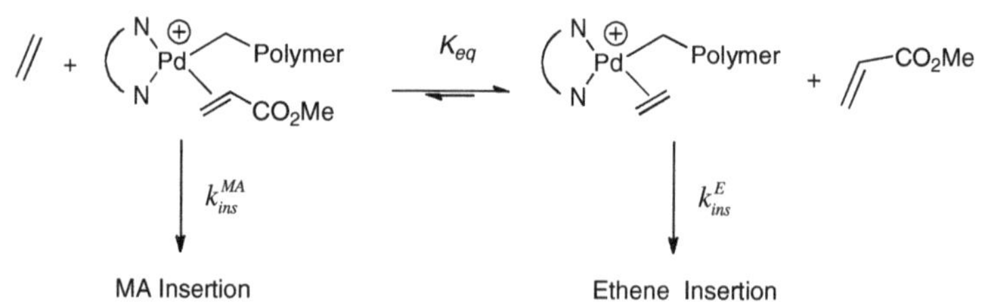

$$\frac{\text{Acrylate Incorp.}}{\text{Ethene Incorp.}} = K_{eq}\frac{k_{ins}^{MA}}{k_{ins}^{E}} \times \frac{[MA]}{[Ethene]}$$

Scheme 4.8 Curtin–Hammett control of the polar comonomer incorporation ratio adding more comonomer causes the concentration of the cyclic species **B3′** to rise, slowing and eventually shutting down the polymerization process. In practice, this limits comonomer incorporation in this system to about 12 mol%. As discussed later in this chapter, the utility of this simple model goes beyond of a simple mechanistic test, since it can be used to interpret computational results and deduce some interesting predictions on the performance of different catalysts.

4.4 Difficulties Met in the Copolymerization of Polar and Non-polar Olefins

Achieving a controlled copolymerization of polar and non-polar monomers by catalysts based on coordination-insertion mechanisms has proven to be an extremely challenging objective. A very intense research devoted to this task for almost two decades and, as a result, a clearer image of the obstacles involved is emerging. This knowledge is essential in order to propose possible solutions. This section will summarize the origin of the difficulties, while the next will review the main innovations devised for overcoming them.

4.4.1 Catalyst Poisoning by the Polar Monomer

Monomers containing polar substituents have different coordination modes, π and σ (Scheme 4.9). In order to undergo migratory insertion, first the monomer has to bind to the metal in the π mode. Thus, the formation of a σ coordination complex is unproductive, blocking (or poisoning) the active center and preventing the polymerization to progress. Strongly Lewis acidic metal centers (such as those in conventional Ziegler–Natta systems) are irreversibly poisoned by polar monomers, but less oxophilic late transition metals are less sensitive to them. Carboxylates, esters, ethers, and other oxygen functional groups bind reversibly to late metal centers. The resulting σ-complexes can readily isomerize to π mode, or can be displaced by the non-polar olefin, allowing the copolymerization to proceed. Such reversible poisoning decreases catalytic activity, but does not prevent the copolymerization process nor influences the comonomer incorporation ratio.

The preference of polar monomers for π or σ coordination has been addressed in a number of theoretical papers using DFT methods [106, 108, 113–115]. These

Scheme 4.9 σ- and π coordination modes of polar monomers

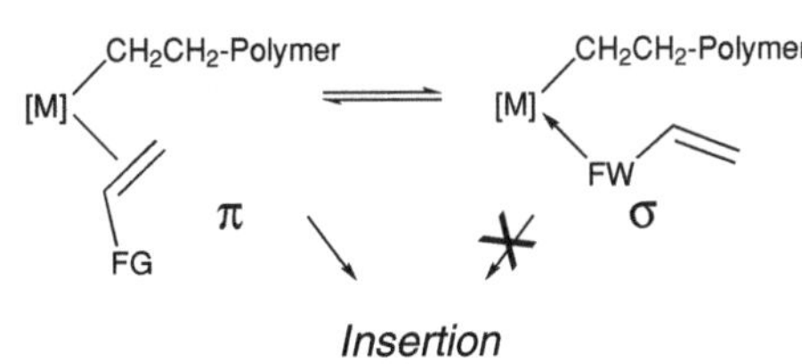

have shown that three main factors control the balance between both coordination modes: the electronic structure of the monomer [111, 112], the nature of the metal centre [106, 113, 114] and the electric charge of the catalyst [109, 113, 114]. Table 4.1 collects relative energies of the π and σ coordination modes calculated for selected monomers and catalysts.

The preference of the monomer for π or σ binding mode is defined, to an important extent, by the shape and energy of its frontier molecular orbitals (FMO), which are made up from those in the C=C and the polar fragments. The FMO energies of several functionalized vinyl monomers are plotted in Fig. 4.5 [113, 114]. Most olefin polymerization catalysts are electron-poor species, therefore their interaction with the monomer essentially involves electron donation to the metal from the HOMO or other high energy filled orbital of the monomer. Hence, π coordination will be more favorable for electron-rich olefins such as vinyl ethers and amines which have high-energy π C=C orbitals. Conversely, the electron donor capability of the π bond decreases when electron-withdrawing groups are directly attached to the C=C bond. For example, in MA the energy of the π C=C bond falls below that of the lone electron pair sitting on the carbonyl oxygen atom. Clearly, this is a factor that favors σ coordination. On the other hand, the lower donor capability of the electron-poor C=C bond is partially compensated by its extra ability to accept electron back-donation from the metal center into low energy empty π^* orbitals [106, 108, 113]. As a result, methyl acrylate is a borderline case that favors σ coordination in some cases and π in others, depending on the properties of the metal center. Thus, it is difficult to draw a simple relationship between the energies of the FMO in the monomers and their binding preferences. A complete picture of monomer bonding can only be obtained after considering another important factor, the capability of the available orbitals to achieve a good overlapping. This is determined by the shape and extension of the orbitals and varies greatly from one case to another [109]. For example, lone pairs on nitrogen tend to be very good donors to metal atoms, therefore the σ coordination mode is

Table 4.1 Energy differences between π and σ coordination modes calculated for selected monomers and catalysts

Catalyst[a]	Monomer	$E(\sigma \rightarrow \pi)$[b]	Ref.
$(\alpha\text{-diim})Ni\text{-}R^+$	MA	+4.0	[113]
$(\alpha\text{-diim})Pd\text{-}R^+$	MA	−3.4	[113]
(sal)Ni-R	MA	−7.5	[113]
(sal)Pd-R	MA	−14.6	[113]
$(\alpha\text{-diim})Ni\text{-}R^+$	VA	+0.7	[113]
$(\alpha\text{-diim})Pd\text{-}R^+$	VA	−5.2	[113]
(sal)Ni-R	VA	−7.2	[113]
$(\alpha\text{-diim})Ni\text{-}R^+$	AN	+16.9	[114]
$(\alpha\text{-diim})Pd\text{-}R^+$	AN	+12.8	[114]
(sal)Pd-R	AN	+3.9	[114]
$(\alpha\text{-diim})Pd\text{-}R^+$	VCl	−8.1	[108]

[a] α-diim, HN=HCH=NH; sal, HN=CH-*o*-$C_6H_4O^-$; R, *n*-Pr or Et
[b] Energy differences, kcal/mol, negative values favor π coordination

particularly favored for monomers containing nitrogen functional groups such as nitriles. On the contrary, non-bonding pairs on halogenated hydrocarbons are very internal and interact only weakly with the transition elements. Accordingly, calculations predict that AN has a strong preference for σ coordination [109, 114], while VCl will be forming exclusively π complexes [106, 108, 109] despite the fact that the nitrogen lone pair of AN is lying at much lower energy than either its π-C=C orbital, or the lone pair orbitals in any other monomers shown in Fig. 4.5, including VCl.

The nature of the metal center also has a critical role in the coordination mode of the monomer. π Coordination is usually more favorable for Pd(II) based catalysts than for their Ni(II) analogues. In general, η^2-olefin complexes in high oxidation state are much more common for second and third row transition elements than for the first row congeners. A simple explanation for this trend is that the *s*, *p* and *d* orbitals have similar radial extensions in the cations of the first transition row elements, which causes unfavorable repulsions between the ligand and metal filled orbitals, while the more extended *d* orbitals in the heavier 2nd and 3rd transition series allow better ligand/metal π-*d* overlapping [114, 115]. DFT calculations have confirmed this trend for a variety of Ni and Pd catalysts. Interestingly, detailed wavefunction analyses suggest that the lower tendency of Ni complexes to bind polar monomers in π mode arises from the electrostatic component of the bond energy, rather than from unfavorable electron–electron (or Pauli) repulsions or from reduced orbital overlapping [113]. These calculations point out the important role of the electric charge: π-coordination is favored by electrically neutral catalysts, such as salicylaldiminato [113] or anilinotroponato complexes [116], due to their stronger capability to back-donate electron density into the empty π* orbitals of the ligand, as compared to more electron-deficient cationic α-diimine derivatives. In an interesting computational experiment, Jordan and Ziegler have shown that π-coordination becomes increasingly favorable in a series of isostructural Pd complexes as their net charge decreases from +1 to −1 [109]. This is in good agreement with calculations on neutral Pd-phosphinosulfonato complexes, which indicate that the π bonding mode of MA is

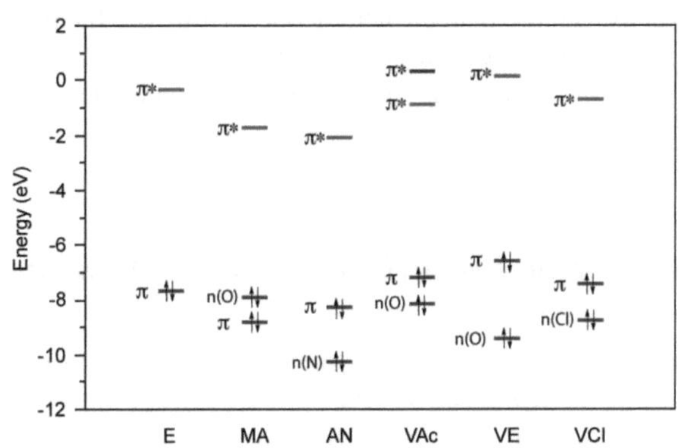

Fig. 4.5 Energies of the frontier orbitals for some polar monomers calculated at the B3LYP/6-311G* level

strongly favored (by ca. 3 kcal/mol) [117]. On the other hand, calculations for α-diimine complexes indicate that steric effects play only a minor role on determining the monomer coordination mode [113].

Coordination of functional monomers to alkyl-Pd and alkyl-Ni moieties has also been investigated experimentally. The unstable alkene complexes were generated at low temperature and characterized by NMR in situ. In agreement with DFT calculations, the preference of MA or VA to bind in π fashion to cationic α-diimine complexes is higher for Pd than for the analogous Ni complexes [29, 96]. Palladium forms exclusively π-complexes with MA and VA, while Ni forms a 9:1 equilibrium mixture of the π- and the σ (*O*-bound) complexes with VA [96]. The latter equilibrium ratio corresponds to a free energy difference of 0.7 kcal/mol that is in excellent agreement with DFT predictions (see Table 4.1) [113]. Also in line with theoretical predictions, polar monomers with poor σ-donor functionalities such as vinyl ethers [118] and halides [110, 119] form exclusively π-complexes, while acrylonitrile invariably coordinates in σ mode to palladium both in cationic [97, 98] and neutral [100] complexes. Equilibrium mixtures of the σ and π complexes have been observed only in the above-mentioned case of a Ni α-diimine complex with VA [96]. Here the individual NMR spectra of the σ/π isomers are resolved only below −120 °C, as exchange of the coordination mode is extremely fast. Increasing the temperature above −50 °C leads to insertion of VA into the Ni–R bond. Exchange between the σ/π modes is probably facile even when only one of the possible coordination isomers can be observed. Thus, the very stable σ-complexes formed by the reaction of acrylonitrile with palladium alkyls evolve gradually to afford insertion products, presumably via undetected π-complexes [97, 98, 100].

Although much importance has been attributed to σ-coordination of the polar comonomer as one of the main problems for the copolymerization of polar comonomers [113], its real impact is reduced by the facile σ/π exchange at Ni and Pd centers. Poisoning by the monomer only becomes a significant problem when the polar group is an exceptionally good σ donor, for example AN. Although this monomer undergoes stoichiometric insertion reactions with Pd α-diimine complexes, it suppresses all catalytic activity if added in excess [100]. Both experimental and theoretical methods provide clear indication of the relative importance of the poisoning effect of polar monomers: Nickel catalysts are more sensitive than Pd ones, and positively charged catalysts more than the neutral ones. However, the most important factor for catalyst poisoning of the catalyst is the nature of the polar comonomer itself. The compatibility of polar monomers with π coordination increases in the order acrylonitrile < acrylates < vinyl esters < vinyl ethers and halides.

4.4.2 Low Reactivity of the Polar Monomer

A typical feature of copolymers formed by coordination-insertion mechanisms is the low incorporation level of the polar comonomer. This indicates that polar monomers are in general much less reactive towards the catalyst than non-polar olefins, especially ethene. In extreme cases, ethene homopolymerization can take place in the presence of polar olefins without these being actually incorporated in the polymer. As described in Sect. 4.3, the comonomer incorporation ratio can be modeled by a classic Curtin–Hammett model, which considers the binding equilibrium of the monomers and their characteristic insertion rates. Migratory insertion is faster for electron-poor alkenes, such as MA, which, at the same time, coordinate weakly to the metal center. Although the two effects are opposite and tend to compensate mutually, the binding capacity of ethene is much higher than any other olefin, including α-olefins [96] and some electron-rich alkenes such as vinyl ethers [118] that are better π-electron donors. Thus, the dominant role of the coordination equilibrium determines the relatively low reactivity of polar monomers, at least in copolymerization with ethene.

Brookhart has determined relevant kinetic and thermodynamic parameters for the Pd(α-diimine) system [29, 96]. Figure 4.6 displays free energy diagrams for the competitive insertion of ethene with MA, VA and vinyl trifluoroacetate (VA^f), measured for two slightly different α-diimine complexes. In either case, the most stable π-complex is formed with ethene. This complex can either undergo migratory insertion with a barrier $\Delta G^{\ddagger}_{E}$, or exchange the ethene ligand to produce the polar comonomer π-complex, which can then evolve to the corresponding insertion product. The energy barrier for the latter process is the sum of the free energy of the exchange preequilibrium, $\Delta G_{E/comon.}$, and the "intrinsic" kinetic barrier for the comonomer insertion, $\Delta G^{\ddagger}_{comon}$. The ratio between the polar monomer and ethene incorporation rates is a function of the difference between their respective insertion barriers, $\Delta\Delta G^{\ddagger}_{ins}$ (see Fig. 4.6 A, below), as well as the relative monomer concentrations in the feed. For MA, $\Delta\Delta G^{\ddagger}_{ins}$ is 3.4 kcal/mol (at 34 °C), which is consistent with 5% comonomer incorporation under realistic copolymerization conditions [29]. The barrier $\Delta G^{\ddagger}_{comon.}$, for VA is ca. 1 kcal/mol larger than $\Delta G^{\ddagger}_{E}$. However, since VA is a better electron donor than MA, the formation of the VA π complex is more favorable. Both factors compensate mutually, and the difference $\Delta\Delta G^{\ddagger}_{ins}$ for the insertion of VA and ethene is 3.5 kcal/mol, i.e., nearly the same than for MA. Thus, the failure of Pd α-diimine systems to copolymerize VA cannot be attributed to difficulties in the comonomer insertion. However, the $\Delta\Delta G^{\ddagger}_{ins}$ term is somewhat higher for VA^f, 4.1 kcal/mol, suggesting that the tendency for comonomer incorporation would be lower in this case. This helps to understand why, although the α-diimine catalyst homopolymerizes ethene in the presence of VA^f, it does not incorporate this comonomer [96].

There is much less information available on the relative insertion rates of ethene and polar co-monomers in Ni complexes. However, DFT calculations for cationic α-diimine catalysts suggest that, while energy barriers to migratory insertion are in general lower for Ni than Pd, the difference between the ethene and MA insertion rates is smaller for the former, suggesting that MA incorporation would be lower [120]. This seems to be confirmed experimentally, since high-pressure MA/E copolymerizations with nickel α-diimine catalysts typically lead to less than 1 mol% incorporation [37, 38].

Although the accuracy of the data required for a reliable calculation of the comonomer incorporation rate is probably beyond the limits of present-time DFT methods, it is illustrative examining the calculations results in the light of the Curtin–Hammett model for comonomer incorporation. Table 4.2 shows π-binding

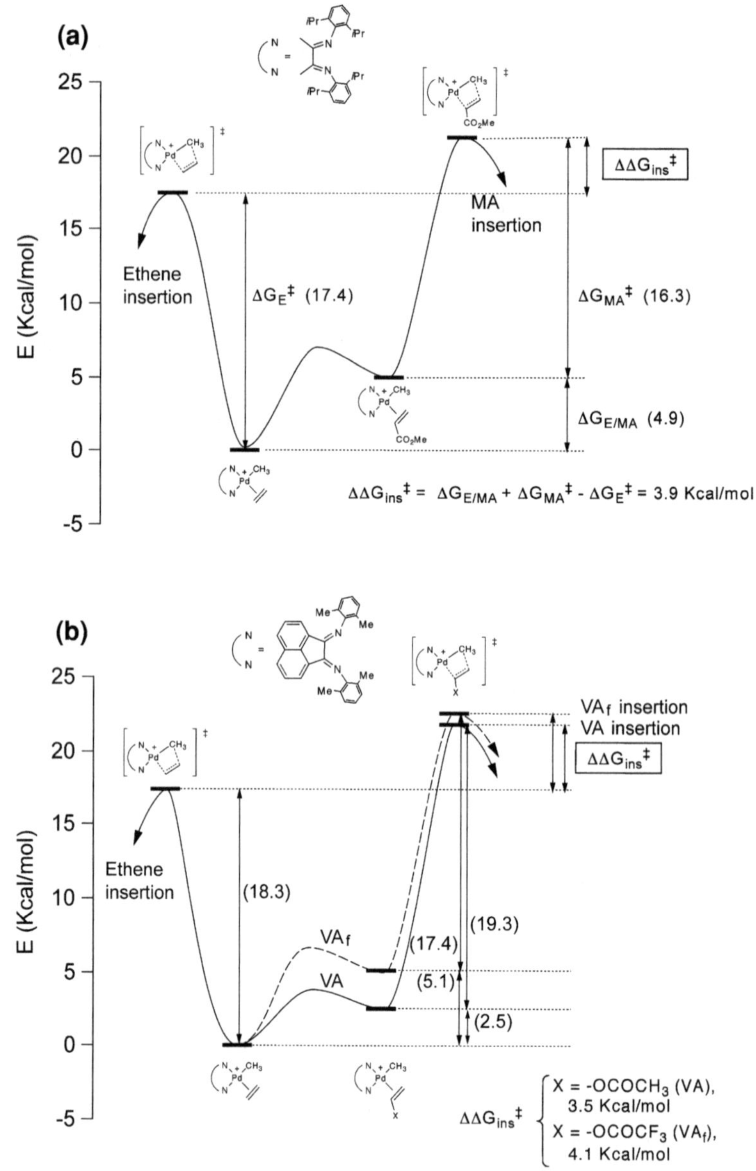

Fig. 4.6 Free energy diagrams for competitive insertion of ethene and methyl acrylate (**a**) and vinyl acetate or trifluoroacetate (**b**) into methylpalladium α-diimine complexes

Table 4.2 Calculated monomer π-binding energies (kcal/mol) to the fragment $[Pd(HN{=}CH{-}CH{=}NH)(Et)]^+$ and insertion barriers for several monomers (data taken from Chen et al. [102])

Comon.	ΔE_{comon} [a]	$\Delta E^{\ddagger}_{comon}$ [b]	$\Delta\Delta E^{\ddagger}_{ins}$ [c]
E	0.0	18.7	0
MA	4.6	17.3	3.2
VA	3.7	18.7	3.7
VCl	3.8	18.1	3.2
AN	8.4	15.4	5.1

[a] π-Binding energy of the monomer relative to ethene (more positive values mean less favorable π binding)
[b] Insertion barrier for each monomer, relative to the corresponding π complex
[c] $\Delta\Delta E^{\ddagger}_{ins} = (\Delta E_{comon} + \Delta E^{\ddagger}_{comon}) - \Delta E^{\ddagger}_{E}$

and insertion barrier energies calculated by Goddard et al. for the simplified catalyst model $[Pd(HN{=}CCHCH{=}NH)(Et)]^+$, which include corrections for solvent effects [108]. In spite of its simplicity and the neglect of the entropic terms, the model provides an excellent description of the real system at a semiquantitative level, including the relative stability of the π-olefin complexes with ethene, MA and VA, and the olefin insertion barriers. The differences between global insertion barrier for polar comonomer and ethene ($\Delta\Delta E^{\ddagger}_{ins}$) are remarkably similar to the experimental $\Delta\Delta G^{\ddagger}_{ins}$ values. One of its most important features is that the model reproduces the comparatively high stability of the ethene π-complex.

This calculation confirms that $\Delta\Delta E^{\ddagger}_{ins}$ has similar values for the monomers MA, VA and VCl, which is not too surprising since, as previously discussed, the influence of electronic effects on the stability of the π-complex and on the size of the insertion barrier are of opposite sign and tend to cancel out [112]. For acrylonitrile, the exceptional weakness of the π-complex is only compensated in part by the easier insertion, and $\Delta\Delta E^{\ddagger}_{ins}$ is significantly higher. This suggests that, even if the catalyst were not poisoned by acrylonitrile, the incorporation rate would be negligible for this comonomer.

The energy balance for the incorporation of electron-deficient olefins can be improved by decreasing the positive charge at the metal fragment. In order to investigate this effect, Jordan and Ziegler have carried out DFT calculations on a hypothetical set of isostructural Pd catalysts containing ligands in which the electric charge is varied from +1 to −1 by adding anionic hydroborate fragments to the α-diimine ligands (Fig. 4.7) [109]. Table 4.3 collects the results obtained for ethene and acrylonitrile. As can be seen, the $\Delta\Delta E^{\ddagger}_{ins}$ parameter becomes more favorable for AN in the neutral catalyst, and even negative in the anionic one, suggesting that such catalyst would discriminate the polar monomer over ethene. This is due to the dramatic increase of the π-bonding capability of acrylonitrile to neutral and anionic molecules, while the effect of electric charge on the corresponding insertion barriers is small. At first sight, this result appears promising, but ethene insertion becomes increasingly difficult on passing from cationic to anionic catalysts. In the latter case, the ethene insertion barrier is over 24 kcal/mol, suggesting that this complex would be too unreactive to display any significant catalytic activity. Thus, it can be concluded that, although decreasing the

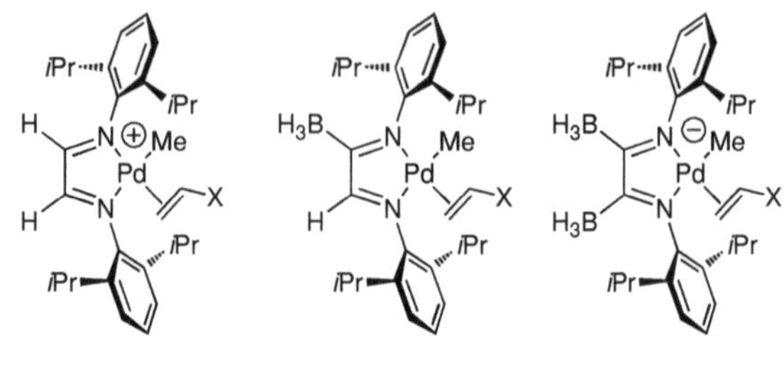

Fig. 4.7 Computational models for cationic, neutral and anionic palladium catalysts

Table 4.3 π-binding energy and insertion barriers for AN and E calculated for cationic, neutral and anionic Pd(α-diimine) catalysts (kcal/mol)

Catalyst	E_{AN} [a]	$\Delta E^{\ddagger}_{AN}$ [b]	$\Delta E^{\ddagger}_{E}$ [c]	$\Delta\Delta E^{\ddagger}_{ins}$ [d]
Cationic	5.8	16.3	16.3	5.8
Neutral	−0.4	20.5	21.6	1.5
Anionic	−6.1	19.3	24.4	−11.6

[a] π Binding energy for acrylonitrile (more negative values mean more favorable π-binding)
[b] Acrylonitrile insertion barrier
[c] Ethene insertion barrier
[d] $(E_{AN} + \Delta E^{\ddagger}_{AN}) - \Delta E^{\ddagger}_{E}$. All energies relative to ethene complex + acrylonitrile

electrophilicity of the catalyst would improve its tolerance to polar functionalities, at the same time it causes a serious decrease of the catalytic activity.

4.4.3 Formation of Less Reactive or "Dormant" Propagating Species

Incorporation of a polar comonomer unit into the growing polymer chain has deep effects on the catalyst behavior. In some cases, the interaction of the metal center with the polar group could cause irreversible catalyst deactivation (see next section). In other cases, however, the polar functionality simply decreases the reactivity of the propagating species, allowing the catalytic process to continue albeit at a slower pace. One of the causes for such activity drop is the propensity of the polar group attached to the polymer chain to coordinate to the metal center, blocking the incorporation of new monomer units. It has already been mentioned (Scheme 4.7) that ethene/MA copolymerization with palladium α-diimine catalysts involves the formation of 6-membered cyclic chelates. However, the formation of such chelates is not the only possible reason for the decrease of the catalyst activity.

Insertion of the polar monomer in 2,1 fashion leads to a metal alkyl species that bears a functional group on the metal-bound carbon atom. Such intermediates usually display lower reactivity. When the functional group is a carbonyl fragment (e.g., -COOR or -COR), the insertion product is a metal enolate. Enolates can bind to the metal either through carbon (*C*-enolate) or oxygen (*O*-enolate), as shown in Scheme 4.10. If the metal center is oxophilic, it will migrate from the carbon atom

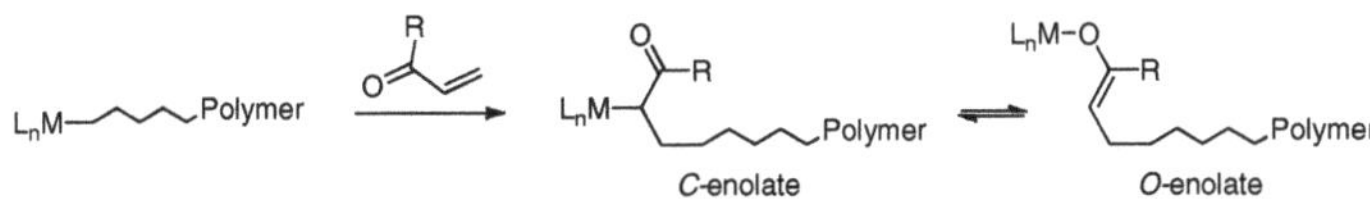

Scheme 4.10 *C*- and *O*-binding modes of enolates generated after the insertion of a carbonyl-substituted olefin

Fig. 4.8 Monomer insertion into α-functionalized metal–carbon bonds

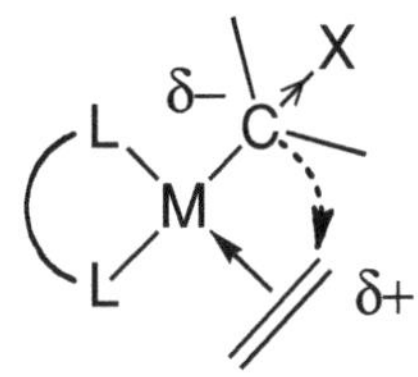

to the oxygen. Although *O*-enolates catalyze the homopolymerization of the polar monomers by a coordination-addition mechanism [121], they are inactive in olefin insertion polymerization. While early transition metals have a strong preference for *O*-coordination [3, 4], "soft" metal centers, such as Pd, almost invariably favor the *C*-coordination mode [122]. The small Ni(II) cation is midway between these two situations, favoring *O*-coordination in some cases (e.g., ketone enolates [123]), and *C*- in others (ester or nitrile derivatives [124]). However, enolate formation will not deactivate a Ni catalyst because the preference for the *C*– or *O*– coordination modes is small and these are readily exchanged.

Aside from the problem of enolate coordination, an electronegative group at the metal-bound carbon atom tends to stabilize the M–C linkage, decreasing its reactivity towards new monomer units. An intuitive way of understanding this effect is to look at olefin migratory insertion as an internal nucleophilic attack of the metal-bound alkyl on the coordinated alkene, as shown in Fig. 4.8. Removing electron density from the α-carbon lowers its nucleophilicity and causes the energy barrier for the insertion to rise.

Acrylonitrile provides a good example of catalyst deactivation by the electron-withdrawing effect of α-functional groups. Typical palladium polymerization/oligomerization catalysts, such as cationic alkyls containing α-diimines [98], salicylaldiminato [100] or other bidentate nitrogen ligands [97] undergo 2,1-insertion of acrylonitrile, affording catalytically inert oligomeric products bridged by the α-cyano group. One of these compounds is the bis(imidazolyl)methane derivative shown in Scheme 4.11 [97]. This oligomer is not reactive towards ethene or AN but is cleaved by carbon monoxide affording monomeric α-cyanoalkyl-carbonyl species that very slowly undergo CO insertion over a period of several days. This demonstrates that the lack of reactivity of such complexes cannot be entirely due to the aggregation in oligomers, but also to the low migratory capability of the α-cyano alkyl group. For comparison, CO insertion in analogous methylpalladium species takes less than 1 min. Analogous Pd complexes containing α-chloroalkyl groups also display lower reactivity towards CO or ethene than their normal alkylic counterparts [125].

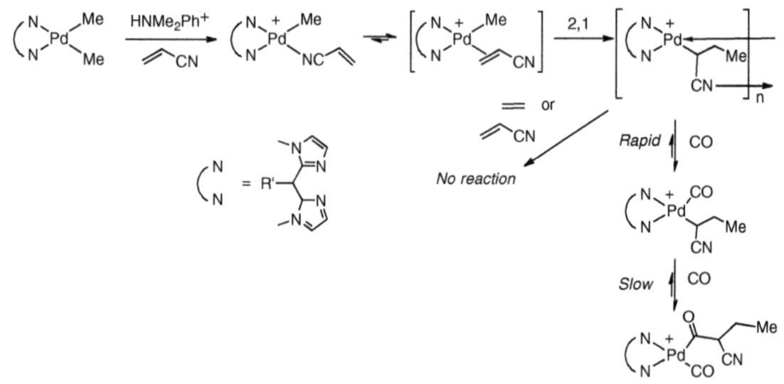

Scheme 4.11 Insertion of acrylonitrile into Pd–Me bonds leads to unreactive α-cyanoalkyl species

Unreactive α-functional alkyls are not formed exclusively by 2,1 insertion reactions, since the *chain walking* mechanism facilitates the migration of the metal center to the vicinity of the polar group also when the insertion takes place with the opposite regioselectivity (Scheme 4.12). Thus, the insertion of vinyl ethers into the Pd–Me bond of cationic α-diimine complexes proceeds with 1,2 regioselectivity to afford β-alkoxyalkyl products, which isomerize into α-alkoxyalkyl species [102, 118, 126]. Analogous reactions with α-olefins containing a *N*-carbazolyl group in their ω-positions also afford α-functionalized alkyls, following successive chain-walking events [104]. Palladium complexes with α-alkoxo or α-amino functionalization become further stabilized by the formation of three-membered chelate complexes. At least in some special cases, the formation of these intermediates does not prevent the copolymerization of the polar monomers. The resulting copolymers have branched structures with the $OSiR_3$ [118] or carbazolyl groups [104] placed at the end of side branches and separated from the main polymer by at least two consecutive CH_2–CH_2-units. This suggests that, even though the Pd-C(X) bond is too unreactive to undergo new monomer insertions, the Pd unit can shift away from the funcitonal group by (*chain walking*) in order to resume the chain growth process.

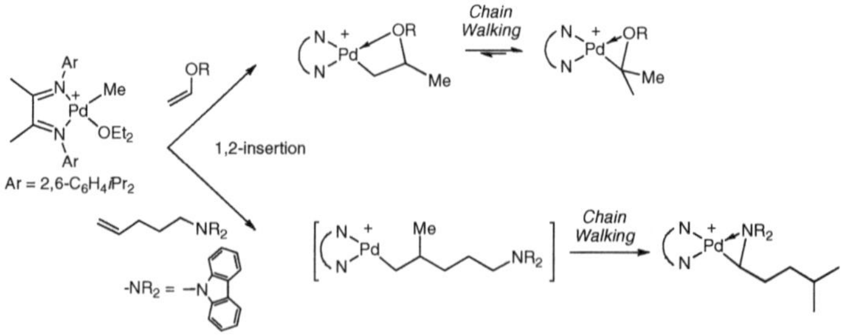

Scheme 4.12 Generation of α-functionalized alkyls by chain waking

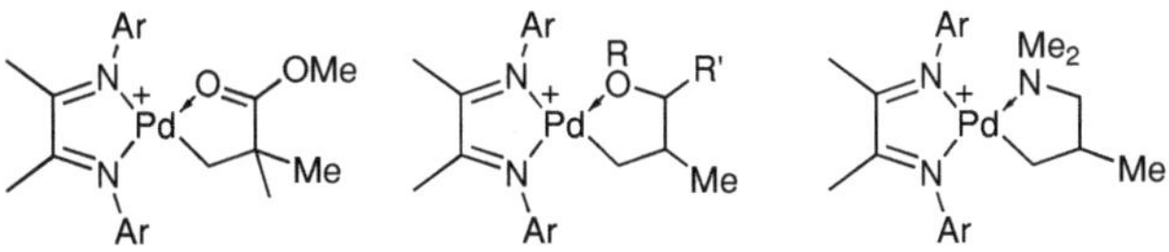

B5 B6 B7

R, R' = OMe
R = Et, R' = H

Fig. 4.9 Stable chelate species isolated from the reaction of α-diimine Pd–Me complexes with MMA (**B5**), allyl ethers (**B6**) and allyldimethylamine (**B7**)

Formation of cyclic chelates is probably the general situation in polar comonomer copolymerization with Ni and Pd α-diimine catalysts. Five-membered chelates **B5**, **B6** and **B7** (Fig. 4.9) have been isolated from the reaction of α-diimine Pd methyl complexes with MMA [105], allyl ethers (acrolein dimethylacetal, allyl ethyl ether) [103], and allyldimethylamine [104], respectively. The Pd-oxygen interaction in chelates **B5** and **B6** can be displaced by ethene, and these compounds can initiate ethene homo- and copolymerization reactions. However, the strong donor properties of the NMe_2 group render the chelate ring of **B7** too stable to undergo any further olefin insertion. Stable chelate complexes are probably a more serious problem in the case of Ni catalysts, but they have been less studied. Johnson has briefly reported that cationic Ni α-diimine alkyls insert MA at −40 °C to afford a relatively stable 4-membered chelate, **C4**, which slowly rearranges to 6-membered chelates **C3** at room temperature [37] (Scheme 4.13). The stability of **C4**, attributed to the strength of the dative Ni–O bond, was invoked to explain the difficulty of Ni catalysts for copolymerizing MA. DFT calculations confirmed that the metal–oxygen dative interaction is stronger for Ni than for Pd [120]. However, according to these calculations, when the catalyst bears bulky *N*-aryl substituents the stability of the Ni and Pd chelates is not so different because the Ni derivatives are more destabilized by steric effects than their Pd analogues. Thus, the causes of the low activity of Ni complexes in ethene/MA copolymerization are not fully understood and could be attributed either to the formation of stable chelates or to catalyst poisoning by the comonomer.

In contrast to MA, VA and VA^f react with Pd α-diimine alkyls affording products with a thermodynamic preference to form 5-membered chelate rings. As shown in Scheme 4.14, VA or VA^f insert selectively in 2,1 fashion to give the isolable chelates **B8** and **B8′** [96]. Similar 5-membered chelates complexes are also formed from ω-alkenyl acetates such as allyl acetate or butenyl acetate, by means of *chain walking* rearrangement of the alkyl chain. Complex **B8** has been

Scheme 4.13 Formation of Ni chelates from MA

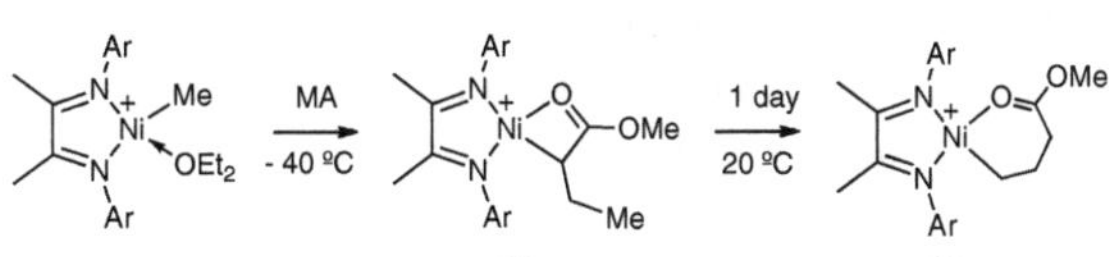

C4 C3

Scheme 4.14 Five-membered chelates from VA and VA[f]

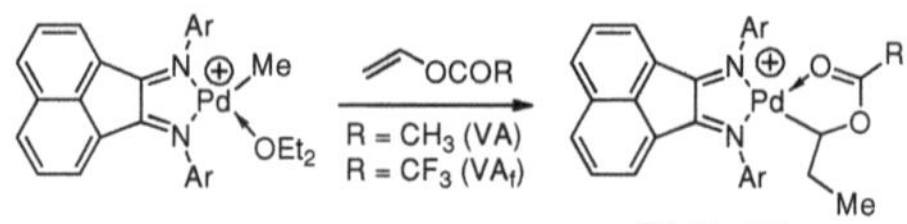

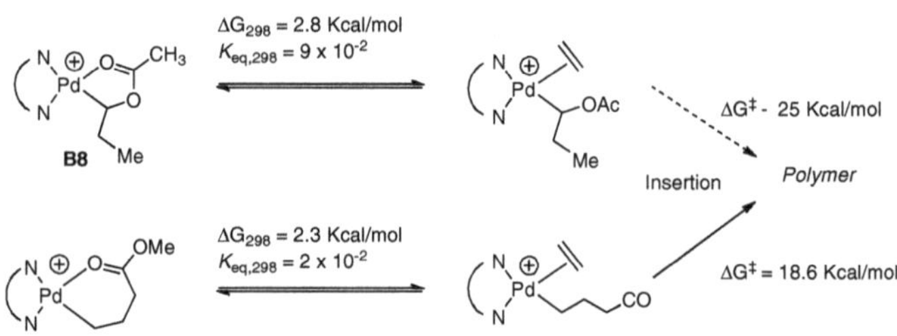

Scheme 4.15 Comparative reactivity of chelate complexes arising from VA (*up*) and MA (*bottom*) insertion

isolated and structurally characterized, but **B8′** is thermally unstable (see next section).

Since Pd α-diimine complexes copolymerize MA and ethene, but become deactivated by VA, it is instructive to compare the behavior of chelates **B3** and **B8** towards ethene (Scheme 4.15). In contrast with **B3**, complex **B8** does not initiate ethene homopolymerization, even under forcing conditions. The low reactivity of this complex is the prime cause of the failure of the α-diimine catalysts to copolymerize ethene and VA. However, NMR studies showed that, like **B3**, **B8** reacts reversibly with ethene at low temperature (below −70 °C) to form the corresponding π-olefin complex. At room temperature, the free energy change for the displacement of the chelate ring by ethene (extrapolated from low temperature measurements) is only 0.5 kcal/mol higher for **B8** than for **B3**. Although this means that the VA chelate is somewhat more difficult to open, the difference is not large enough to explain why this monomer shuts down the Pd(α-diimine) catalyst, while MA is successfully copolymerized. The different behavior of the two monomers is due to the fact, that subsequent ethene insertion step is much more difficult for **B8** than for **B3**. Note that in the case of MA, the insertion takes place in a "normal" Pd–CH_2 bond, and therefore the energy barrier is essentially the same than in Pd alkyls (ca. 18.5 kcal/mol). However, the π-ethene species arising from **B8** is less reactive due to the presence of an electronegative OAc substituent in the metal-bound carbon. Ethene insertion has not been experimentally observed in this case, but DFT calculations by Goddard point to a high-energy barrier of 25.1 kcal/mol. The latter value is in good agreement for the experimental value measured with the more reactive chelate **B8′** obtained from VA[f] (21. 5 kcal/mol). Thus, it can be concluded that the deactivation of the Pd(α-diimine) complex by

Scheme 4.16 4-Membered chelate from MA insertion in a neutral Ni-salicylaldiminato polymerization catalyst

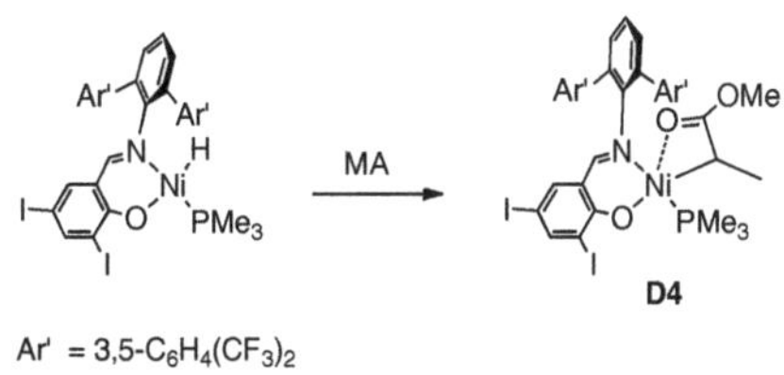

VA is due to the deactivating effect of the α-acetoxy group rather than to the stability of the chelates of type **B8**. It is interesting to consider that, although the formation of chelate **B3** in the MA system significantly decreases the polymerization rate, it probably has a protective effect preventing the formation of "dormant" species with a carboxylate substituent sitting on the α-carbon.

The role of chelated intermediates seems to be more important for cationic than for neutral complexes, although for the latter there is less information available. Mecking has found that the products arising from the reaction of MA or VA with nickel salicylaldiminato complexes are 5-coordinated, with a weak secondary Ni–O interaction (Scheme 4.16) [127]. DFT calculations suggest that 2,1 insertion of MA in Pd-phosphinosulfonato catalysts leads to reactive 4- or 5-membered chelates [117]. However, such chelates have not been detected experimentally, because the Pd···O=C interaction is so weak that is displaced by the auxiliary coligand present in the reaction, even if this is an extremely weak ligand such as dmso [67]. In spite of this, kinetic measurements of the ethene/MA copolymerization rate suggest that, similarly to the α-diimine system, the limiting step is the monomer insertion subsequent to that of MA. Since the stabilization of the MA insertion intermediate by chelate interaction is not important, the limiting factor is the intrinsically low reactivity of the Pd–CH(CO_2Me)R bond. The same argument could also explain the low polymerization and incorporation rates observed in this system with other polar comonomers that are not expected to produce strong intramolecular chelates, such as VF [64].

4.4.4 *Irreversible Catalyst Deactivation*

As a general rule, the activity of olefin polymerization catalysts decays as a result of side reactions leading to catalytically inert products. In certain cases, polar comonomers can promote such deactivation processes or induce specific reactions causing the irreversible deactivation of the catalyst.

Neutral polymerization catalysts such as nickel phosphinoenolato [26, 27], anilinotroponato [128, 129] or anilinoperinaphthenato [130] derivatives are prone to deactivate through the formation of inactive bis(ligand) complexes [Ni(O–N)$_2$]. Although nickel salicylaldiminato catalysts are rather stable and compatible with many polar substances, they are deactivated by MA [131]. Grubbs has shown that the salicylaldiminato catalyst precursor **D1**, containing PPh_3 as stabilizing coligand, reacts with MA to afford a mixture of methyl cinnamate and methyl

3-phenylpropionate (Scheme 4.17, up) [42]. The inactive bis-ligand complex **D2** was detected as the main nickel-containing byproduct. While methyl cinnamate is a Heck-type product arising from MA insertion followed by β-H elimination, the origin of the saturated ester is less evident. Isotopic labeling experiments have demonstrated that, under rigorously anhydrous conditions, the extra hydrogen in the latter comes exclusively from MA. Moisture traces can also act as a proton source, producing saturated products (same scheme, bottom) [42, 132]. It is believed that the deactivation of neutral nickel catalysts is related to the tendency of intermediate hydrides to transfer hydrogen by one of the two pathways shown in Scheme 4.18. One of them (*Path A*) involves reductive O–H coupling and elimination of the ligand in protonated form. If the ligand is acidic enough, it can cleave the Ni–C bond of a second catalyst molecule affording the saturated R–H product [129]. The same products can be formed by direct H transfer from the

Scheme 4.17 Reaction of salicylaldiminato catalyst precursors with MA

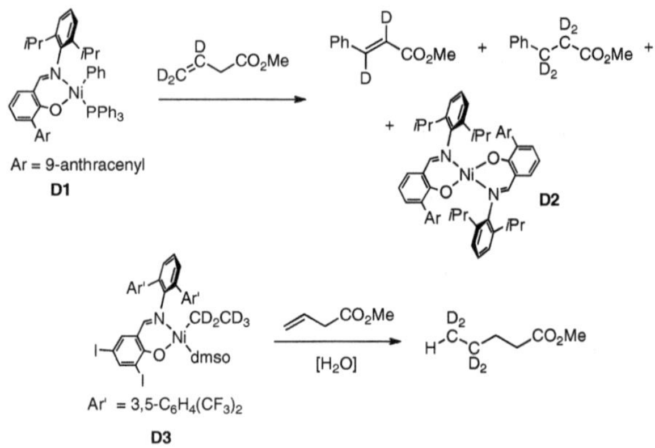

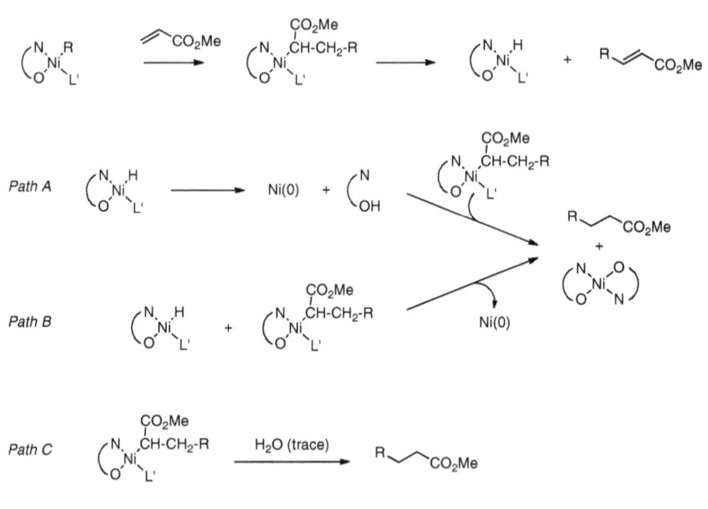

Scheme 4.18 Deactivation pathways of neutral nickel polymerization catalysts

hydride to the enolate complex arising from the acrylate insertion (*Path B*) [132]. Finally, the enolate complex could undergo hydrolysis by moisture traces (*Path C*).

Using salicylaldiminato complexes stabilized by PMe_3, Mecking has shown that MA insertion products decay by a bimolecular mechanism involving β-hydrogen elimination and hydride transfer (i.e., the mechanism depicted by *Path B*) [127]. However, as previously noted by Grubbs [42], the hydrolytic deactivation pathway *C* becomes important when the PMe_3 coligand is replaced by PPh_3. Mecking showed that the hydrolytic stability of the acrylate insertion products is closely related to the stabilizing capability of the ancillary ligand present in the catalyst: water readily cleaves the enolate intermediates containing PPh_3, but reacts more slowly with species containing the strongly donor ligand PMe_3. Thus, reaction of the highly reactive dmso-stabilized ethyl complex **D3** with MA exclusively produces methyl pentanoate because the enolate intermediate is trapped by water traces faster than it decomposes by β-hydrogen elimination (Scheme 4.17, bottom) [127].

The ultimate causes of the irreversible decomposition of nickel salicylaldiminato complexes by MA seem to lie in a combination of low reactivity of the MA insertion products towards ethene insertion, and low stability of hydride species. In this scenario, β-hydrogen elimination followed by deactivation processes A–C become dominant. If the hydride intermediates were more stable, MA insertion would not lead to catalyst deactivation but only to chain transfer, and polyolefins capped with polar units would be produced. Tomov and Gibson have observed this type of reactivity using nickel phosphinoenolato catalysts (type **A**, Fig. 4.2) modified with bulky substituents [133].

A different deactivation pathway appears when the polar monomer bears an electronegative group X directly bound to the vinyl group, for example, vinyl chlorides, ethers or esters (X = -Cl, -OR or -OCOR) (Scheme 4.19). Insertion of such monomers transfers the substituent X to the propagating species, in a way that

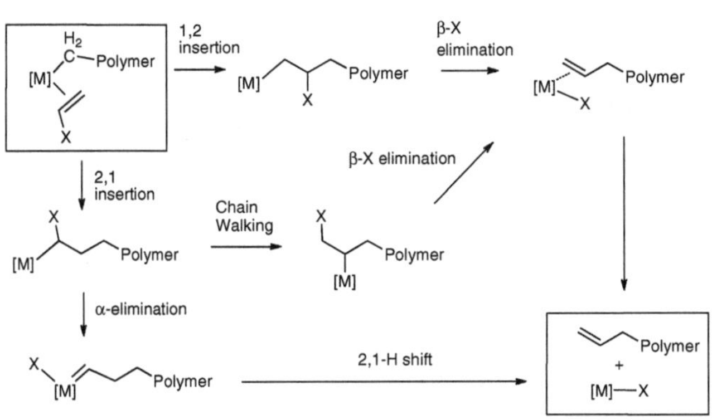

Scheme 4.19 Mechanisms of irreversible heteroatom cleavage following 1,2 or 2,1 insertion of functionalized vinyls

facilitates its abstraction by the metal center in a process that is formally similar to β-hydrogen elimination. However, in contrast to β-H elimination, abstraction of the electronegative X fragment is irreversible and causes the M–C bond to be replaced by a catalytically inert M-X fragment. Vinyl halides are potent venoms for olefin polymerization catalysts, because β-halogen abstraction and formation of strong M-halide bonds are thermodynamically very favorable. Analogous OAc and OR abstraction reactions have been detected in the reaction of polymerization catalysts with alkenyl acetates [93, 127, 134] and phenyl vinyl ether [118].

The most direct route for heteroatom elimination is 1,2-insertion followed by β-X elimination. However, 1,2 insertion is not essential for catalyst deactivation. As shown in Scheme 4.19, the same products formed after 1,2 insertion can be equally produced when the monomer inserts with 2,1 regioselectivity, by two alternative mechanisms. The α-haloalkyl complex formed in the 2,1 insertion step could experience α-X elimination to an unstable alkylidene complex, which evolves into the final products by 2,1-hydrogen shift. Otherwise, chain-walking migration of the metal fragment allows the rearrangement of the 2,1-insertion product into a β-haloalkyl complex, which directly undergoes β-X elimination.

Insertion of VCl into Zr–Me bond of zirconocene complexes produces propene and chlorozirconium species. Since chain-walking isomerization does not occur in d^0 systems, the reaction most likely involves 1,2 insertion of VCl followed by β-Cl elimination [135]. DFT calculations for the reaction of VCl with transition metal hydrides provide support for the 1,2-insertion regioselectivity [136, 137]. Iron diiminopyridine complexes activated with MAO also react with VCl producing propene [119]. The mechanism of this reaction is probably the same as for the metallocene alkyls, but the Fe–Cl complex is recycled by reaction with the alumoxane (Scheme 4.20, up). Recycling the Fe catalyst allows ethene polymerization to proceed in the presence of VCl or VA, apparently without comonomer incorporation. However, when a mixture of ethene and deuterated VCl or VA is oligomerized with a $FeCl_2$(diiminopyridine)/MAO catalyst, the resulting polyolefins have deuterated vinyl terminal groups, indicating that VCl or VA insertion takes place and it is immediately followed by halide abstraction (Scheme 4.20, bottom) [134].

Scheme 4.20 Reaction of VCl or VA with Fe diiminopyridine terminates polyethylene chain propagation

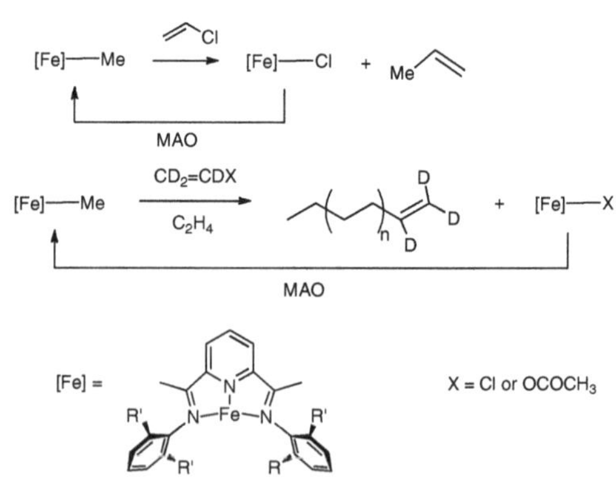

Jordan [99, 119] and Sen [111] have studied the mechanism of the insertion of vinyl halides into Pd–Me bonds. Vinyl chloride or bromide complexes were generated at low temperature and their evolution monitored by NMR. As observed for the Zr and Fe catalysts, the Pd π-complexes evolve affording propene and different halogenated metal products, but the initial insertion products were not detected. DFT calculations for Pd α-diimine complexes suggest that VCl insertion should favor the 2,1 regioselectivity [99, 107–109]. This has been recently confirmed by an isotopic labeling experiment involving the complex $[Pd(Me)(OEt_2)(ArN{=}CC(Me)C(Me)C{=}NAr)]^+$ (Ar = 2,6-diispropylphenyl) and VCl selectively deuterated at the positions 1 or 2 [99]. The distribution of deuterium in the propene product, shown in Scheme 4.21, is consistent with 2,1-insertion followed either by isomerization to a β-chloroalkyl intermediate or by α-halide abstraction and 1,2-hydrogen shift. The first of these two options appears as more likely, since chain-walking is precisely one of the main features of Pd(α-diimine) alkyls and, on the other hand, α-chloroalkyl complexes are usually stable [125]. DFT calculations provided additional support for this conclusion [99].

Palladium α-diimine catalysts copolymerize non-polar olefins with certain monomers containing OR substituents, including some allyl ethers (acrolein dimethylacetal/ethene) [103] and silyl vinyl ethers ($Ph_3SiOCH{=}CH_2$/1-hexene) [118]. One of the difficulties met in the copolymerization of this kind of monomers is the formation of catalytically inactive η^3-allylcomplexes. It is believed that the process involves OR group abstraction leading to highly reactive alkoxo-π-alkene intermediates. These evolve with C–H abstraction and alcohol elimination, as shown in Scheme 4.22. The tendency to undergo this deactivation step depends strongly on the nature of the OR group of the comonomer. Elimination of aliphatic alkoxo (-OR) or siloxane ($-OSiR_3$) moieties is relatively difficult, which allows the above-mentioned copolymerizations. However, elimination of phenoxo (-OPh) from phenyl allyl ether is a rapid and very favorable process. Alkoxo abstraction is not the only problem in the copolymerization of vinyl ethers, since cationic Pd catalyst readily trigger facile cationic homopolymerization reactions (see next section).

Insertion of VA in the Ni-Me bond of the nickel α-diimine complex, shown in Scheme 4.23, leads to a mixture of 5- and 6-membered metallacyclic chelates arising from 2,1 and 1,2 insertion, respectively. The 1,2-insertion product, having

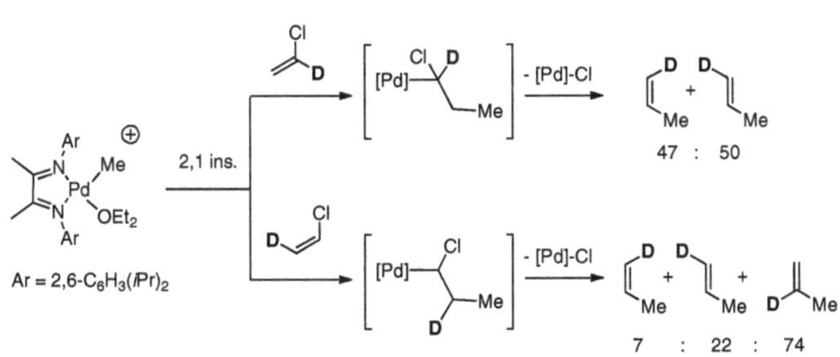

Scheme 4.21 Deuterium labeling experiment confirming the 2,1 regioselectivity of VCl insertion in Pα-diimine catalysts

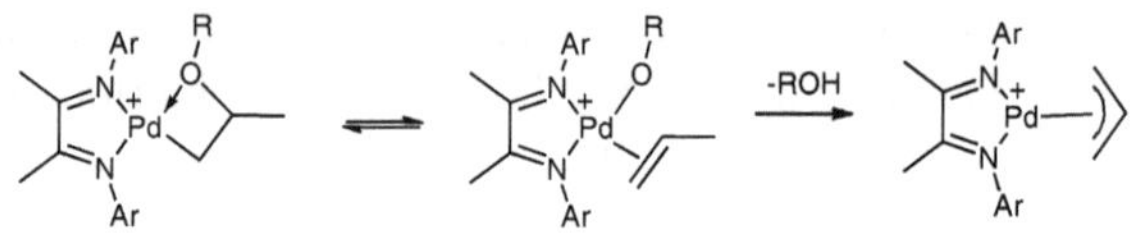

Scheme 4.22 Deactivation of palladium catalysts by ether monomers

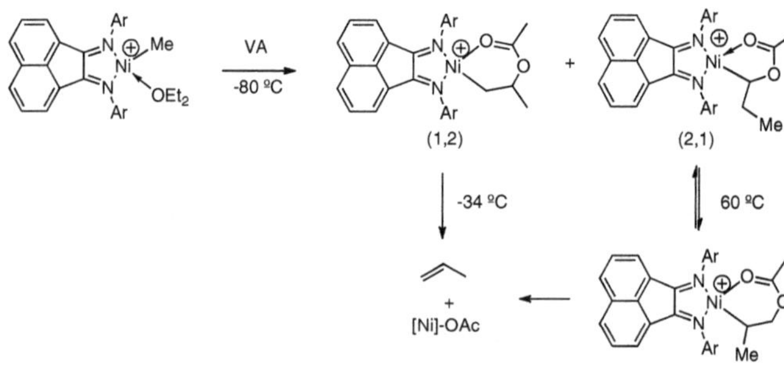

Scheme 4.23 Thermal decomposition of chelates arising from 1,2 or 2,1 insertion of VA in α-diimine complexes

the acetate group in the β position, is thermally unstable and decomposes at low temperature affording propene and paramagnetic Ni acetate complexes. On the other hand, the regioisomer resulting from 2,1 insertion is thermally robust (decomposes at 70 °C). As discussed in the preceding section, analogous Pd complexes arising from 2,1 insertion of VA are also very stable. The decomposition of 2,1-VA insertion product requires previous opening of the 5-membered chelate ring and isomerization to the corresponding β-acetoxy alkyls by the chain walking mechanism. Similar chelates arising from the reaction of VA^f with both Ni and Pd complexes are not as stable due to the better migratory properties of the electronegative trifluoroacetate group. Interestingly, the Ni VA^f insertion product is more stable than its Pd analogue, probably because opening the chelate is more difficult in the former case. The β-carboxylate abstraction process is probably responsible for the deactivation of Ni and Pd catalysts by VA^f. However, the available data are not conclusive on whether acetate abstraction is one of the causes that prevents the copolymerization of ethene and VA with Pd α-diimine catalysts [96].

4.4.5 Undesired Competing Polymerization Processes

An important problem in the copolymerization of polar olefins is the occurrence of polymerization mechanisms other than the coordination-insertion process, which lead to products that are markedly different from the intended copolymers [5]. Highly reactive olefins such as acrylates, vinyl halides or vinyl ethers are prone to polymerize by radical or cationic mechanisms. Such processes can also occur during copolymerization reactions and may be difficult to suppress, particularly if the copolymerization process is slow. It is a well-known fact that due to their redox reactivity many transition metal complexes can readily initiate many chemical transformations involving free radicals, including polymerizations [138]. Palladium complexes are no exception, and it was soon realized that some complexes with structures which are typical for polymerization catalysts, like the pyrrole-imine complex **N1** [54] or the salicylaldiminato **D5** [55] (Fig. 4.10) initiate ethene/acrylate copolymerization reactions by free-radical rather than insertion mechanisms.

This ability of some Pd alkyl complexes to initiate radical polymerization suggests that they behave as sources of free radicals. Sen has shown that neutral (halo)alkyl Pd α-diimine complexes are unstable with respect to homolytic Pd–C fission [139]. Treatment of the cyclic chelate **B3** with tetralkylamonium halides causes the opening of the chelate ring to afford a linear alkyl, which rapidly transposes by chain walking into a thermodynamically favored enolate species (Scheme 4.24). These species decompose spontaneously producing free radicals, which decay mainly to methyl crotonate, together with smaller amounts of methyl butyrate and dimethyl suberate. If the reaction is carried out in the presence of MA, these free radicals trigger its polymerization. These results pose some doubt on whether the decomposition of MA insertion products in the salicylaldiminato system reported by Grubbs and Mecking discussed in the previous section, could arise by a bond homolysis mechanism as well. These Pd *C*-enolate complexes are probably more prone to homolysis than other Pd alkyl species because the resulting radicals are particularly stable.

Sen and Espinet have shown that simple palladium perfluorophenyl complexes with PPh_3 ligands can polymerize MA and copolymerize MA with 1-hexene producing MA rich copolymers [56]. This system has some properties that are unusual for a radical polymerization mechanism. For instance, the catalyst homopolymerizes MA but not MMA, a monomer that is even more easily polymerized by free radical methods. In fact, the polymerization of MA does not take place in the presence of MMA. However, the poly-MA and poly-MA-co-1-hexene produced by this catalyst are very similar to the materials obtained with a classic radical initiator such as AIBN. A non-conventional radical mechanism involving reversible Pd-C(CO_2Me)R homolysis/recombination was proposed to explain these observations.

Establishing that olefin/polar olefin copolymerization takes place by a coordination-insertion mechanism, and does not involve competition by free radical processes is not a trivial task, especially when only small amounts of comonomer are incorporated. The composition of the copolymer can be a good indicator,

L = PPh_3, PMe_3, Py
N1

L = N(R)=C(H)(CH_2Ph)
D5

Ar = 2,6-C_6H_3*i*Pr_2

Fig. 4.10 Pd catalysts that polymerize MA by free radical mechanisms

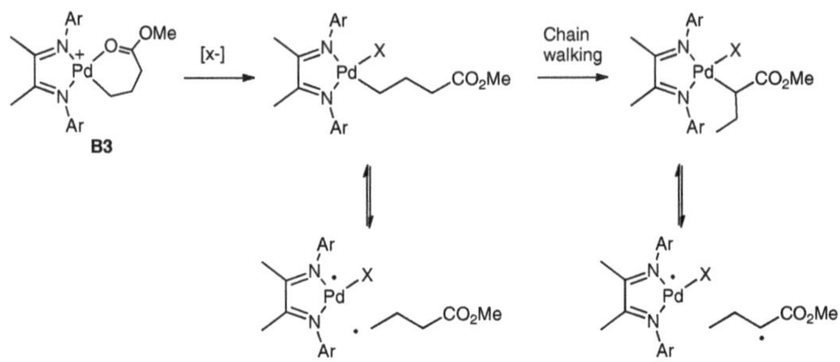

Scheme 4.24 Pd–C homolysis of neutral α-diimine alkyls

because those produced by radical mechanisms have high polar comonomer contents, while coordination-insertion mechanisms lead to products that are higher in the less polar component. However, a mixture of homopolymers might resemble a typical copolymer with low polar comonomer content. GPC analysis with two types of detectors (e.g., refractometric and UV) detectors is a very useful characterization technique in such cases, but separation of the components of the mixture by selective extraction and precipitation is sometimes the only reliable tool for a correct characterization.

Classic tests for radical reactions such as polymerization inhibition with radical traps can be misleading, as such reagents can react with hydride complexes that are key intermediates in the insertion polymerization mechanism [140]. Mild radical traps such as phenols are usually inefficient inhibitors of radical polymerizations [56], but more efficient reagents such as TEMPO and galvinoxyl can produce false negative results as well, because they react with aluminum alkyls such as alumoxanes commonly used as cocatalysts in polymerization reactions.

Electron-rich olefins such like vinyl ethers vinyl-*t*-butylether or vinyl-methylether are easily polymerized through non-selective cationic propagation mechanisms. These processes can be initiated by cationic complexes such as Pd(α-diimine) derivatives, which prevents their copolymerization with non-polar comonomers with these catalysts [118]. This problem can be avoided by using electrically neutral catalysts, such as Pd(phosphinosulfonato) complexes [60].

4.5 Overcoming the Difficulties: Strategies and Breakthroughs

4.5.1 Strategies Based on the Comonomer

One of the most direct ways to improve the compatibility between catalysts and polar comonomers is to adapt the structure of the latter or to protect their functional groups. For example, bulky acrylates (e.g., *t*-butyl acrylate) or fluorinated esters such as VA^f are tolerated more readily by polymerization catalysts because their polar functionalities have less capacity to interact with the metal center [96, 113]. In some cases, both the polar and non-polar monomers can be tuned simultaneously to facilitate their copolymerization with a certain catalyst. Jordan's work on the copolymerization of vinyl ethers by palladium α-diimine complexes provides a good example of this [126]. As mentioned in the preceding section, electron-rich monomers such as methyl or *t*-butyl vinyl ethers are very prone to cationic polymerization, a process that can be initiated by cationic complexes such as Pd α-diimine catalysts. This leads to vinyl ether homopolymers and prevents their copolymerization with non-polar olefins. If the strongly electron-donor -OR group of the monomer is replaced by the more electronegative–OPh fragment, the cationic polymerization pathway is suppressed, but the catalyst is then readily deactivated through facile β-OPh elimination reactions. Vinyl siloxanes, R_3Si-$OCH{=}CH_2$ provide a compromise with electronic properties midway between the two types of ether functionalities. Cationic polymerization is slow with the trimethylsilyl derivative, and does not take place with the bulkier triphenylsilyl ether. However, the steric hindrance of the Ph_3SiO- group decreases significantly the ability of the monomer to bind to the sterically hindered Pd center. In order to increase its capability to compete with the non-polar component, the less reactive olefin 1-hexene was used instead of ethene. This particular selection of monomers is successful and allows the Pd catalyst to copolymerize affording branched products containing up to 20 mol% comonomer.

Some other strategies based on monomer modification by masking this functionality with a protective group or alkyl reagents have already been mentioned in Sect. 4.2. These seek to minimize the interaction of the polar group with the active center, disrupting their capability to poison the metal center or to form "dormant" chelate intermediates once the insertion has taken place (see Scheme 4.2). Selecting suitable monomers, with separated vinyl and functional fragments, can also avoid the unfavorable interactions of the metal center and the polar groups.

From a practical point of view, the main drawback of strategies based on the monomer is their high economic cost. Using stoichiometric amounts of masking agents is expensive, and special comonomers are much more costly than simple substances such as MA or VA. However, these methods can be useful for the production of specialty polymers with interesting properties. A couple of examples will illustrate this concept.

The first example is the preparation of nanostructured polyolefin copolymers with functionalized norbornene monomers. Neutral nickel catalysts, e.g. salicylaldiminato or iminocarboxamidato alkyl complexes (**D** and **G**, respectively, Fig. 4.2), have excellent compatibility with this type of monomers and catalyze their copolymerization with ethene. Bazan has used the iminocarboxamidato catalysts **G** to develop interesting ethene/norbornenyl acetate copolymers [45]. The copolymerization reactions exhibit a quasi-living behavior affording high molecular weight copolymers with very narrow molecular weight distributions. The comonomer contents can be accurately controlled through the ethene/comonomer ratio in the feed. Changes in this ratio during the polymerization process are translated into the relative rates of comonomer insertion, and therefore become "recorded" in the polymer molecules as local differences in the copolymer richness. For example, an ethene pressure jump from 50 to 1,100 psi produces a pseudoblock copolymer with two segments of different composition. A technically easier approach for this strategy is to continuously feed ethene while allowing the depletion of the comonomer. This leads to a "tapered" comonomer distribution. If the comonomer concentration is restored at regular times, peaks are produced in the monomer/comonomer distribution along the polymer chains (Scheme 4.25). The pseudomultiblock structure of the resulting copolymers leads to phase separation and to peculiar layered microscopic structures that are similar to those observed in typical block copolymers [48, 49].

Late transition metal catalysts enclose an undeveloped potential for the synthesis of advanced functional materials. Guan has shown how the unique properties of Pd α-diimine catalysts can be exploited in a straightforward manner in the synthesis of advanced functional materials [30]. As discussed in Sect. 4.3, the topology of branched polyethylenes produced with these catalysts can be controlled by the ethene concentration during the synthesis [30–32]. Low ethene pressure favors the formation of polymers with a dendritic morphology (i.e., branches in branches), while higher pressures leads to less entangled structures with a larger number of terminal branches. When MA is introduced in the system, the same branching schemes can be obtained, but with many of these branches terminated with the functional group. This leads to the production of interesting copolymers with tailored topologies (Scheme 4.26). A problem with this is that the molecules of EMA copolymers produced with Pd α-diimine catalysts are too small to allow a detailed investigation by light scattering methods.

Guan has developed the special comonomers **O–U** in order to increase the molecular weight of the copolymers, and to allow the introduction of different functional groups. For example, inserting a polymethylene spacer between the

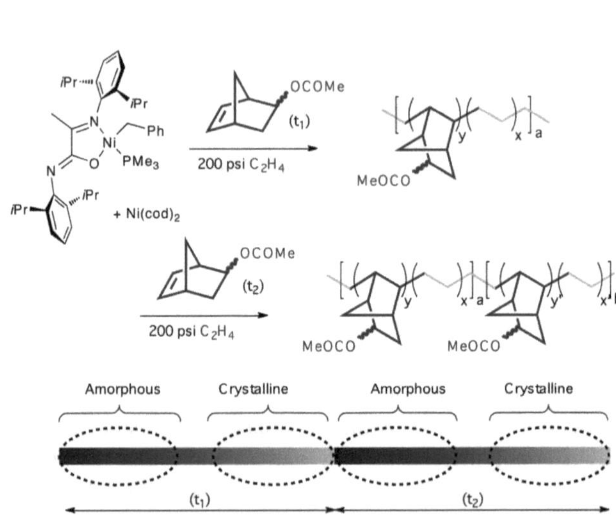

Scheme 4.25 Synthesis of ethene-norbornenyl acetate pseudoblock copolymers

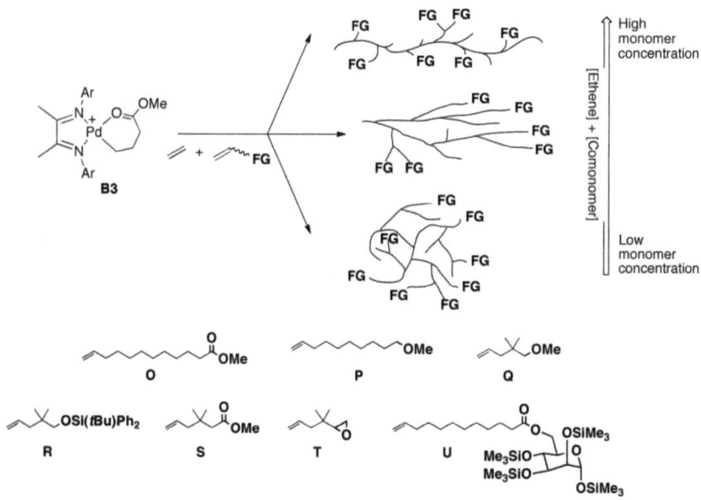

Scheme 4.26 Functional copolymers with well-defined branching topologies

vinyl and the carboxylate fragments of the monomer (**O**) increases the polymerization degree. This simple approach does not work with other functional monomers. For example olefins with terminal ether groups (**P**) are not copolymerized, presumably because the Pd atom migrates to the ether group and the catalyst is deactivated by β-OR abstraction. This problem is solved by inserting a quaternary C atom which prevents *chain walking* of the metal center to the OR group (comonomer **Q**). Consistently with Jordan's findings, the catalyst performance also improves when the OR group is replaced by the bulky siloxy group (**R**). Noteworthy, this comonomer does not induce the typical decrease of the catalyst activity observed in almost every polar/non polar copolymerizations, and the copolymers have essentially the same molecular weight than ethene homopolymers produced with this Pd catalyst. Using a quaternary-C blocked spacer, highly reactive functionalities such as epoxides can be introduced without perturbing the active center (**S**, **T**). Even silyl-protected carbohydrate moieties can be copolymerized, such as the mannose derivative **U**. Post-polymerization deprotection of the carbohydrate moiety leads to polyethylene functionalized with biologically active monosaccharide units. Although the precise location of the functional groups is not established, it is assumed that most likely these are placed at the branch terminations, as observed in MA or vinyl ether copolymers [29, 118].

4.5.2 Catalyst Activation Methods

Lewis acids such as aluminum or boron derivatives commonly used as co-catalysts in polymerization reactions react with polar vinyl monomers, initiating their polymerization [121, 138]. In order to prevent undesired side polymerization processes, catalysts intended for copolymerization of polar monomers are usually designed to avoid the use of such conventional co-catalysts. A simple approach consists of generating the catalyst in situ from free ligands and suitable precursors. A classic example is the ethene oligomerization system employed in the SHOP process, where a nickel phosphinoenolato complex is directly produced from $Ni(cod)_2$ and a phosphorusylide [141]. More recently, Pd-phosphinosulfonato catalysts have been generated from free phosphinosulfonic acids and simple Pd compounds [58, 71, 72]. One drawback of this methodology is that only a fraction of the ligand or the metal precursors are converted effectively into active species, which limits catalytic efficiency. Thus, using well-defined organometallic species as single-component catalysts is a more efficient strategy.

A typical single-component polymerization catalyst is a metal alkyl stabilized by the main ligand and a secondary ligand or "base". The purpose of the base is to stabilize the metal complex by occupying the coordination position required for the catalytic process. In order to activate the catalyst, the free base has to be released under the catalysis conditions. Scheme 4.27 summarizes the main methods available for the synthesis of base-stabilized catalyst precursors. Cationic complexes of Ni and Pd α-diimine alkyl complexes developed by Brookhart, are classic examples [10, 29, 96]. Cationic α-diimine complexes are strongly electrophilic, and only the weakest bases are labile enough to be displaced by ethene or other olefins. Extremely reactive alkyls containing Et_2O as auxiliary ligand are obtained by reacting dialkyl precursors $M(R)_2$(α-diimine) with suitable protic acids (method A). Cationic Pd complexes containing nitrile co-ligands can also be prepared by halide abstraction from readily available (halo)alkyl precursors (method B). A key feature of cationic catalysts is the use of low coordination capability counteranions with very small tendency to occupy the free coordination site. For a soft Lewis acid center such as Pd(II), many anions of reduced coordinating capability, such as PF_6^-, BF_4^-, etc., fulfill this condition [142]. However, some of these are small anions basic enough to interact with a harder Lewis-acid center such as Ni(II). Brookhart's choice of the very stable fluorinated tetraarylborate anion BAr_4^{f-} proves very satisfactory for its stability and high solubility in non-polar hydrocarbon solvents. This anion has been adopted by many other researchers and is widely used in the development of late-transition metal catalysts for olefin polymerization, as well as in other applications in homogeneous catalysis.

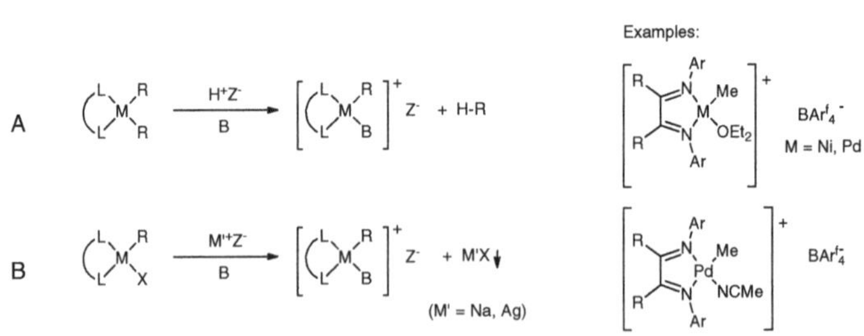

Scheme 4.27 Synthesis of base-stabilized cationic catalysts

An acceptable replacement for BAr_4^{f-} is SbF_6^-, which can be advantageous for isolation and characterization purposes [29].

In contrast with cationic catalysts, electrically neutral catalysts do not require special non-coordinating anions. Compared to their cationic counterparts, neutral catalysts are milder Lewis acids, and better π-donors. However, displacement of coordinated coligands by the π-acceptor olefin monomers is more favorable, allowing the use of stronger ligands as bases. The main methods developed for the synthesis of base-stabilized neutral alkyl complexes are shown in Scheme 4.28. A classic synthesis of nickel phosphinoenolato complexes involves the oxidative addition of phosphine ylide precursors to $Ni(cod)_2$ in the presence of phosphines or other stabilizing bases [26, 27] (method C). However, a more general approach to the synthesis of neutral catalysts involves methathetical halide exchange between well-known organometallic precursors, such as $[NiPh(Br)(PPh_3)_2]$ [39, 52] or $[Ni(CH_2Ph)((Cl)(PMe_3)_2]$ [47] with alkaline salts of the anionic ligand (D). Method E uses the ligand in neutral form as a protic acid for cleaving one M–C bond of a suitable dialkyl precursor such as $[PdMe_2(Py)_2]$ [63]. A clever extension of the latter methodology resorts to the use of dialkyl precursors $[MMe_2(tmda)]$, containing the *N*,*N*,*N*′,*N*′-tetramethylethylenediamine ligand (tmda), as shown in F.

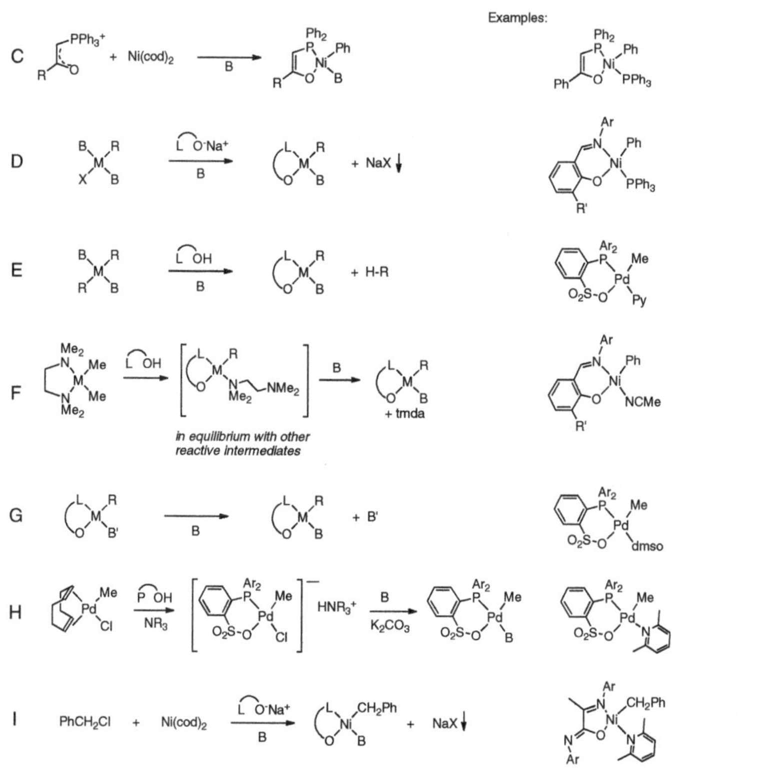

Scheme 4.28 Syntheses of base-stabilized neutral catalysts

These dialkyls are stable and easily handled precursors because the tmda ligand forms very stable chelates. However, irreversible cleavage of one of the M–Me bonds by the acidic ligand forces the formation of highly reactive intermediates, from which tmda is readily displaced [143]. This method allows direct access to reactive salicylaldiminato complexes stabilized by acetonitrile [131, 142], pyridine [62, 74, 144] or other weak coordinating bases [41]. Extremely reactive salicylaldiminato and phosphinosulfonato catalysts containing dmso as stabilized base have been prepared by either methods F [67] or G [132]. Nozaki used chloride as the stabilizing base. The resulting anionic complexes can be used for the synthesis of more reactive catalysts containing sterically crowded bases such as 2,6-lutidine, that can be more readily dissociated from the metal center (method H) [59, 61]. Bazan has also described the preparation of a iminocarboxamidato catalyst containing 2,6-lutidine by the oxidative addition reaction outlined in equation I. In contrast with PMe_3-ligated analogues, the 2,6-lutidine complex is active in the absence of ligand scavengers (see below) [145].

The role of the base used in the catalyst design may go beyond a mere protection of the coordination vacancy. In some cases, the presence of the base in the reaction medium can be beneficial for the catalyst stability [40]. But, in general, the base competes with the monomer for the open coordination site of the catalyst, reducing its activity and/or the molecular weight of the products. The effect can be negligible when the base is very weak (e.g., Et_2O), but catalyst precursors stabilized by such weak ligands are sensitive compounds, which may be not convenient for routine use. These problems have prompted a search for different catalyst precursors, preferably base-free derivatives, which combine high reactivity and easy use and storage (Fig. 4.11). For example, reactive chelates of type **B3** are very stable and can be used as convenient catalyst sources. The synthesis of such precursors has been simplified and can be accomplished by treating readily available chloro(methyl) α-diimine complexes with MA and the sodium salt of the adequate counteranion (e.g., $NaBAr^f_4$ or $NaSbF_6$) [29]. Related metallacycles such as **B5** or **B6** (see Fig. 4.9), are similarly obtained from different monomers (MMA [105], acroleine dimethyl acetal or ethyl vinyl ether [103]) and can also be applied as useful catalyst precursors. In contrast, similar base-free precursors for nickel α-diimine catalysts has still to be developed. Although the Ni analogue of **B3** has been detected (compound **C3**, see Scheme 4.13) [38, 96], its use as catalyst precursors has not been investigated. Easily prepared Ni and Pd η^3-allyl complexes may be an alternative to these chelate precursors, since the organometallic ligand can release a coordination position by changing its hapticity from η^3 to η^1. Nickel allyls containing α-diimine ligands (Fig. 4.11) have been used as ethene polymerization catalysts [146–148]. However, η^3 coordination of the allyl ligand is too stable and their performance is rather poor unless they are combined with alkylaluminum activators. The allyl complex **C1'** was applied for the copolymerization of ethene with methyl acrylate under rather rigorous conditions. In order to achieve significant activities, this catalyst has to be activated with large loads of $B(C_6F_5)_3$ and $NaBAr^f_4$ [37, 38]. Related η^3-benzyl complexes are expected to be comparatively more reactive than the allyl precursors. Zwitterionic Ni η^3-benzyl complexes

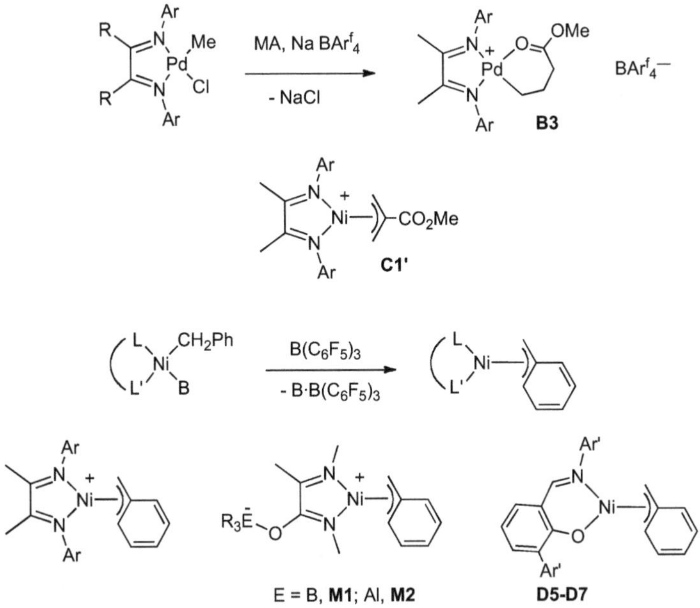

Fig. 4.11 Base-free cationic, zwitterionic and neutral catalysts

M, prepared by ligand abstraction with strong Lewis acids such as $B(C_6F_5)_3$ are highly active catalysts for ethene homopolymerization [44]. Some other η^3-benzylnickel derivatives, such as the cationic $[Ni(\eta^3\text{-}CH_2Ph)(\alpha\text{-diimine})]^+$ [149], have also been prepared by similar phosphine abstraction reactions. However, the use of η^3-benzyl derivatives of Ni as base-free catalysts compatible with polar comonomers has not been widely explored, and it has only been demonstrated with the neutral salicylaldiminate derivatives **D5-D7** (see Sect. 4.5.4). In addition to η^3-allyl and η^3-benzyl derivatives, several base-free complexes of composition "M(R)(L-O)" with L-O = phosphinoenolato [27] and phosphinosulfonato ligands [63, 69] have been isolated. These compounds are dimers stabilized by bridging metal–oxygen interactions. In spite of their ability to catalyze ethene polymerization, their possible use as base-free catalysts for the incorporation of polar comonomers has received little attention.

The activity of base-stabilized catalysts can be promoted by using suitable "ligand scavenging" reagents. A typical example is the activation of phosphine-containing complexes with $Ni(cod)_2$ [27, 40, 44–49]. The latter removes the phosphine from the reaction mixture forming stable $Ni(PR_3)_n$ complexes. Addition of $Ni(cod)_2$ or other "phosphine sponges" is facultative when the catalyst contains a readily dissociable ligand such as PPh_3 [27, 40, 43, 45], but strictly necessary for strongly binding phosphine such as PMe_3 [46–49]. Nozaki showed that halide precipitating salts ($NaBAr^f_4$ better than AgOTf) boost the catalytic activity of anionic $[Pd(Me)(Cl)(P\text{–}O)]^-$ phosphinosulfonato complexes in ethene-MA copolymerization [59].

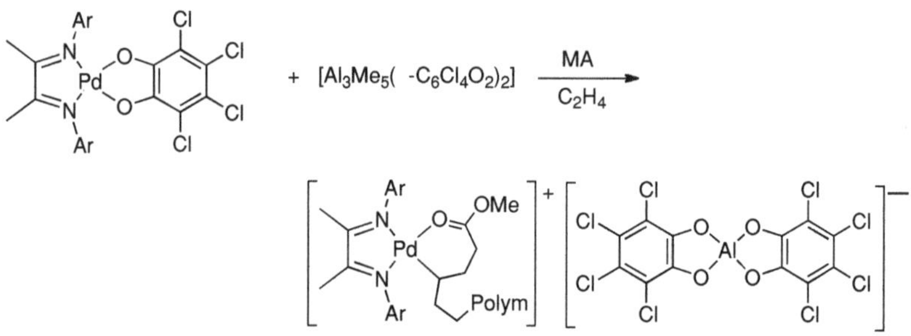

Scheme 4.29 Generation of active α-diimine catalysts by exchange of bidentate catecholate ligands

The use of "soft" activating agents as phosphine sponges or halide precipitating salts brings back the classic "two components" methodologies to the polar comonomer polymerization field, with the corresponding advantages in terms of catalyst preparation and handling, avoiding the use of aluminum or boron co-catalysts. However, some recent publications have described the use of alumoxanes to activate Ni [52, 53] and Ti [78] complexes for the copolymerization of ethene and polar comonomers. In order to minimize the impact of the aluminum reagents in the copolymerization system the cocatalysts are used in relatively small excess, well below the dose of the polar comonomer. The dose of the activating aluminum reagent can be further diminished when the catalyst and co-catalyst are designed to achieve a good match. A combination of α-diimine palladium catecholate complexes and alkylaluminum catecholates in 1:1 molar ratio leads to the formation of ion pairs containing weakly coordinating catecholaluminate anions that are effective for ethene polymerization [150] (Scheme 4.29). The aluminum catecholate complex interacts only weakly with MA, which makes this system very appropriate for copolymerization reactions. Indeed, at low MA concentrations this combination produces essentially the same type of EMA copolymers obtained with the classic tetraarylborate-based system (Brasse et al., unpublished results). However, under high MA concentrations, the aluminum cocatalyst triggers radical homopolymerization reactions, resulting in the formation of EMA and polyMA mixtures. As mentioned above, the possibility of polar comonomer copolymerization reactions with catalysts generated by readily available two components systems instead of sophisticated organometallic complexes represents an bonus for practical application and deserves further attention.

4.5.3 Selecting the Metal

Doubtless, one of the key aspects of catalyst design is the choice of the metal. It is reasonable to assume that a good catalyst for copolymerization of non-polar and polar monomers should also be an *excellent* catalyst for the homopolymerization of ethene and other non-polar olefins. After the discovery of Ni and Pd-based olefin polymerization catalysts, most other transition elements have been subjected to a meticulous scrutiny to find out which ones produce other active catalysts. It is now known that nearly all elements from the first transition row from Ti to Ni, with the exception of Mn, are good candidates for this purpose, as are some heavier elements from groups 4 (Zr, Hf), and 10 (Pd). The second aspect to consider is the compatibility of the catalyst with polar functionalities. Leaving aside the polymerization of protected or masked comonomers, this condition apparently rules out catalysts based on early-transition metals. However, Fujita has demonstrated recently that even Ti complexes with a suitable catalyst design can be applied for this purpose [78] (see Chap. 1, this volume).

Iron 2,6-bisiminopyridine complexes are relatively tolerant to polar functionalities. They polymerize ethene in the presence of MA [79], and they have also been used to hydrogenate functionalized olefins [151]. The failure to catalyze copolymerization could be associated with specific monomers rather than a general incompatibility with polar functionalities [79, 132]. As mentioned in Sect. 4.2, several copper catalysts show some capacity for ethene polymerization and, at the same time, can copolymerize ethene and acrylates, which poses the question of whether these copolymerization reactions involve true insertion mechanisms. However, the acrylate content of copolymers produced with copper cannot be lowered below 40 mol%, which argues in favor of a radical-type mechanism [82–85].

The success of Fujita's Ti catalysts suggests that the horizon of polar comonomer copolymerization could expand to other transition elements beyond Ni and Pd. However, in the present state of the art there is little alternative to these two elements. The available evidence indicates that Pd is superior to Ni for polar comonomer incorporation, essentially because π-coordination of the monomer is more favorable with the former, and σ-interactions, which poison the catalyst are stronger for the latter (see discussion in Sect. 4.3). Thus, nickel catalysts are less successful in the incorporation of industrially relevant polar vinyl monomers, but some catalysts capable of performing such copolymerizations have been described very recently [52, 53, 152]. Developing the potential of Ni catalysts for polar/non polar monomer copolymerization is important for practical application, since the high cost of Pd prevents its application in the large-scale production of polymeric materials.

4.5.4 Ligand Design

The aim of this section is to discuss a few examples in which specific modifications of a known catalyst design leads to improvements on the ability to incorporate polar comonomers. This line of work has achieved very significant advances, but predicting how a given modification will alter the catalyst performance is still very difficult.

Since most of our current knowledge regarding the copolymerization of polar and non-polar olefin comes from the study of the α-diimine-based catalysts, many of the proposals concern the modification of these ligands or closely related ones containing N and O donor atoms. One of the important advances in the field has been the introduction of palladium phosphinosulfonate catalysts, which are not nitrogen-based. Due to the focus of this book, this section first addresses the progress of the nitrogen-based catalyst designs, and concludes by examining some of the main features of palladium phosphinosulfonato catalysts.

4.5.4.1 Steric Effects

Steric effects are probably the most prominent aspect of ligand design for olefin polymerization catalysts. As discussed in Sect. 4.3, the steric effects exerted by bulky aryl substituents play a key role in the control of the activity and the molecular weight of the polymers produced by catalysts based on α-diimines and other related nitrogen-based ligands [7]. These effects are also important for the performance of these catalysts in copolymerization reactions. Incorporation of MA by Pd-α-diimine catalysts increases when the *ortho* substituents of the aryl groups are relatively small, because they oppose less steric hindrance to monomer coordination, hence discrimination by size becomes less important. For example going from 2,6-di*iso*propyl to 2,6-dimethyl changes the MA incorporation level from ca. 4 to 15–20% [29]. At the same time, decreasing the size of the *ortho* substituents causes the molecular weight of the polymer to drop. Thus, the improvement in comonomer incorporation is gained at the cost of significant decrease of the polymer molecular weight.

Although steric effects have a significant influence on the activity of α-diimine catalysts in ethene homopolymerization, they do not affect their catalytic activity in ethene-MA copolymerization. Large aryl substituents increase the monomer insertion rate, but at the same time they hinder the displacement of the chelate ring formed upon MA insertion, which is believed to control the overall reaction rate.

The steric hindrance of the α-diimine ligand influences the regioselectivity of the comonomer insertion. DFT calculations show that encumbered ligands decrease the preference of MA for 2,1 insertion by destabilizing its transition state with regard to that of the competing 1,2 insertion pathway [106]. Accordingly, Sen reasoned that such destabilization should be even more pronounced for a bulkier ligand, such as MMA. Overriding the preference for 2,1 insertion selectivity have some positive effects, such as preventing the formation of unreactive α-functionalized intermediates and avoiding homolytic decomposition which leads to the formation of free radicals. Indeed, MMA reacts with the Pd catalyst selectively affording the 1,2 insertion product **B5** (Fig. 4.9) but the reaction is very slow. Compared with ethene, MMA is too unreactive and does not incorporate. Probably, the same steric effects responsible for the inversion of the regioselectivity are also the cause of the low reactivity of this monomer [105].

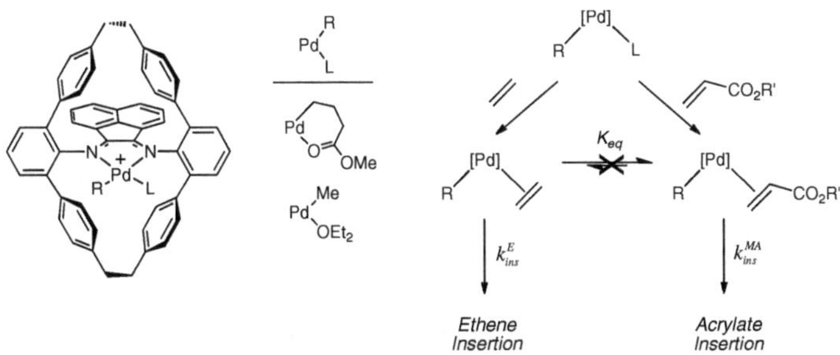

Scheme 4.30 Suppression of the Curtin–Hammett monomer insertion control by cyclophane-based α-diimine ligands

The cyclophane-based catalyst **B7** (Scheme 4.30) provides a very interesting example of how an appropriate ligand design provides control on certain aspects of the copolymerization process [153]. In spite of being more sterically crowded, this catalyst achieves up to 20 mol% incorporation of methyl or *t*-butyl acrylate, significantly more than the analogous catalyst with 2,6-diisopropylphenyl substituents. This might seem paradoxical on the light of the previously commented effect of the ligand size on MA incorporation, but it has been explained by the disruption of the Curtin–Hammett mechanism of monomer discrimination described in Sect. 4.3 (see Scheme 4.8). The cyclophane ligand encapsulates the metal center, preventing the associative exchange of coordinated monomers [154]. The exchange of ethene and acrylate monomers in the coordination sphere of Pd becomes so slow that it cannot be detected by NMR spin recovery experiments. Thus, monomer discrimination ceases to be under thermodynamic control, and higher incorporation of the weaker binding acrylate monomer takes place. Kinetic measurements have shown that relative insertion rates of ethene and MA in the Pd–C bond catalysts are nearly the same in the cyclophane and 2,6-diisopropylphenyl substituted catalysts, demonstrating that the differences in the MA incorporation values emanate exclusively from the disruption of the associative monomer exchange equilibrium.

4.5.4.2 Electronic Effects

Electronic effects exert an important influence on the activity of olefin polymerization catalysts, and particularly on their ability to incorporate polar comonomers. The electric charge of the catalyst has a deep effect on the metal electrophilicity and is one of the most important factors that determine their tolerance to polar functionalities. As discussed in Sect. 4.2, neutral catalysts are expected to facilitate polar comonomer incorporation. However, decreasing the positive charge at the metal center has a negative side, as the benefits on the improved tolerance to polar monomers are counteracted by poorer performance as polymerization catalysts. Specially designed ligands that behave as electron buffers may help to regulate electron density on the metal center, optimizing the balance between tolerance polar groups and high catalytic activity (Fig. 4.12). Zwitterionic iminocarboxamidato catalysts (**M**) constitute an interesting approach to this concept (see also Chap. 2, Vol. 35). These are very active catalysts for ethene polymerization [44], but their capacity for the incorporation of polar comonomers seems to be very limited. For $ER_3 = AlMe_3$ these catalysts polymerize ethene in the presence of 10-undecen-1-ol, incorporating very small amounts of the comonomer [155]. PymNox catalysts (**I**) represent another case in which the electric charge is delocalized in the ligand [53]. This class of complexes relate to the neutral salicylaldiminato catalysts by replacement of the anionic phenolato fragment by a neutral pyridine-*N*-oxide unit. However, the two types of ligands probably have very similar donor capabilities because aromatic *N*-oxides have a strongly dipolar electronic structure, with a large fraction of negative charge sitting at the oxygen atom. The active species involved in polymerization with PymNox catalysts are cationic, but as a result of the peculiar electronic structure of the ligand, PymNox catalysts are best described as a resonance hybrid with a delocalized positive charge. As a result, it can be expected that the PymNox complexes should exhibit intermediate properties between those of typical cationic and neutral catalysts. Accordingly, they are one order of magnitude more active than the analogous salicylaldiminato catalysts in ethene polymerization, but at the same time they show good tolerance towards polar molecules and one of them ($R' = Me$) copolymerizes ethene and MA with ca. 1 mol% incorporation.

Pd-α-diimine complexes bearing electron donor or withdrawing substituents in the *para* position of the aryl substituents show different capability for MA incorporation [29]. Guan has carried out a detailed study of the electronic effects on the catalysts represented in Scheme 4.31 [156]. As can be seen, MA incorporation increases as the substituents become more electron donor. Catalysts with stronger electron-donor ligands maintain higher catalytic activity in the presence of the polar comonomer, while those with electron-withdrawing groups become readily deactivated at sufficiently high comonomer concentrations. This is

Fig. 4.12 Shaping the charge distribution in the catalyst with non-innocent ligands

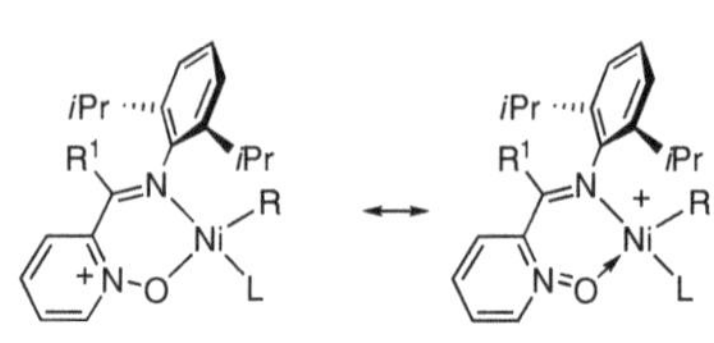

I

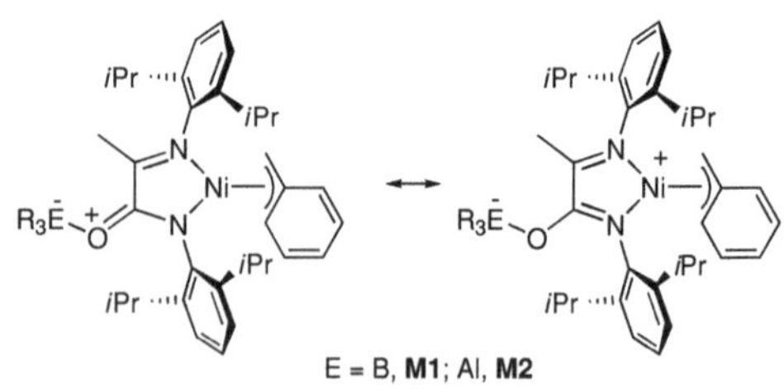

E = B, **M1**; Al, **M2**

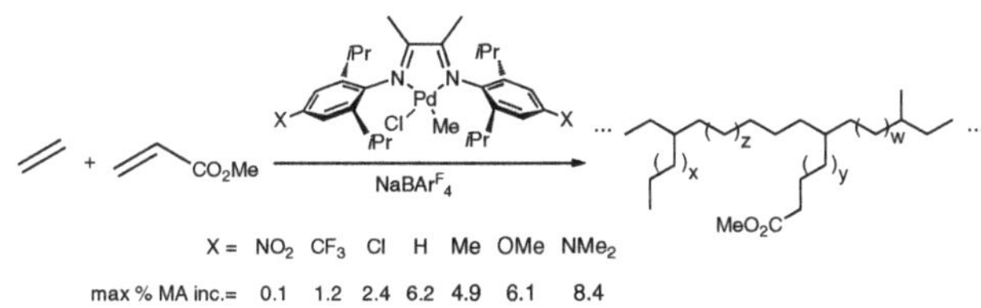

Scheme 4.31 Electronic effects in ethylene–MA copolymerization with Pd(α-diimine) catalysts

consistent with the idea that a somewhat less electrophilic metal center should be more adequate for polar comonomer incorporation. It also suggests that fine-tuning of the donor capacity of the ligand can significantly improve the polar comonomner incorporation capability.

4.5.4.3 Cooperative Effects

A very promising strategy for enhancing the performance of olefin polymerization catalysts consists in building binuclear structures where complex cooperative interactions between metal centers can emerge. Marks has studied extensively the effects of nuclearity and proximity of metal centers in group 4 catalysts containing two metal units. The forced proximity of the metal centers has a significant impact on both catalytic activity and monomer enchainment [157]. The same approach has been applied to late transition metal catalysts. Binuclear complexes formed by tethered phosphinoenolato [158–160] or salicylaldiminato [161–163] units have been prepared. Compared to similar monomeric catalysts, many of these catalysts show significant differences on activity, tolerance to polar solvents or the branching topology of the resulting polymers. It has been demonstrated that appropriate design of binuclear salicylaldiminato complexes leads to remarkable enhancements on their capability to incorporate polar comonomers (Fig. 4.13). In 2005, Lee reported that the binuclear complexes **D5–D7** incorporate functionalized norbornenes two to three times more efficiently than similar mononuclear catalysts [164]. Interestingly, the amount of comonomer incorporated by these catalysts does not show a linear dependency on the feed composition, as it increases more sharply at higher comonomer concentrations. The effectiveness of the binuclear arrangement depends on the size of the linker, increasing as **D6** < **D5** < **D7**. More recently, Marks has advanced further on this concept, developing a new bimetallic catalyst design, **D8**, which incorporates both nickel centers in a rigid salicylaldiminato-type structure based on a central naphthalene core [152, 165]. In addition to the substantial effects on the catalyst activity and branching degree observed in ethene homopolymerization, the capability of these catalysts to incorporate functionalized norbornenes is four times of the related mononuclear catalysts. Furthermore, the binuclear catalyst exhibits a notable capability for incorporating acrylate comonomers, up 8–9% MMA and 11% MA, which is a rather unique property considering that classic salicylaldiminato complexes are rapidly deactivated by this monomer (see Sect. 4.4).

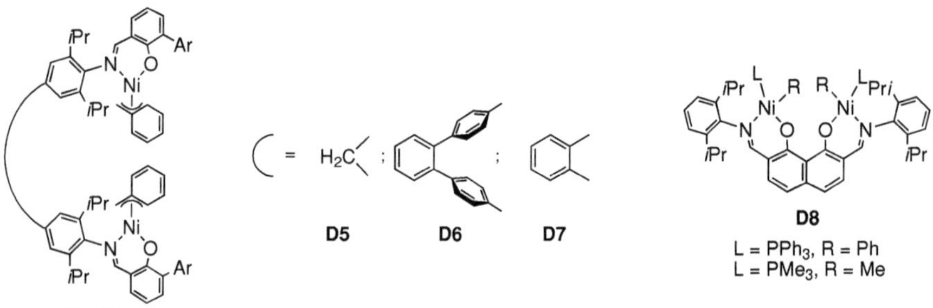

D5 - D7
Ar = 9-antrhacenyl

Fig. 4.13 Bimetallic catalysts for copolymerization of ethene with functionalized norbornenes (**D5–D8**) and acrylates (**D8**)

It is believed that the enhancement of the comonomer incorporation capability of bimetallic catalysts is due to the ability of the two active centers to cooperate binding both the polar and the vinyl groups, hence facilitating the migratory insertion pathway (Scheme 4.32) [152, 157, 164]. According to this explanation, cooperativity would be maximized when the separation between the metal centers matches the size of the polar comonomer.

Molecular models of complexes **D5–D7** indicate that the distance between metal centers in **D7**, ca. 7.3 Å, is similar to the size of the functional norbornene monomers, while the metal–metal separations in **D5** and **D6** are too long or too short, respectively. This could justify why **D7** is the best catalyst of the group. The metal–metal distance in the naphthalene derivative **D8** is shorter, and therefore is best suited for the incorporation of the smaller acrylate monomers.

4.5.4.4 Palladium Phosphinosulfonato Catalysts, a System Without Nitrogen-Based Ligands

To conclude this section, palladium phosphinosulfonato complexes will be examined and compared with other group 10 catalysts containing nitrogen-based ligands.

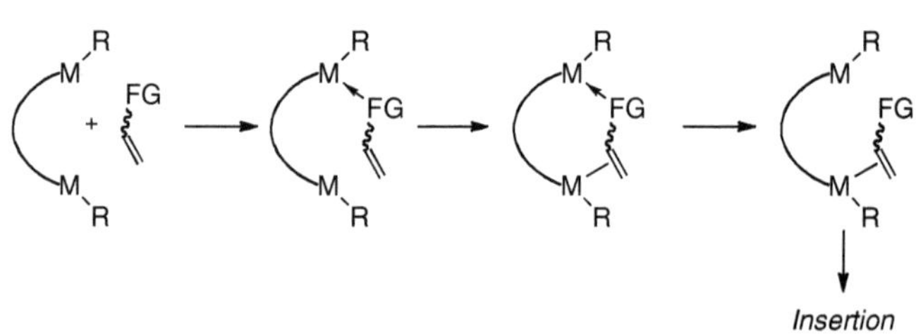

Scheme 4.32 Cooperative mechanism for the incorporation of polar comonomers by bimetallic catalysts

Phosphinosulfonato catalysts combine two features that help to increase catalyst tolerance to polar monomers: they are palladium-based and, in addition, electrically neutral. This combination is unusual amongst polymerization catalysts, probably because the intermediate alkyl or hydride complexes readily decompose by reductive coupling reactions that are common in Pd chemistry. The very low nucleophilicity and basicity of the sulfonate group is probably fundamental to prevent such processes [8].

One of the most obvious differences between catalysts containing N- (either N,N or N,O donors) and P,O based ligands (phosphinosulfonato, phosphinoenolato and other related ligands) is that bulky 2,6-disubstituted aryls are not required to produce polyethylene in the latter case. This points out important differences in the mechanisms regulating chain growth in the two families of catalysts. As discussed in Sect. 4.3, β-H elimination is faster for palladium α-diimine catalysts than chain propagation; thus, molecular weight control depends on the ability of bulky aryl substituents to hinder associative chain displacement by blocking the axial positions of the complex. In contrast, theoretical calculations for phosphinosulfonato catalysts [117, 166] have shown that the slower rate of chain transfer relative to chain propagation is due to the comparatively difficult β-H elimination. Of course, steric effects can add to this factor, and the molecular weight of the polymers is further increased if the phosphinosulfonato ligand is supplemented with bulky aryl substituents on the P atom [62]. The magnitude of the energy barriers for β-H elimination and ethene insertion are similar, therefore chain propagation prevails at high ethene concentration (giving rise to high molecular weight, linear polymers), while chain transfer and chain walking become dominant at low ethene pressure [153, 166].[2] That palladium phosphinosulfonato catalysts have chain-walking capability is evidenced when [Pd(Me)(Py)(P–O)] (P–O: phosphinosulfonato) reacts with 6-chloro-1-hexene in the absence of ethene. This reaction produces a mixture of chlorinated olefin oligomers and the corresponding chlorocomplex [Pd(Cl)(Py)(P–O)], as a result of β-chloride abstraction that requires chain walking migration of the metal to the vicinity of the chlorine atom [63].

Another important characteristic of the phosphinosulfonato ligands is the extreme disparity of the electronic properties of their two donor centers, a strongly σ donor and moderate π acceptor phosphine fragment and a weakly σ-donor sulfonato group. It should be recalled here that the mechanism of olefin polymerization by square-planar complexes containing non-symmetrical ligands involves two sets of *cis–trans* isomeric intermediates that alternate along the reaction pathway (see Sect. 4.3). Isomers having the alkyl chain in *trans* to the sulfonato group (the weaker σ-donor) are thermodynamically more stable, while those with the alkyl in *trans* to the phosphine group are more reactive

[2] Note the contrast with the Pd α-diimine system: in this case the rate of insertion is independent on the monomer concentration. This is due to the fact that here the resting state is the π-ethene complex and the rate-limiting step is the intramolecular monomer insertion.

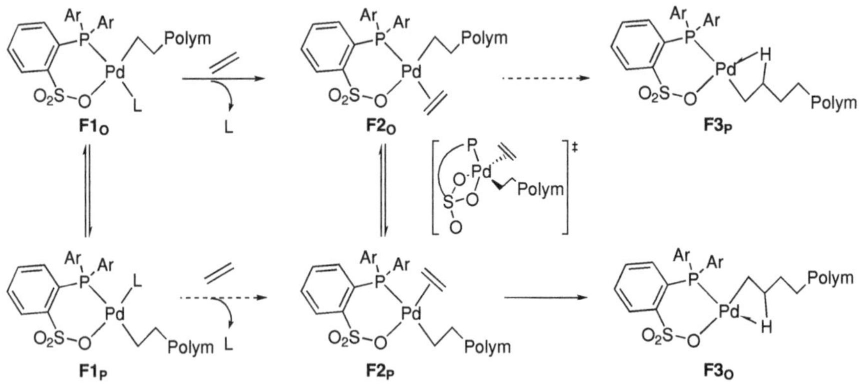

Scheme 4.33 Propagation mechanism for Pd phosphinosulfonato catalysts

(Scheme 4.33). DFT calculations show that ethene insertion will be much more favorable for the less stable complex **F2P** than for the stable isomer **F2O**. The energy difference between the two insertion barriers is very large in this system (10.7 vs. 31.3 kcal/mol), therefore *cis/trans* isomerization of the π-olefin isomers constitutes the only feasible mechanism for chain propagation, as discussed previously for the Ni salicylaldiminato system. Normally, *cis–trans* isomerization is considerably more difficult for Pd because the energy of tetrahedral intermediates is prohibitively high. However, DFT calculations by Morokuma have disclosed another remarkable property of the phosphinosulfonato ligand: it provides a low-energy path for *cis–trans* isomerization. Simultaneous interaction of Pd with two oxygen atoms of the SO_3 group enables Berry's pseudorotation, which is a process typical for five-coordinated species that exchanges the relative positions of the ligands with a low energy barrier [166]. It is possible that the lack of such an isomerization mechanism is the reason why other Pd complexes with non-symmetrical P,O ligands are inactive in olefin polymerization, remaining frozen in their most stable thermodynamic configurations [167]. Ziegler has calculated that the effective insertion barrier for ethene in phosphinosulfonato catalysts is 18.2 kcal/mol, very close to the value calculated by the same author for the Pd-α-diimine system [117].

Some of the properties of phosphinosulfonato catalysts can be interpreted on the basis of the low symmetry of the ligand and the strong labilizing *trans* effect of the P atom. Thus, the energy barrier for migratory insertion is relatively low in this system because, in the reactive isomers **F2P**, the P atom labilizes the alkyl chain placed in *trans*, increasing its migratory ability. In the stable form of the catalyst (**F1O**, **F2O**), the alkyl group is *cis* to the P donor. Therefore, the P atom directs its strong *trans effect* to the position available for the monomer, which decreases the tendency of σ donors to coordinate in this site. Very likely, this effect contributes to improve the tolerance of the system to polar substances and favors the

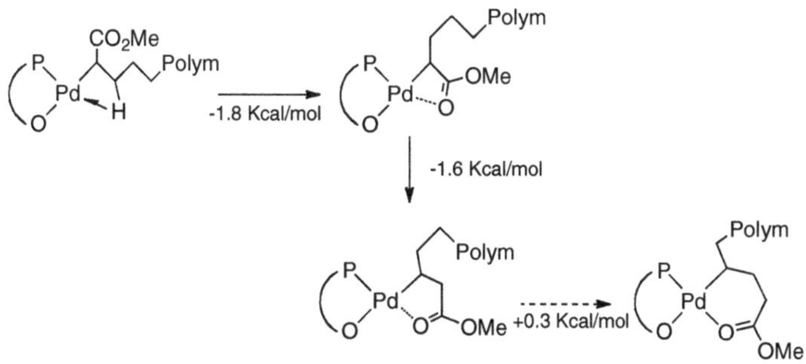

Scheme 4.34 Thermodynamic preferences for chelate formation with neutral phosphinosulfonato catalysts

productive binding of polar monomers in π fashion. Similarly, the *trans effect* increases the lability of the chelates formed upon insertion of polar monomers. Based on the linear structure and random distribution of the MA units in the EMA copolymer produced by phosphinosulfonato catalysts, Pugh and Drent deduced that, in contrast to the Pd α-diimine system, strong 6-membered chelated intermediates are not formed with these catalysts [58]. DFT calculations provide additional support to this deduction [117], showing that coordination of the carbonyl oxygen to Pd is weaker in this system, and that the thermodynamic driving force for the expansion of the chelate ring is very small (Scheme 4.34). Mecking has also shown that any chelate formed by MA insertion is displaced even by extremely weak ligands such as dmso [67]. Hence, if the rate-limiting step in the ethene-MA copolymerization in the phosphinosulfonato system is the insertion following MA incorporation, this is due to the intrinsically low reactivity of the Pd–C bond in the resulting α-carbonyl substituted species. For comparison, the reactivity of Ni–C bonds formed by MA in the salicylaldiminato system is too low to undergo any further insertion (see Sect. 4.4). The weakness of the chelate coordination allows multiple MA insertions when the MA/ethene ratio in the feed is high enough [67, 70], in contrast with the α-diimine system, where the chelate ring can be displaced only by ethene but not by the weaker binding MA comonomer.

In summary, the stability and catalytic activity of neutral palladium phosphinosulfonato ligands are enhanced by their special features: (1) the intrinsically high barrier to β-H elimination, producing high molecular weight and linear polymers; (2) the strongly non-symmetrical configuration and the labilizing influence of the P atom lowers the effective energy barrier for monomer insertion and, at the same time, enhances the tolerance of the system by disfavoring *trans* σ-coordination of monomers or chelate formation; and (3) the sulfonato donor group plays an essential role for the success of this class of catalysts providing a low energy isomerization path between *cis–trans* isomers, and preventing reductive coupling reactions that destabilize other neutral palladium complexes with anionic ligands.

4.6 Conclusion and Outlook

Development of olefin polymerization catalysts that are not only compatible with polar comonomers, but capable of incorporating them into polyolefin molecules by coordination-insertion mechanisms is an extremely challenging goal. After 15 years of intensive research, a good understanding of the difficulties associated with the incorporation of polar comonomers has been gained. Early concerns, like poisoning of the catalyst by the comonomer in its σ-coordination mode are presently considered as a relatively minor problem. In contrast, other difficulties that had not been clearly foreseen, such as the low reactivity of functionalized intermediates are currently deemed crucial to understand how polar monomers affect catalyst activity and why certain systems undergo undesired radical polymerization processes. Gaining such knowledge would not have been possible without the extraordinary mechanistic research carried out by Brookhart and other authors, and the invaluable contribution of theoretical methods. The improved understanding of the problems associated with the introduction of polar substances in a coordination-insertion polymerization system has stimulated many researchers to devise the means to solve them. Many creative proposals have been made, and success has been achieved where the difficulties have appeared impossible to overcome. Remarkable cases illustrate this statement. One of them is the development of palladium phosphinosulfonato system, which demonstrated that copolymerizing ethene in a controlled manner with important functionalized vinyls, including acrylates, VA, VF, vinyl sulfones, etc., with catalysts based on coordination-insertion mechanisms is a real possibility. Perhaps the success of these catalysts containing P,O ligands suggests that classic catalyst designs based on nitrogen ligands devised by Brookhart or Grubbs in the last decade of the past century could be, after all, not so well-suited for compatibility with polar monomers as initially thought. However, very recent results have shown how cleverly designed bimetallic catalysts based on nitrogen ligands can supply the deficiencies of a single metal center with cooperative interaction of multiple active units.

So far, relatively few systems have been successful in the incorporation of polar comonomers, and the recorded catalytic activities are still too low for practical application in large-scale processes. Furthermore, using precious elements such as Pd for production of bulk materials is not realistic, but this problem could be alleviated in the future by developing efficient catalysts based on Ni or other abundant first-row transition series metals. Fujita's work with Ti phenoxiimine catalysts appears particularly promising on this regard.

There is still a long way to go before olefin copolymerization catalysts capable of controlled incorporation of MA, MMA or other polar comonomers become attractive for industrial uses. However, Pd- and Ni-based catalysts are fulfilling some of the most ambitious expectations in terms of control of the copolymerization process. In contrast with classic radical copolymerization techniques, these catalysts offer for the first time the possibility of controlling the copolymer molecular weight, microstructure and comonomer incorporation rate. In addition, it is now possible to control the distribution of the comonomer units, placing the polar groups in branch terminations, randomly dispersed over the polymer molecule, or generating controlled monomer/comonomer gradients. Work by Guan on Pd α-diimine catalysts has demonstrated that it is possible to produce copolymers with controlled topologies spanning from open structures to hyperbranched molecules. It is possible now to dream of polyfunctional materials prepared by relatively simple olefin copolymerization reactions.

Acknowledgment Financial support from the DGI, Junta de Andalucía and European Union (Projects CTQ2009-11721 Project FQM5074 and FEDER funds) and CSIC (PIF 08-017-1) is gratefully acknowledged.

References

1. Fink G, Mulhaupt M, Britzinger HH (eds) (1995) Ziegler catalysts: recent scientific innovations and technological improvements. Springer, Heidelberg
2. Patil AO, Schulz DN, Novak BM (eds) (1997) Functional polymers. ACS, Washington
3. Padwa AR (1989) Functionally substituted poly (α-olefins). Prog Polym Sci 14(6):811–813
4. Boffa LS, Novak BM (2000) Copolymerization of polar monomers with olefins using transition-metal complexes. Chem Rev 100(4):1479–1494
5. Sen A, Borkar S (2007) Perspective on metal-mediated polar monomer/alkene copolymerization. J Organomet Chem 692:3291–3299
6. Novák I, Borsig F, Hrèková L, Fiedlerová A, Kleinová A, Pollak V (2007) Study of surface and adhesive properties of polypropylene grafted by maleic anhydride. Polym Eng Sci 47(8):1207–1212
7. Ittel SD, Johnson LK (2000) Late-metal catalysts for ethene homo- and copolymerization. Chem Rev 100(4):1169–1204
8. Goodall BL (2009) Late transition metal catalysts for the copolymerization of olefins and polar monomers. In: Guan Z (ed) Metal catalysis in olefin polymerization. Topics in organometallic chemistry, vol 26. Springer, Weinheim, pp 159–178
9. Nakamura A, Ito S, Nozaki K (2009) Coordination-insertion copolymerization of fundamental polar monomers. Chem Rev 109:5215–5244
10. Johnson LK, Killian CM, Brookhart M (1995) New Pd(II)- and Ni(II)-based catalysts for polymerization of ethene and a-olefins. J Am Chem Soc 117:6414–6415
11. Killian CM, Tempel DJ, Johnson LK, Brookhart M (1996) Living polymerization of α-olefins using Ni(II)-α-diimine catalysts. Synthesis of new block polymers based on -olefins. J Am Chem Soc 118:11664–11665
12. Johnson LK, Mecking S, Brookhart M (1996) Copolymerization of ethene and propylene with functionalized vinyl monomers by palladium(II) catalysts. J Am Chem Soc 118:267–268
13. Scheirs J, Kaminsky W (eds) (1999) Metallocene-based polyolefins, preparation, properties and technology. Wiley, Weinheim
14. Berkefeld A, Mecking S (2008) Coordination copolymerization of polar vinyl monomers H_2CCHX. Angew Chem Int Ed 47:2538–2542
15. Imuta J, Kashiwa N (2002) Catalytic regioselective introduction of allyl alcohol into the nonpolar polyolefins: development of one-pot synthesis of hydroxyl-capped polyolefins mediated by a new metallocene IF catalyst. J Am Chem Soc 124:1176–1177
16. Haghiara H, Murata M, Uozumi T (2001) Alternating copolymerization of ethene and 5-Hexen-1-ol with [ethene(1-indenyl)(9-fluorenyl)]-zirconium dichloride/methylaluminoxane as the catalyst. Macromol Rapid Commun 22:353–357

17. Hakala K, Helaja T, Löfgren B (2000) Metallocene/methylaluminoxane-catalyzed copolymerizations of oxygen-functionalized long-chain olefins with ethene. J Polym Sci Ser A Polym Chem 38:1966–1971
18. Schneider MJ, Schäfer R, Mülhaupt R (1997) Aminofunctional linear low density polyethene via metallocene-catalysed ethene copolymerization with N, N-bis(trimethylsilyl)-1-amino-10-undecene. Polymer 38:2455–2459
19. Bruzaud S, Cramail H, Duvugnac L, Deffeux A (1997) ω-Chloro-α-olefins as co- and termonomers for the synthesis of functional polyolefins. Macromol Chem Phys 198:291–303
20. Marques MM, Correia SG, Ascenso JR, Ribeiro AFG, Gomes PT, Dias AR, Foster P, Rausch MD, Chien JCW (1999) Polymerization with TMA-protected polar vinyl comonomers. I. Catalyzed by group 4 metal complexes with η5-type ligands. J Polym Sci Ser A Polym Chem 37:2457–2469
21. Correia SG, Marques MM, Ascenso JR, Ribeiro AFG, Gomes PT, Dias AR, Blais M, Rausch MD, Chien JCW (1999) Polymerization with TMA-protected polar vinyl comonomers. II. Catalyzed by nickel complexes containing α-diimine-type ligands. J Polym Sci Ser A Polym Chem 37:2471–2480
22. Aaltonen P, Fink G, Löfgren B, Seppala J (1996) Synthesis of hydroxyl group containing polyolefins with metallocene/methylaluminoxane catalysts. Macromolecules 29:5255–5260
23. Aaltonen P, Löfgren B (1995) Synthesis of functional polyethenes with soluble metallocene/methylaluminoxane catalyst. Macromolecules 28:5353–5357
24. Chung TC, Lu HL, Li CL (1995) Functionalization of polyethene using borane reagents and metallocene catalysts. Polym Intl 37:197–205
25. Yang HX, Liu CR, Wang C, Sun XL, Guo YH, Wang XK, Wang Z, Xie ZW, Tang Y (2009) [O–NSR]$TiCl_3$-catalyzed copolymerization of ethene with functionalized olefins. Angew Chem Int Ed 48:8099–8012
26. Klabunde U, Mulhaupt R, Herskovitz T, Janowicz AH, Calabrese SD (1987) Ethene homopolymerization with P, O-chelated nickel catalysts. J Polym Sci Ser A Polym Chem 1987:1989–2003
27. Klabunde U, Ittel SD (1987) Nickel catalysts for ethene homo- and co-polymerization. J Mol Catal 41:123–134
28. Soula R, Saillard B, Spitz R, Claverie J, Llaurro MF, Monnet C (2002) Catalytic copolymerization of ethene and polar and nonpolar a-olefins in emulsion. Macromolecules 35:1513–1523
29. Mecking S, Johnson LK, Wang L, Brookhart M (1998) Mechanistic studies of the palladium-catalyzed copolymerization of ethylene and α-olefins with methyl acrylate. J Am Chem Soc 120:888–899
30. Chen G, Ma SX, Guan Z (2003) Synthesis of functional olefin copolymers with controllable topologies using a chain-walking catalyst. J Am Chem Soc 125:6697–6704
31. Guan Z, Cotts PM, McCord EF, McLain S (1999) Chain walking: a new strategy to control polymer topology. Science 283:2059–2062
32. Guan Z (2002) Control of polymer topology by chain-walking catalysts. Chem Eur J 8:3087–3092
33. Marques MM, Fernandes S, Correia SG, Ascenso JR, Caroço S, Gomes PT, Mano J, Pereira SG, Nunes T, Dias AR, Rausch MD, Chien JCW (2000) Synthesis of acrylamide/olefin copolymers by a diimine nickel catalyst. Macromol Chem Phys 201:2464–2468
34. Marques MM, Fernandes S, Correia SG, Caroço S, Gomes PT, Dias AR, Mano J, Rausch MD, Chien JCW (2001) Synthesis of polar vinyl monomer α-olefin copolymers by -diimine nickel catalyst. Polym Int 50:579–587
35. Fernandes S, Soares A, Lemos F, Lemos MANDA, Mano JF, Maldanis RJ, Rausch MD, Chien JCW (2005) Copolymerization of ethylne/unsaturated alcohols using nickel catalysts: effect of the ligand on the activity and comonomer incorporation. J Organomet Chem 690:895–909

36. Fernandes S, Ascenso JR, Gomes PT, Costa SI, Silva LC, Chien JCW, Marques MM (2005) Synthesis of acrylamide end-functionalised poly(1-hexene) using an alpha-diimine nickel catalyst. Polym Int 54:249–255
37. Johnson LK, Bennett A, Dobbs K, Hauptman E, Ionkin A, Ittel SD, McCord E, McLain S, Radzewich C, Yin Z, Wang L, Wang Y, Brookhart M (2002) Polym Mater Sci Eng 86:319–322
38. Johnson LK, Wang L, McLain S, Bennett A, Dobbs K, Hauptman E, Ionkin A, Ittel SD, Kutnisky K, Marshall W, McCord E, Radzewich C, Rinehart A, Sweetman KJ, Wang Y, Yin Z, Brookhart M (2003) Copolymerization of ethene and acrylates by nickel catalysts. In: Patil AO, Hlatky GG (eds) Beyond metallocenes, vol 857. American Chemical Society, Washington, DC, pp 131–142 ACS Symposium Series
39. Wang CM, Friedrich S, Younkin TR, Li RT, Grubbs RH, Bansleben DA, Day MW (1998) Neutral nickel(II)-based catalysts for ethene polymerization. Organometallics 17:3149–3151
40. Younkin TR, Connor EF, Henderson JI, Friedrich SK, Grubbs RH, Bansleben DA (2000) Neutral, single-component nickel (II) polyolefin catalysts that tolerate heteroatoms. Science 287:460–462
41. Korthals B, Göttker-Schnetmann I, Mecking S (2007) Nickel(II)-methyl complexes with water-soluble ligands L [(salicylaldiminato-κ^2-(N, O)NiMe(L)] and their catalytic properties in disperse aqueous systems. Organometallics 26:1311–1316 , and references therein
42. Waltman AW, Younkin TR, Grubbs RH (2004) Insights into the deactivation of neutral nickel ethene polymerization catalysts in the presence of functionalized olefins. Organometallics 23:5121–5123
43. Connor EF, Younkin TR, Henderson JI, Hwang S, Grubbs RH, Roberts WP, Litzau JJ (2002) Linear functionalized polyethene prepared with highly active neutral Ni(II) complexes. J Polym Sci Ser A Polym Chem 40:2842–2854
44. Sun J, Shan Y, Cui Y, Schumann H, Hummert M (2004) Novel cyclohexyl-substituted salicylaldiminato-nickel(II) complex as a catalyst for ethylene homopolymerization and copolymerization. J Polym Sci Ser A Polym Chem 42:6071–6080
45. Boardman BM, Bazan GC (2009) α-Iminocarboxamidato nickel complexes. Acc Chem Res 42:1597–1606
46. Diamanti SJ, Ghosh P, Shimizu F, Bazan GC (2003) Ethene homopolymerization and copolymerization with functionalized 5-Norbornen-2-yl monomers by a novel nickel catalyst system. Macromolecules 36:9731–9735
47. Rojas RS, Wasilike J-C, Wu G, Ziller JW, Bazan GC (2005) α-Iminocarboxamide nickel complexes: synthesis and uses in ethene polymerization. Organometallics 24:5644–5653
48. Diamanti SJ, Khanna V, Hotta A, Coffin RC, Yamakawa D, Kramer EJ, Fredrickson GH, Bazan GC (2006) Tapered block copolymers containing ethene and a functionalized comonomer. Macromolecules 39:3270–3274
49. Coffin RC, Diamanti SJ, Hotta A, Khanna V, Kramer EJ, Fredrickson GH, Bazan GC (2007) Pseudo-tetrablock copolymers with ethene and a functionalized comonomer. Chem Commun 3350–3352
50. Wang L, Hauptman E, Johnson LK, Marshall W, McCord E, Wang Y, Ittel SD, Radzewich C, Kutninsky K, Ionkin AS (2002) Novel nickel catalysts for ethylene/polar monomer polymerization. Polym Mater Sci Eng 86:322
51. Guo CY, Pulecke N, Basvani K, Kindermann MK, Heinicke J (2010) 2-Phosphinophenolato nickel catalysts: formation of ethene copolymers with isolated sec-alkyl, aryl, and functionally substituted alkyl groups. Macromolecules 43:1416–1424
52. Li X-F, Li Y-G, Li Y-S, Chen Y-X, Hu N-H (2005) Copolymerization of ethene with methyl methacrylate with neutral nickel(II) complexes bearing-ketoiminato chelate ligands. Organometallics 2005:2502–2510
53. Brasse M, Cámpora J, Palma P, Álvarez E, Cruz V, Ramos J, Reyes ML (2008) Nickel 2-iminopyridine N-oxide (PymNox) complexes: cationic counterparts of salicylaldiminato-based neutral ethene polymerization catalysts. Organometallics 27:4711–4723

54. Tian G, Boone HW, Novak BM (2001) Neutral palladium complexes as catalysts for olefin–methyl acrylate copolymerization: a cautionary tale. Organometallics 34:7656–7663
55. Kang M, Sen A (2005) Neutral N-O chelated palladium(II) complexes: syntheses, characterization, and reactivity. Organometallics 14:3508–3515
56. Elia C, Elyashiv-Barad S, Sen A, López-Fernández R, Albéniz AC, Espinet P (2002) Palladium-based system for the polymerization of acrylates. Scope and mechanism. Organometallics 20:4249
57. Berkefeld A, Mecking S (2008) Coordination copolymerization of polar vinyl monomers H_2CCHX. Angew Chem Int Ed 47:2548–2542
58. Drent E, van Dijk R, van Ginkel R, van Oort B, Pugh RI (2003) Palladium catalysed copolymerisation of ethene with alkylacrylates: polar comonomer built into the linear polymer chain. Chem Commun 1:744–745
59. Kochi T, Yoshimura K, Nozaki K (2005) Synthesis of anionic methylpalladium complexes with phosphine–sulfonate ligands and their activities for olefin polymerization. Dalton Trans 25–27
60. Luo S, Vela J, Lief GR, Jordan RF (2007) Copolymerization of ethene and alkyl vinyl ethers by a (phosphine-sulfonate)PdMe catalyst. J Am Chem Soc 129:8946–8947
61. Kochi T, Noda S, Yoshimura K, Nozaki K (2007) Formation of linear copolymers of ethene and acrylonitrile catalyzed by phosphine sulfonate palladium complexes. J Am Chem Soc 129:8948–8949
62. Skupov KM, Marella PR, Simard M, Yap GA, Allen N, Conner D, Goodall BL, Claverie J (2007) Palladium aryl sulfonate phosphine catalysts for the copolymerization of acrylates with ethene. Macromol Chem Phys 28:2033–2038
63. Vela J, Lief GR, Shen Z, Jordan RF (2007) Ethene polymerization by palladium alkyl complexes containing Bis(aryl)phosphino-toluenesulfonate ligands. Organometallics 26:6624–6635
64. Weng W, Shen Z, Jordan RF (2007) Copolymerization of ethene and vinyl fluoride by (phosphine-sulfonate)Pd(Me)(py) catalysts. Organometallics 129:15450–15451
65. Skupov KM, Piche L, Claverie J (2008) Linear polyethene with tunable surface properties by catalytic copolymerization of ethene with N-Vinyl-2-pyrrolidinone and N-Isopropylacrylamide. Macromolecules 41:2309–2310
66. Ito S, Munakata K, Nakamura A, Nozaki K (2009) Copolymerization of vinyl acetate with ethene by palladium/alkylphosphine—sulfonate catalysts. J Am Chem Soc 131:1406–1407
67. Guironnet D, Roesle P, Rünzi T, Göttker-Schnetmann I, Mecking S (2009) Insertion polymerization of acrylate. J Am Chem Soc 131:422–423
68. Bouilhac C, Rünzi T, Mecking S (2010) Catalytic copolymerization of ethene with vinyl sulfones. Macromolecules 43:3589–3590
69. Bettucci L, Bianchini C, Claver C, García-Suárez EJ, Ruíz A, Meli A, Oberhauser W (2007) Ligand effects in the non-alternating CO–ethene copolymerization by palladium(II) catalysis. Dalton Trans 5590–5602
70. Guironnet D, Caporaso L, Neuwald B, Göttker-Schnetmann I, Cavallo L, Mecking S (2010) Mechanistic insights on acrylate insertion polymerization. J Am Chem Soc 132:4418–4426
71. Liu S, Borkar S, Newsham D, Yennawar H, Sen A (2007) Synthesis of palladium complexes with an anionic po chelate and their use in copolymerization of ethene with functionalized norbornene derivatives: unusual functionality tolerance. Organometallics 26:210–216
72. Drent E, van Dijk R, van Ginkel R, van Oort B, Pugh RI (2002) The first example of palladium catalysed non-perfectly alternating copolymerisation of ethene and carbon monoxide. Chem Commun 964–965
73. Newsham D, Borkar S, Sen A, Conner D, Goodall BL (2007) Inhibitory role of carbon monoxide in palladium(II)-catalyzed nonalternating ethene/carbon monoxide copolymerizations and the synthesis of polyethene-*block*-poly(ethene-*alt*-carbon monoxide). Organometallics 27:3636–3638

74. Guironnet D, Rünzi T, Göttker-Schnetmann I, Mecking S (2008) Control of molecular weight in Ni(II)-catalyzed polymerization via the reaction medium. Chem Commun 4695–4697
75. Zhou X, Bontemps S, Jordan RF (2008) Base-free phosphine-sulfonate nickel benzyl complexes. Organometallics 27:4821–4824
76. Noda S, Kochi T, Nozaki K (2009) Synthesis of allylnickel complexes with phosphine sulfonate ligands and their application for olefin polymerization without activators. Organometallics 28:656–658
77. Zhang D, Wang JY, Yue Q (2010) Synthesis, characterization and catalytic behaviors of water-soluble phosphine-sulfonato nickel methyl complexes bearing PEG-amine labile ligand. J Organomet Chem 695:903–908
78. Terao H, Ishii S, Mitani M, Tanaka H, Fujita T (2008) Ethylene/polar monomer copolymerization behavior of Bis(phenoxy–imine)Ti complexes: formation of polar monomer copolymers. J Am Chem Soc 130:17636–17637
79. Britovsek GJP, Gibson VC, Spitzmesser SK, Tellmann KP, White AJP, Williams DJ (2002) Cationic 2, 6-bis(imino)pyridine iron and cobalt complexes: synthesis, structures, ethylene polymerisation and ethylene/polar monomer co-polymerisation studies. J Chem Soc Dalton Trans 1159–1171
80. Castro PM, Lapplainen K, Ahlgren M, Leskela M, Repo T (2003) Iron-based catalysts bearing Bis(imido)–pyridine ligands for the polymerization of tert-butyl acrylate. J Polym Sci Ser A Polym Chem 41:1380–1389
81. Fullana MJ, Miri MJ, Vadhavkar SS, Kolhaktar N, Delis AC (2008) Polymerization of methyl acrylate and as comonomer with ethylene using a bis(imino)pyridine iron(II) chloride/methylaluminoxane catalyst. J Polym Sci Ser A Polym Chem 46:5542–5558
82. Stibrany RT, Schulz DN, Kacker S, Patil AO, Baugh LS, Rucker SP, Zushma S, Berluche E, Sissano JA (2003) Polymerization and copolymerization of olefins and acrylates by Bis(benzimidazole) copper catalysts. Macromolecules 36:8584–8586
83. Baugh LS, Sissano JA, Kacker S, Berluche E, Stibrany RT, Schulz DN, Rucker SP (2006) Fluorinated and ring-substituted bisbenzimidazole copper complexes for ethylene/acrylate copolymerization. J Polym Sci Ser A Polym Chem 44:1817–1840
84. Raspolli Galletti AM, Carlini C, Giaiacopi S, Martinelli M, Sbrana G (2007) Bis(salicylaldiminato)copper(II)/methylaluminoxane catalysts for homo- and copolymerizations of ethylene and methyl methacrylate. J Polym Sci Ser A Polym Chem 45:1134–1142
85. Nagel M, Paxton WF, Sen A (2004) Metal-mediated polymerization of acrylates: relevance of radical traps? Macromolecules 37:9305–9307
86. Guan Z, Popeney C (2009) Recent progress in late transition metal a-diimine catalysts for olefin polymerization. In: Metal catalysts in olefin polymerization. Topics on organometallic chemistry, vol 26. Springer, Berlin
87. Svejda SA, Johnson LK, Brookhart M (1999) Low-temperature spectroscopic observation of chain growth and migratory insertion barriers in (α-Diimine)Ni(II) olefin polymerization catalysts. J Am Chem Soc 121:10364–10365
88. Tempel DJ, Johnson LK, Huff RL, White PS, Brookhart M (2000) Mechanistic studies of Pd(II)-α-Diimine-catalyzed olefin polymerizations. J Am Chem Soc 122:6686–6700
89. Shulz LH, Tempel DJ, Brookhart M (2001) Palladium(II) β-agostic alkyl cations and alkyl ethylene complexes: investigation of polymer chain isomerization mechanisms. J Am Chem Soc 123:11539–11555
90. Shulz LH, Brookhart M (2001) Measurement of the barrier to β-hydride elimination in a β-agostic palladium-ethyl complex: a model for the energetics of chain-walking in (α-Diimine)PdR+olefin polymerization catalysts. Organometallics 20:3975–3982
91. Leatherman MD, Svejda SA, Johnson LK, Brookhart M (2003) Mechanistic studies of Nickel(II) alkyl agostic cations and alkyl ethylene complexes: investigations of chain propagation and isomerization in (α-diimine)Ni(II)-catalyzed ethylene polymerization. J Am Chem Soc 125:3068–3081

92. Deng L, Woo T, Cavallo L, Margl PM, Ziegler T (1997) The role of bulky substituents in Brookhart-type Ni(II) Diimine catalyzed olefin polymerization: a combined density functional theory and molecular mechanics study. J Am Chem Soc 119:6177–6189
93. Gates DP, Svejda SA, Oñate E, Kilian CM, Johnson LK, White PS, Brookhart M (2000) Synthesis of branched polyethylene using (R-Diimine)nickel(II) catalysts: influence of temperature, ethylene pressure, and ligand structure on polymer properties. Macromolecules 33:2320–2334
94. Chan MSW, Deng L, Ziegler T (2000) Density functional study of neutral salicylaldiminatoi nickel(II) complexes as olefin polymerization catalysts. Organometallics 19:2741–2750
95. Ramos J, Cruz VL, Martínez-Salazar J, Brasse M, Palma P, Cámpora J (2010) Density functional study for the polymerization of ethylene monomer using a new nickel catalyst. J Polym Sci Ser A Polym Chem 49:1160–1165
96. Williams BS, Leatherman MD, White AJP, Brookhart M (2005) Reactions of vinyl acetate and vinyl trifluoroacetate with cationic Diimine Pd(II) and Ni(II) alkyl complexes: identification of problems connected with copolymerizations of these monomers with ethylene. J Am Chem Soc 127:5132–5146
97. Wu F, Foley SR, Burns CT, Jordan RF (2005) Acrylonitrile insertion reactions of cationic palladium alkyl complexes. J Am Chem Soc 127:1841–1853
98. Stojcevic G, Propokchuk EM, Baird MC (2005) Coordination insertion reactions of acrylonitrile into Pd-H and Pd-methyl bonds in α-diimine-palladium(II) system. J Organomet Chem 690:4349–4355
99. Kilyanek SM, Stoebenau EJ, Vinayavekhin N, Jordan RF (2010) Mechanism of the reaction of vinyl chloride with (α-diimine)$PdMe^{+}$ species. Organometallics 29:1750–1760
100. Groux LF, Weiss T, Reddy DN, Chase PA, Piers WE, Ziegler T, Parvez M, Benet-Buchholz J (2005) Insertion of acrylonitrile into palladium methyl bonds in neutral and anionic Pd(II) complexes. J Am Chem Soc 127:1854–1869
101. Reddy KR, Chen C-L, Liu Y-H, Peng S-M, Chen J-T, Liu S-T (1999) New imine-phosphine palladium complexes catalyze copolymerization of CO-ethylene and CO-norbornylene and provide well-characterized stepwise insertion intermediates of various unsaturated substrates. Organometallics 18:2574–2576
102. Chen C, Luo S, Jordan RF (2008) Multiple insertion of a silyl vinyl ether by (α-Diimine)PdMe+species. J Am Chem Soc 130:12892–12893
103. Li W, Zhang X, Meetsm A, Hessen B (2004) Palladium-catalyzed copolymerization of ethene with acrolein dimethyl acetal: catalyst action and deactivation. J Am Chem Soc 126:12246–12247
104. Li W, Zhang X, Meetsma A, Hessen B (2008) Reactions of cationic palladium α-Diimine complexes with nitrogen-containing olefins: branched polyethylene with carbazole functionalities. Organometallics 27:2052–2057
105. Borkar S, Yennawar H, Sen A (2007) Methacrylate insertion into cationic diimine palladium(II)–alkyl complexes and the synthesis of poly(alkene-block-alkene/carbon monoxide) copolymers. Organometallics 26:4711–4714
106. Michalak A, Ziegler T (2001) DFT studies on the copolymerization of α-olefins with polar monomers: ethylene–methyl acrylate copolymerization catalyzed by a Pd-based Diimine catalyst. J Am Chem Soc 123:12266–12278
107. von Schenck H, Strömberg S, Zetterberg K, Ludwig M, Åckermark B (2001) Insertion aptitudes and insertion regiochemistry of various alkenes coordinated to cationic (σ-R)(diimine)palladium(II) ($R = –CH_3, –C_6H_5$). A theoretical study. Organometallics 20:2813–2819
108. Philipp DM, Muller RP, Goddard WA, Storer J, McAdon M, Mullins M (2002) Computational insights on the challenges for polymerizing polar monomers. J Am Chem Soc 124:10198–10120
109. Szabo MJ, Jordan RF, Michalak A, Piers WE, Weiss T, Yang S-Y, Ziegler T (2004) Polar copolymerization by a palladium-diimine-based catalyst. Influence of the catalyst charge

and polar substituent on catalyst poisoning and polymerization activity. A density functional theory study. Organometallics 23:5565–5572
110. Deubel DV, Ziegler T (2002) Challenge of the copolymerization of olefins with N-containing polar monomers. Systematic screening of nickel(II) and palladium(II) catalysts with Brookhart and Grubbs ligands. 2. Chain-propagation barriers, intrinsic regioselectivity, and Curtin-Hammett reactivity. Organometallics 41:4432–4441
111. Kang M, Sen A, Zakharov L, Rheingold AL (2002) Diametrically opposite trends in alkene insertion in late and early transition metal compounds: relevance to transition-metal-catalyzed polymerization of polar vinyl monomers. J Am Chem Soc 2002:12080–12081
112. Rix FC, Brookhart M, White PS (1998) Electronic effects on the β-alkyl migratory insertion reaction of para-substituted styrene methyl palladium complexes. J Am Chem Soc 118:2436–2448
113. Michalak A, Ziegler T (2001) DFT studies on the copolymerization of -olefins with polar monomers: comonomer binding by nickel- and palladium-based catalysts with Brookhart and Grubbs ligands. Organometallics 20:1521–1532
114. Deubel DV, Michalak A, Ziegler T (2002) DFT studies on the copolymerization of -olefins with polar monomers: comonomer binding by nickel- and palladium-based catalysts with Brookhart and Grubbs ligands. Organometallics 21:1603–1611
115. Frenking G, Fröhlich N (2000) The nature of the bonding in transition-metal compounds. Chem Rev 100:717–774
116. Michalak A, Ziegler T (2003) Polymerization of ethylene catalyzed by a nickel(+2) anilinotropone-based catalyst: DFT and stochastic studies on the elementary reactions and the mechanism of polyethylene branching. Organometallics 22:2069–2079
117. Haras A, Anderson GDW, Michalak A, Rieger B, Ziegler T (2006) Computational insight into catalytic control of poly(ethylene-methyl acrylate) topology. Organometallics 25:4491–4497
118. Luo S, Jordan RF (2006) Copolymerization of silyl vinyl ethers with olefins by (-diimine)PdR+. J Am Chem Soc 128:12072–12073
119. Foley SR, Stockland RA, Jordan RF (2003) Reaction of vinyl chloride with late transition metal olefin polymerization catalysts. J Am Chem Soc 125:4350–4361
120. Michalak A, Ziegler T (2003) A comparison of Ni- and Pd-Diimine complexes as catalysts for ethylene/methyl acrylate copolymerization. A static and dynamic density functional theory study. Organometallics 22:2660–2669
121. Chen EXY (2009) Coordination polymerization of polar vinyl monomers by single-site metal catalysis. Chem Rev 110:5157–5214
122. Vicente J, Arcas A, Fernández-Hernández JM, Bautista D (2008) New acetonyl Pd(II) complexes. Organometallics 27:6155–6169
123. Cámpora J, Maya CM, Palma P, Carmona E, Gutiérrez E, Ruiz C, Graiff C, Tiripicchio A (2005) Cyclic enolates of Ni and Pd: equilibrium between C- and O-bound tautomers and reactivity studies. Chem Eur J 11:6889–6904
124. Cámpora J, Matas I, Palma P, Álvarez E, Graiff C, Tiripicchio A (2007) Acid-base reactions of methylnickel hydroxo, alkoxo, and amide complexes with carbon acids. Studies on the reactivity of noncyclic nickel enolates. Organometallics 26:5712–5721
125. Foley SR, Shen H, Qadeer AA, Jordan RF (2007) Generation and insertion reactivity of cationic palladium complexes that contain halogenated alkyl ligands. Organometallics 23:600–609
126. Chen C, Luo S, Jordan RF (2010) Cationic polymerization and insertion chemistry in the reactions of vinyl ethers with (α-Diimine)PdMe+species. J Am Chem Soc 132:5273–5284
127. Berkefeld A, Drexler M, Miler HM, Mecking S (2009) Mechanistic insights on the copolymerization of polar vinyl monomers with neutral Ni(II) catalysts. J Am Chem Soc 131:12613–12622
128. Hicks FA, Jenkins JC, Brookhart M (2003) Synthesis and ethylene polymerization activity of a series of 2-anilinotropone-based neutral nickel(II) catalysts. Organometallics 22:3533–3545

129. Jenkins JC, Brookhart M (2004) A mechanistic investigation of the polymerization of ethylene catalyzed by neutral Ni(II) complexes derived from bulky anilinotropone ligands. J Am Chem Soc 126:5827–5842
130. Jenkins JC, Brookhart M (2003) A highly active anilinoperinaphthenone-based neutral nickel(II) catalyst for ethylene polymerization. Organometallics 22:250–256
131. Connor EF, Younkin TR, Henderson JI, Waltman AW, Grubbs RH (2003) Synthesis of neutral nickel catalysts for ethylene polymerization–the influence of ligand size on catalyst stability. Chem Commun 2272–2273
132. Berkefeld A, Mecking S (2009) Deactivation pathways of neutral Ni(II) polymerization catalysts. J Am Chem Soc 131:1565–1574
133. Gibson VC, Tomov A (2001) Functionalised polyolefin synthesis using [P, O]Ni catalysts. Chem Commun 1964–1965
134. Boone HW, Athey PS, Mullins MJ, Philipp D, Muller RP, Goddard WA (2002) Copolymerization studies of vinyl chloride and vinyl acetate with ethylene using a transition-metal catalyst. J Am Chem Soc 124:8790–8791
135. Stockland RA, Foley SR, Jordan RF (2003) Reaction of vinyl chloride with group 4 metal olefin polymerization catalysts. J Am Chem Soc 125:796–809
136. Watson LA, Yandulov D, Caulton KG (2001) C–D0 (D0 = π-donor, F) cleavage in $H_2CCH(D0)$ by $(Cp_2ZrHCl)n$: mechanism, agostic fluorines, and a carbene of Zr(IV). J Am Chem Soc 123:603–601
137. Straizsar S, Wolczanski PT (2001) Insertion of H_2CCHX (X = F, Cl, Br, OiPr) into $(tBu_3SiO)_3TaH_2$ and β-X-elimination from $(tBu_3SiO)_3HTaCH_2CH_2X$ (X = OR): relevance to Ziegler–Natta copolymerizations. J Am Chem Soc 123:4728–4740
138. di Lena F (2010) Matyjaszewski: transition metal catalysts for controlled radical polymerization. Prog Polym Sci 35:959–1021
139. Nagel M, Sen A (2006) Intermediacy of radicals in rearrangement and decomposition of metal–alkyl species: relevance to metal-mediated polymerization of polar vinyl monomers. Organometallics 2006(25):4726–4724
140. Albéniz AC, Espinet P, López-Fernández R, Sen A (2002) A warning on the use of radical traps as a test for radical mechanisms: they react with palladium hydrido complexes. J Am Chem Soc 124:11278–11279
141. Keim W (1990) Nickel—an element with wide application in industrial homogeneous catalysis. Angew Chem Int Ed 29:235–244
142. Mecking S (2000) Cationic nickel and palladium complexes with bidentate ligands for the C—C linkage of olefins. Coord Chem Rev 203:325–351
143. Berkefeld A, Möller HM, Mecking S (2009) Unusual reactivity of N, N, N′, N′-tetramethylethylenediamine-coordinated neutral nickel(II) polymerization catalysts. Organometallics 28:4048–4055
144. Zuidveld MA, Wehrmann P, Röhr C, Mecking S (2004) Remote substituents controlling catalytic polymerization by very active and robust neutral nickel(II) complexes. Angew Chem Int Ed 40:869–873
145. Rojas RS, Barrera Galland G, Wu G, Bazan GC (2007) Single-component α-iminocarboxamide nickel ethylene polymerization and copolymerization initiators. Organometallics 876:5339–5345
146. Ionkin A, Marshall WJ (2004) ortho-5-Methylfuran- and benzofuran-substituted 3-Allyl(-diimine)nickel(II) complexes: syntheses, structural characterization, and the first polymerization results. Organometallics 23:3276–3283
147. Ionkin A, Marshall W (2004) Synthesis and structural characterization of eta(3_)allyl(alpha-diimine)nickel(II) complexes bearing trimethylsilyl groups. J Organomet Chem 689:1057–1063
148. Strauch JW, Erker G, Kehr G, Frohlich R (2002) Formation of a butadienenickel-based zwitterionic single-component catalyst for ethylene polymerization: an alternative activation pathway for homogeneous Ziegler-Natta catalysts of late transition metals. Angew Chem Int Ed 41:2543–2546

149. Fessina V, Ramminger C, Seferin M, Mauler RS, de Souza RF, Monteiro AL (2003) η^3-Benzyl-nickel-Diimine complex: synthesis and catalytic properties in ethylene polymerization. Macrom Rapid Commun 24:667–670
150. Brasse M, Cámpora J, Davies M, Teuma E, Palma P, Álvarez E, Sanz E, Reyes ML (2007) Integrating catalyst and co-catalyst design in olefin polymerization catalysis: transferable dianionic ligands for the activation of late transition metal polymerization catalysts. Adv Synth Catal 349:2111–2120
151. Trovitch RJ, Lobkovsky E, Bill E, Chirik PJ (2008) Functional group tolerance and substrate scope in Bis(imino)pyridine iron catalyzed alkene hydrogenation. Organometallics 27:1470–1478
152. Rodriguez B, Delferro M, Marks TJ (2009) Bimetallic effects for enhanced polar comonomer enchainment selectivity in catalytic ethylene polymerization. J Am Chem Soc 131:5902–5919
153. Popeney C, Camacho DH, Guan Z (2007) Efficient incorporation of polar comonomers in copolymerizations with ethylene using a cyclophane-based Pd(II) -Diimine catalyst. J Am Chem Soc 129:10062–10063
154. Popeney C, Guan Z (2009) A mechanistic investigation on copolymerization of ethylene with polar monomers using a cyclophane-based Pd(II) alpha-Diimine catalyst. J Am Chem Soc 131:12384–12393
155. Carone CLP, Bisatto R, Galland GB, Rojas R, Bazan G (2008) Ethylene polymerization and copolymerization with a polar monomer using an α-iminocarboxamide nickel complex activated by trimethylaluminum. J Polym Sci Ser A Polym Chem 46:54–59
156. Popeney C, Guan Z (2005) Ligand electronic effects on late transition metal polymerization catalysts. Organometallics 24:1145–1155
157. Li H, Marks TJ (2006) Nuclearity and cooperativity effects in binuclear catalysts and cocatalysts for olefin polymerization. Proc Nac Academ Sci 103:15925–15302
158. Kurtev K, Tomov A (1994) Ethene polymerization by binuclear nickel-ylide complexes. J Mol Catal 88:141–150
159. Tomov A, Kurtev K (1995) Binuclear nickel-ylide complexes as effective ethylene oligomerization/polymerization catalysts. J Mol Catal A Chem 103:95–103
160. Tomov A, Broyer JP, Spitz R (2000) Emulsion polymerization of ethylene in water medium catalysed by organotransition metal complexes. Macromol Symp 150:53–58
161. Hu T, Tang L-M, Li X-F, Hu N-H (2005) Synthesis and ethylene polymerization activity of a novel, highly active single-component binuclear neutral nickel(II) catalyst. Organometallics 24:2628–2632
162. Chen Q, Yu J, Huang J (2007) Arene-bridged salicylaldimine-based binuclear neutral nickel(II) complexes: synthesis and ethylene polymerization activities. Organometallics 26:617–625
163. Whermann P, Mecking S (2008) Highly active binuclear neutral nickel(II) catalysts affording high molecular weight polyethylene. Organometallics 27:1399–1408
164. Sujith S, Joe DJ, Na SJ, Park Y-W, Choi CH, Lee BY (2005) Ethylene/polar norbornene copolymerizations by bimetallic salicylaldimine-nickel catalyst. Macromolecules 38:10027–10033
165. Rodriguez B, Delferro M, Marks TJ (2008) Neutral bimetallic nickel(II) phenoxyiminato catalysts for highly branched polyethylenes and ethylene-norbornene copolymerizations. Organometallics 27:2166–2168
166. Noda S, Nakamura A, Kochi T, Chung LW, Morokuma K, Nozaki K (2009) Mechanistic studies on the formation of linear polyethylene chain catalyzed by palladium phosphine–sulfonate complexes: experiment and theoretical studies. J Am Chem Soc 121:14088–14100
167. Malinoski JM, White PS, Brookhart M (2003) Structural characterization of $[2\text{-}(t\text{-}Bu)_2PCH_2C(O)C_6H_5]PdMe(2\text{–}C_2H_4)^+BAr'_4$-: a model for the catalyst resting state for ethylene polymerization. Organometallics 22:621–623

Index

MIX
Papier aus verantwortungsvollen Quellen
Paper from responsible sources
FSC® C105338

If you have any concerns about our products,
you can contact us on
ProductSafety@springernature.com

In case Publisher is established outside the EU,
the EU authorized representative is:
Springer Nature Customer Service Center GmbH
Europaplatz 3, 69115 Heidelberg, Germany

Printed by Libri Plureos GmbH
in Hamburg, Germany